A Mathematical Introduction to Logic

A MATHEMATICAL INTRODUCTION TO LOGIC

HERBERT B. ENDERTON
University of California, Los Angeles

ACADEMIC PRESS New York San Francisco London
A Subsidiary of Harcourt Brace Jovanovich, Publishers

ACADEMIC PRESS, INC.
111 Fifth Avenue, New York, New York 10003

United Kingdom Edition published by
ACADEMIC PRESS, INC. (LONDON) LTD.
24/28 Oval Road, London NW1

LIBRARY OF CONGRESS CATALOG CARD NUMBER: 78-182659

AMS (MOS) 1970 Subject Classifications: 0201, 02B10

PRINTED IN THE UNITED STATES OF AMERICA

for Eric and Bert

Contents

Chapter Two — FIRST-ORDER LOGIC

Chapter Three — UNDECIDABILITY

Chapter Four — SECOND-ORDER LOGIC

Preface

This book gives a mathematical treatment of the basic ideas and results of logic. It is intended to serve as a textbook for an introductory mathematics course in logic at the junior–senior level. The objectives are to present the important concepts and theorems of logic and to explain their significance and their relationship to the reader's other mathematical work.

As a text, the book can be used in courses anywhere from a quarter to a year in length. In one quarter, I generally reach the material on models of first-order theories (Section 2.6). The extra time afforded by a semester would permit some glimpse of undecidability, as in Section 3.0. In a second term, the material of Chapter 3 (on undecidability) can be more adequately covered.

The book is intended for the reader who has not studied logic previously, but who has some experience in mathematical reasoning. There are no specific prerequisites aside from a willingness to function at a certain level of abstraction and rigor. There is the inevitable use of basic set theory. In Chapter 0 is a concise summary of the set theory used. One should not begin the book by studying this chapter; it is instead intended for reference if and when the need arises. The instructor can adjust the amount of set theory employed; for example it is possible to avoid cardinal numbers completely (but some good theorems are then lost). The book contains some examples drawn from abstract algebra. But they are just examples, and are not essential to the exposition. The later chapters (Chapter 3 and 4) tend to be more demanding of the reader than are the earlier chapters.

Induction and recursion are given a more extensive discussion (in Section 1.2) than has been customary. I prefer to give an intuitive account of these subjects in lectures and have a precise version in the book rather than to have the situation reversed.

Exercises (over 150 in all) are given at the end of nearly all the sections. If the exercise bears a boldface numeral, then the results of that exercise are used in the exposition in the text. Exercises of unusual difficulty are marked with an asterisk.

I cheerfully acknowledge my debt to my teachers, a category in which I include also those who have been my colleagues or students. I would be pleased to receive comments and corrections from the users of this book.

List of Symbols

The numbers indicate the pages on which the symbol first occurs.

Introduction

Symbolic logic is a mathematical model of deductive thought. Or at least that was true originally; as with other branches of mathematics it has grown beyond the circumstances of its birth. Symbolic logic is a model in much the same way that modern probability theory is a model for situations involving chance and uncertainty.

How are models constructed? You begin with a real-life object, for example an airplane. Then you select some features of this original object to be represented in the model, for example its shape, and others to be ignored, for example its size. And then you build an object which is like the original in some ways (which you call essential) and unlike it in others (which you call irrelevant). Whether or not the resulting model meets its intended purpose will depend largely on the selection of the properties of the original object to be represented in the model.

Logic is more abstract than airplanes. The real-life objects are certain "logically correct" deductions. For example,

All men are mortal.
Socrates is a man.
Therefore, Socrates is mortal.

The validity of inferring the third sentence (the conclusion) from the first two (the assumptions) does not depend on special idiosyncracies of Socrates. The inference is justified by the form of the sentences rather than by empirical facts about mortality. It is not really important here what "mortal" means; it does matter what "all" means.

> Borogoves are mimsy whenever it is brillig.
> It is now brillig, and this thing is a borogove.
> Hence this thing is mimsy.

Again we can recognize that the third sentence follows from the first two, even without the slightest idea of what a mimsy borogove might look like.

Logically correct deductions are of more interest than the above frivolous examples might suggest. In fact, axiomatic mathematics consists of many such deductions laid end to end. Thus the deductions made by the working mathematician constitute real-life originals whose features are to be mirrored in our model.

The logical correctness of these deductions is due to their form but is independent of their content. This criterion is vague, but it is just this sort of vagueness which prompts us to turn to mathematical models. A major goal will be to give, within a model, a precise version of this criterion. The questions (about our model) we will initially be most concerned with are:

1. What does it mean for one sentence to "follow logically" from certain others?
2. If a sentence does follow logically from certain others, what methods of proof might be necessary to establish this fact?

Actually we will present two models. The first (sentential logic) will be very simple and will be woefully inadequate for interesting deductions. Its inadequacy stems from the fact that it preserves only some crude properties of real-life deductions. The second model (first-order logic) is admirably suited to deductions encountered in mathematics. When a working mathematician asserts that a particular sentence follows from the axioms of set theory, he means that this deduction can be translated to one in our model.

While our model is well suited to mathematics, it must be admitted that other models of deductive thought have been proposed for other purposes. For example, philosophers and mathematicians have considered so-called modal logic and intuitionistic logic, which represent different selections of properties of real-life deductions. In addition, the models have been generalized and studied for their own interest and for purposes unrelated to their original genesis. For example, one interesting topic of current study is logic with infinitely long sentences.

Thus far we have avoided giving away much information about what our model, first-order logic, is like. As brief hints, we now give some examples of the expressiveness of its formal language. First, take the English sentence which asserts the principle of extensionality, "If the same things are members of a first object as are members of a second object, then those objects are the same." This can be translated into our first-order language as

$$\forall x\, \forall y(\forall z(z \in x \leftrightarrow z \in y) \rightarrow x \approx y).$$

As a second example, we can translate "For every positive number ε there is a positive number δ such that for any x whose distance from a is less than δ, the distance between $f(x)$ and b is less than ε" as

$$\forall \varepsilon(\varepsilon > 0 \rightarrow \exists \delta(\delta > 0 \wedge \forall x(dxa < \delta \rightarrow dfxb < \varepsilon))).$$

We have given some hints as to what we intend to do in this book. We should also correct some possible misimpressions by saying what we are not going to do. This book does not propose to teach the reader how to think. The word "logic" is sometimes used to refer to remedial thinking, but not by us.

CHAPTER ZERO

Useful Facts about Sets

We assume that the reader already has some familiarity with normal everyday set-theoretic apparatus. Nonetheless, we give here a brief summary of facts from set theory we will need; this will at least serve to establish the notation. It is suggested that the reader, instead of poring over this chapter at the outset, simply refer to it if and when issues of a set-theoretic nature arise in later chapters.

A set is a collection of things, called its members or elements. As usual, we write "$t \in A$" to say that t is a member of A, and "$t \notin A$" to say that t is not a member of A. We write "$x = y$" to mean that x and y are the same object. That is, the expression "x" on the left of the equals sign is a name for the same object as is named by the other expression "y." If $A = B$, then for any object t it is automatically true that $t \in A$ iff $t \in B$. (The word "iff" means "if and only if.") This is simply because A and B are the same thing. The converse is the principle of extensionality: If A and B are sets such that for every object t,

$$t \in A \quad \text{iff } t \in B,$$

then $A = B$.

A useful operation is that of adjoining one extra object to a set. For a set A, let $A\,;t$ be the set whose members are (i) the members of A, plus (ii) the (possibly new) member t. Here t may or may not already belong to A, and we have

$$t \in A \qquad \text{iff } A\,;t = A.$$

One special set is the empty set $\varnothing$, which has no members at all. For any object x there is the singleton set $\{x\}$ whose only member is x. More generally, for any finite number $x_1, \ldots, x_n$ of objects there is the set $\{x_1, \ldots, x_n\}$ whose members are exactly those objects. Observe that $\{x, y\} = \{y, x\}$, as both sets have exactly the same members. We have only used different expressions to denote the set.

This notation will be stretched to cover some simple infinite cases. For example, $\{0, 1, 2, \ldots\}$ is the set $\mathbb{N}$ of natural numbers, and $\{\ldots, -2, -1, 0, 1, 2, \ldots\}$ is the set $\mathbb{Z}$ of all integers.

We write "$\{x : \_x\_\}$" for the set of all objects x such that $\_x\_$. We will take considerable liberty with this notation. For example, $\{\langle m, n\rangle : m < n \text{ in } \mathbb{N}\}$ is the set of all ordered pairs of natural numbers for which the first component is smaller than the second. And $\{x \in A : \_x\_\}$ is the set of all elements in A such that $\_x\_$.

If A is a set all of whose members are also members of B, then A is a *subset* of B, abbreviated "$A \subseteq B$." Note that any set is a subset of itself. Also, $\varnothing$ is a subset of every set. ("$\varnothing \subseteq A$" is "vacuously true," since the task of verifying, for every member of $\varnothing$, that it also belongs to A, requires doing nothing at all. Or from another point of view, "$A \subseteq B$" can be false only if some member of A fails to belong to B. If $A = \varnothing$, this is impossible.) From the set A we can form a new set, the *power set* $\mathscr{P}A$ of A, whose members are the subsets of A. Thus

$$\mathscr{P}A = \{x : x \subseteq A\}.$$

For example,

$$\mathscr{P}\varnothing = \{\varnothing\},$$

$$\mathscr{P}\{\varnothing\} = \{\varnothing, \{\varnothing\}\}.$$

The *union* of A and B, $A \cup B$, is the set of all things which are members of A or B (or both). For example, $A\,;t = A \cup \{t\}$. Similarly, the *intersection* of A and B, $A \cap B$, is the set of all things which are members of both A and B. A and B are *disjoint* iff their intersection is empty. A collection of sets is *pairwise disjoint* iff any two members of the collection are disjoint.

More generally, consider a set A whose members are themselves sets. The union of A, $\bigcup A$, is the set obtained by dumping all the members of A into a single set:

$$\bigcup A = \{x : x \text{ belongs to some member of } A\}.$$

Similarly,

$$\bigcap A = \{x : x \text{ belongs to all members of } A\}.$$

For example, if

$$A = \{\{0, 1, 5\}, \{1, 6\}, \{1, 5\}\},$$

then

$$\bigcup A = \{0, 1, 5, 6\},$$
$$\bigcap A = \{1\}.$$

Two other examples are

$$A \cup B = \bigcup\{A, B\},$$
$$\bigcup \mathscr{P}A = A.$$

In cases where we have a set A_n for each natural number n, the union of all these sets, $\bigcup\{A_n : n \in N\}$, is usually denoted "$\bigcup_{n \in N} A_n$" or just "$\bigcup_n A_n$."

The ordered pair $\langle x, y\rangle$ of objects x and y must be defined in such a way that

$$\langle x, y\rangle = \langle u, v\rangle \quad \text{iff } x = u \quad \text{and} \quad y = v.$$

Any definition which has this property will do; the standard one is

$$\langle x, y\rangle = \{\{x\}, \{x, y\}\}.$$

For ordered triples we define

$$\langle x, y, z\rangle = \langle\langle x, y\rangle, z\rangle.$$

More generally we define n-tuples inductively by

$$\langle x_1, \ldots, x_{n+1}\rangle = \langle\langle x_1, \ldots, x_n\rangle, x_{n+1}\rangle$$

for $n > 1$. It is convenient to define also $\langle x\rangle = x$; the above equation then holds also for $n = 1$. S is a *finite sequence* of members of A iff for some positive integer n, we have $S = \langle x_1, \ldots, x_n\rangle$, where each $x_i \in A$. (Finite sequences are often defined to be certain finite functions, but the above definition is slightly more convenient for us.)

A *segment* of the finite sequence $S = \langle x_1, \ldots, x_n \rangle$ is a finite sequence

$$\langle x_k, x_{k+1}, \ldots x_{m-1}, x_m \rangle, \qquad \text{where} \quad 1 \le k \le m \le n.$$

This segment is an *initial segment* iff $k = 1$ and it is *proper* iff it is different from S.

If $\langle x_1, \ldots, x_n \rangle = \langle y_1, \ldots, y_n \rangle$, then it is easy to see that $x_i = y_i$ for $1 \le i \le n$. (The proof uses induction on n and the basic property of ordered pairs.) But if $\langle x_1, \ldots, x_m \rangle = \langle y_1, \ldots, y_n \rangle$, then it does not in general follow that $m = n$. After all, every ordered triple is also an ordered pair. But we claim that m and n can be unequal only if some x_i is itself a finite sequence of y_j's, or the other way around:

Lemma 0A Assume that $\langle x_1, \ldots, x_m \rangle = \langle y_1, \ldots, y_m, \ldots, y_{m+k} \rangle$. Then $x_1 = \langle y_1, \ldots, y_{k+1} \rangle$.

Proof We use induction on m. If $m = 1$, the conclusion is immediate. For the inductive step, assume that $\langle x_1, \ldots, x_m, x_{m+1} \rangle = \langle y_1, \ldots, y_{m+k}, y_{m+1+k} \rangle$. Then the first components of this ordered pair must be equal: $\langle x_1, \ldots, x_m \rangle = \langle y_1, \ldots, y_{m+k} \rangle$. Now apply the inductive hypothesis. ■

For example, suppose that A is a set such that no member of A is a finite sequence of other members. Then if $\langle x_1, \ldots, x_m \rangle = \langle y_1, \ldots, y_n \rangle$ and each x_i and y_j is in A, then by the above lemma $m = n$. Whereupon we have $x_i = y_i$ as well.

From sets A and B we can form the set $A \times B$ of all pairs $\langle x, y \rangle$ for which $x \in A$ and $y \in B$. A^n is the set of all n-tuples of members of A. For example, $A^3 = (A \times A) \times A$.

A *relation* R is a set of ordered pairs. The *domain* of R, dom R, is the set of all objects x such that $\langle x, y \rangle \in R$ for some y. The *range* of R, ran R, is the set of all objects y such that $\langle x, y \rangle \in R$ for some x. The union of dom R and ran R is the *field* of R, fld R.

An *n-ary relation* on A is a subset of A^n. If $n > 1$, it *is* a relation. But a 1-ary (unary) relation on A is just a subset of A. A particularly simple binary relation on A is the equality relation $\{\langle x, x \rangle : x \in A\}$ on A. For an n-ary relation R on A and subset B of A, the *restriction* of R to B is the intersection $R \cap B^n$.

A *function* is a relation F with the property that for each x in dom F there is only one y such that $\langle x, y \rangle \in F$. As usual, this unique y is said to be the value $F(x)$ which F assumes at x. We say that F *maps* A into B and write

$$F : A \to B$$

to mean that F is a function, dom $F = A$, and ran $F \subseteq B$. If in addition ran $F = B$, then F maps A *onto* B. F is *one-to-one* iff for each y in ran F there is only one x such that $\langle x, y\rangle \in F$. If the pair $\langle x, y\rangle$ is in dom F, then we let $F(x, y) = F(\langle x, y\rangle)$. This notation is extended to n-tuples; $F(x_1, \ldots, x_n) = F(\langle x_1, \ldots, x_n\rangle)$.

An *n-ary operation* on A is a function mapping A^n into A. For example, addition is a binary operation on N, whereas the successor operation S (where $S(n) = n + 1$) is a unary operation on N. If f is an n-ary operation on A, then the *restriction* of f to a subset B of A is the function g with domain B^n which agrees with f at each point of B^n. Thus

$$g = f \cap (B^n \times A).$$

This g will be an n-ary operation on B iff B is closed under f, in the sense that $f(b_1, \ldots, b_n) \in B$ whenever each b_i is in B. In this case, $g = f \cap B^{n+1}$, in agreement with our definition of the restriction of a relation.

A particularly simple unary operation on A is the *identity* function i on A, given by the equation

$$i(x) = x \qquad \text{for} \quad x \in A.$$

For a relation R, we define the following:

R is *reflexive* on A iff $\langle x, x\rangle \in R$ for every x in A.

R is *symmetric* iff whenever $\langle x, y\rangle \in R$, then also $\langle y, x\rangle \in R$.

R is *transitive* iff whenever both $\langle x, y\rangle \in R$ and $\langle y, z\rangle \in R$, then also $\langle x, z\rangle \in R$.

R satisfies *trichotomy* on A iff for every x and y in A, exactly one of the three possibilities, $\langle x, y\rangle \in R$, $x = y$, or $\langle y, x\rangle \in R$, holds.

R is an *equivalence relation* on A iff R is a binary relation on A which is reflexive on A, symmetric, and transitive.

R is an *ordering relation* on A iff R is transitive and satisfies trichotomy on A.

For an equivalence relation R on A we define, for $x \in A$, the *equivalence class* $[x]$ of x to be $\{y : \langle x, y\rangle \in R\}$. The equivalence classes then partition A. That is, the equivalence classes are subsets of A such that each member of A belongs to exactly one equivalence class. For x and y in A,

$$[x] = [y] \qquad \text{iff } \langle x, y\rangle \in R.$$

At a few points in the book we will use the axiom of choice. But usually these uses can be eliminated if the theorems in question are restricted to countable languages. Of the many equivalent statements of the axiom of choice, Zorn's lemma is especially useful.

Say that a collection C of sets is a *chain* iff for any elements x and y of C, either $x \subseteq y$ or $y \subseteq x$.

Zorn's Lemma Let A be a nonempty set such that for any chain $C \subseteq A$, the set $\bigcup C$ is in A. Then there is some element $m \in A$ which is maximal in the sense that it is not a subset of any other element of A.

The set N of natural numbers is the set $\{0, 1, 2, \ldots\}$. (It can also be defined set-theoretically, but we will not do so here.) A set A is *finite* iff there is some one-to-one function f mapping (for some natural number n) the set A onto $\{0, 1, \ldots, n-1\}$. (We can think of f as "counting" the members of A.)

A set A is *countable* iff there is some function mapping A one-to-one into N. For example, any finite set is obviously countable. Now consider an infinite countable set A. Then from the given function f mapping a one-to-one into N, we can extract a function f' mapping A one-to-one *onto* N. For some $a_0 \in A$, $f(a_0)$ is the least member of ran f; let $f'(a_0) = 0$. In general there is a unique $a_n \in A$ such that $f(a_n)$ is the $(n+1)$st member of ran f; let $f'(a_n) = n$. Note that $A = \{a_0, a_1, \ldots\}$. (We can also think of f' as "counting" the members of A, only now the counting process is infinite.)

Theorem 0B Let A be a countable set. Then the set of all finite sequences of members of A is also countable.

Proof The set S of all such finite sequences can be characterized by the equation

$$S = \bigcup_{n \in N} A^{n+1}.$$

Since A is countable, we have a function f mapping A one-to-one into N.

The basic idea is to map S one-to-one into N by assigning to $\langle a_0, a_1, \ldots, a_m \rangle$ the number $2^{f(a_0)+1} 3^{f(a_1)+1} \cdot \ldots \cdot p_m^{f(a_m)+1}$, where p_m is the $(m+1)$st prime. This suffers from the defect that this assignment might not be well-defined. For conceivably there could be $\langle a_0, a_1, \ldots, a_m \rangle = \langle b_0, b_1, \ldots, b_n \rangle$, with a_i and b_j in A but with $m \neq n$. But this is not serious; just assign to each member of S the *smallest* number obtainable in the above fashion. This gives us a well-defined map; it is easy to see that it is one-to-one. ∎

At times we will speak of *trees*, which can be useful in providing intuitive pictures of some situations. But our comments on trees will always be informal, as we exclude this concept from the theorems and their proofs. Accordingly, our discussion here of trees will be informal.

For each tree there is an underlying finite partial ordering. We can draw a picture of this partial ordering R; if $\langle a, b\rangle \in R$, then we put a lower than b and connect the points by a line. Two typical pictures of tree orderings are

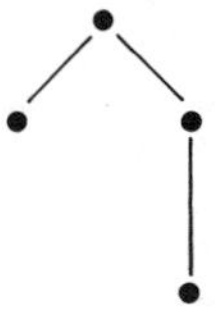

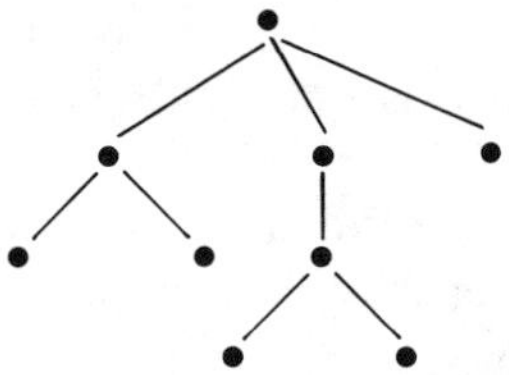

(Perhaps the word "tree" should be replaced by "root" since our branching is downward, not upward.) There is always a highest point in the picture. Furthermore, while branching is permitted below some vertex, the points above any given vertex must lie along a line.

In addition to this underlying finite partial ordering, a tree also has a labeling function whose domain is the set of vertices. For example, one tree, in which the labels are natural numbers, is

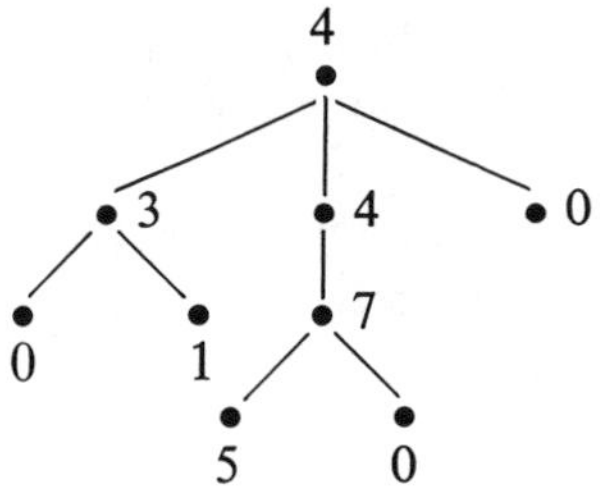

Cardinal numbers

All infinite sets are big, but some are bigger than others. (For example, the set of real numbers is bigger than the set of integers.) Cardinal numbers provide a convenient, although not indispensable, way of talking about the size of sets.

It is natural to say that two sets A and B have the same size iff there is a function which maps A one-to-one onto B. If A and B are finite, then this concept is equivalent to the usual one: If you count the members of A and the members of B, then you get the same number both times. But it is applicable even to infinite sets A and B, where counting is difficult.

Formally, then, say that A and B are *equinumerous* ($A \sim B$) iff there is a one-to-one function mapping A onto B. For example, the set $\mathbb{N}$ of natural numbers and the set $\mathbb{Z}$ of integers are equinumerous. It is easy to see that equinumerosity is reflexive, symmetric, and transitive.

For finite sets we can use natural numbers as measures of size. The same natural number would be assigned to two finite sets (as measures of their size) iff the sets were equinumerous. Cardinal numbers are introduced to enable us to generalize this situation to infinite sets.

To each set A we can assign a certain object, the *cardinal number* (or *cardinality*) of A (card A), in such a way that two sets are assigned the same cardinality iff they are equinumerous:

(K) $$\text{card } A = \text{card } B \qquad \text{iff } A \sim B.$$

There are several ways of accomplishing this; the standard one these days takes card A to be the least ordinal equinumerous with A. (The success of this definition relies on the axiom of choice.) We will not discuss ordinals here, since for our purposes it matters very little what card A actually is, any more than it matters what the number 2 actually is. What matters most is that (K) holds. It is helpful, however, if for a finite set A, card A is the natural number telling how many elements A has. Something is a *cardinal number*, or simply a *cardinal*, iff it is card A for some set A.

(Georg Cantor, who first introduced the concept of cardinal number, characterized in 1895 the cardinal number of a set M as "the general concept which, with the help of our active intelligence, comes from the set M upon abstraction from the nature of its various elements and from the order of their being given.")

Say that A is *dominated* by B ($A \preceq B$) iff A is equinumerous with a subset of B. In other words, $A \preceq B$ iff there is a one-to-one function mapping A into B. The companion concept for cardinals is

$$\text{card } A \leq \text{card } B \qquad \text{iff } A \preceq B.$$

(It is easy to see that $\leq$ is well defined; that is, whether or not $\kappa \leq \lambda$ depends only on the cardinals κ and λ themselves, and not the choice of sets having these cardinalities.) Dominance is reflexive and transitive. A set A is dominated by $\mathbb{N}$ iff A is countable. The following is a standard result in this subject.

Schröder–Bernstein Theorem (a) For any sets A and B, if $A \preceq B$ and $B \preceq A$, then $A \sim B$.

(b) For any cardinal numbers κ and λ, if $\kappa \leq \lambda$ and $\lambda \leq \kappa$, then $\kappa = \lambda$.

Part (b) is a simple restatement of part (a) in terms of cardinal numbers. The following theorem, which happens to be equivalent to the axiom of choice, is stated in the same dual manner.

Theorem 0C (a) For any sets A and B, either $A \preceq B$ or $B \preceq A$.
(b) For any cardinal numbers κ and λ, either $\kappa \leq \lambda$ or $\lambda \leq \kappa$.

Thus of any two cardinals, one is smaller than the other. (In fact, any nonempty set of cardinal numbers contains a smallest member.) The smallest cardinals are those of finite sets: $0, 1, 2, \ldots$. There is next the smallest infinite cardinal, card $\mathbb{N}$, which is given the name $\aleph_0$. Thus we have

$$0, 1, 2, \ldots, \aleph_0, \aleph_1, \ldots,$$

where $\aleph_1$ is the smallest cardinal larger than $\aleph_0$. The cardinality of the real numbers, card $\mathbb{R}$, is called "$2^{\aleph_0}$." Since $\mathbb{R}$ is uncountable, we have $\aleph_0 < 2^{\aleph_0}$.

The operations of addition and multiplication, long familiar for finite cardinals, can be extended to all cardinals. To compute $\kappa + \lambda$ we choose disjoint sets A and B of cardinality κ and λ, respectively. Then

$$\kappa + \lambda = \text{card}\,(A \cup B).$$

This is well defined; i.e., $\kappa + \lambda$ depends only on κ and λ, and not on the choice of the disjoint sets A and B. For multiplication we use

$$\kappa \cdot \lambda = \text{card}(A \times B).$$

Clearly these definitions are correct for finite cardinals. The arithmetic of infinite cardinals is surprisingly simple. The sum or product of two infinite cardinals is just the larger of them:

Cardinal Arithmetic Theorem For cardinal numbers κ and λ, if $\kappa \leq \lambda$ and λ is infinite, then $\kappa + \lambda = \lambda$. Furthermore, if $\kappa \neq 0$, then $\kappa \cdot \lambda = \lambda$.

In particular, for infinite cardinals κ,

$$\aleph_0 \cdot \kappa = \kappa.$$

Theorem 0D For an infinite set A, the set $\bigcup_n A^{n+1}$ of all finite sequences of elements of A has cardinality equal to card A.

We already proved this for the case of a countable A (see Theorem 0B).

Proof Each A^{n+1} has cardinality equal to card A, by the cardinal arithmetic theorem. So we have the union of $\aleph_0$ sets of this size, yielding $\aleph_0 \cdot$ card $A =$ card A points altogether. ■

EXAMPLE It follows that the set of algebraic numbers has cardinality $\aleph_0$. First, we can identify each polynomial (in one variable) over the integers with the sequence of its coefficients. Then by the theorem there are $\aleph_0$ polynomials. Each polynomial has a finite number of roots. To give an extravagant upper bound, note that even if each polynomial had $\aleph_0$ roots, we would then have $\aleph_0 \cdot \aleph_0 = \aleph_0$ algebraic numbers altogether. Since there are at least this many, we are done.

Since there are uncountably many (in fact, $2^{\aleph_0}$) real numbers, it follows that there are uncountably many (in fact, $2^{\aleph_0}$) transcendental numbers.

Throughout the book we will utilize an assortment of standard mathematical abbreviations. We have already used "■" to signify the end of a proof. A sentence "If . . . , then . . ." will sometimes be abbreviated ". . . $\Rightarrow$ . . . ." We also have "$\Leftarrow$" for the converse implication. For "if and only if" we use, in addition to the word "iff," the symbol "$\Leftrightarrow$." For the word "therefore" we have the abbreviation "$\therefore$."

The notational device that extracts "$x \neq y$" as the denial of "$x = y$" and "$x \notin y$" as the denial of "$x \in y$" will be extended to other cases. For example, on page 33 we define "$\Sigma \models \tau$"; then "$\Sigma \not\models \tau$" is its denial.

CHAPTER ONE

Sentential Logic

§ 1.0 INFORMAL REMARKS ON FORMAL LANGUAGES

We are about to construct (in the next section) a language into which we can translate English sentences. Unlike natural languages (like English or Chinese), it will be a formal language, with precise formation rules. But before the precision begins, we will discuss here some of the features we want to incorporate into the language.

As a first example, the English sentence "Traces of potassium were observed" can be translated into the formal language as, say, the symbol **K**. Then for the closely related sentence "Traces of potassium were not observed," we can use **(¬K)**. Here ¬ is our negation symbol, read as "not." One might also think of translating "Traces of potassium were not observed" by some new symbol, e.g., **J**, but we will prefer instead to break such a sentence down into atomic parts as much as possible. For an unrelated sentence, "The sample contained chlorine," we choose, say, the symbol **C**. Then the following compound English sentences can be translated as the formulas shown at the right:

If traces of potassium were observed, then the sample did not contain chlorine. **(K → (¬C))**

The sample contained chlorine, and traces of potassium were observed. **(C ∧ K)**

In the second example we use our conjunction symbol ∧ to translate "and." The first example uses the more familiar arrow to translate "if . . . , then . . . ." In the following example the disjunction symbol ∨ is used to translate "or":

Either no traces of potassium were observed, or the sample did not contain chlorine. **((¬K) ∨ (¬C))**

The sample neither contained chlorine nor were traces of potassium observed. **(¬(C ∨ K))** or **((¬C) ∧ (¬K))**

In this last example we have given two alternative translations. The relationship between them will be discussed later.

One important aspect of the decompositions we will make of compound sentences is that whenever we are given the truth or falsity of the atomic parts, we can then immediately calculate the truth or falsity of the compound. Suppose, for example, that the chemist emerges from his laboratory and announces that he observed traces of potassium but that the sample contained no chlorine. We then know that the four above sentences are true, false, true, and false, respectively. In fact, we can construct in advance a table giving the four possible experimental results (Table I). We will return to the discussion of such tables in Section 1.3.

TABLE I

K	C	(¬(C ∨ K))	((¬C) ∧ (¬K))
F	*F*	*T*	*T*
F	*T*	*F*	*F*
T	*F*	*F*	*F*
T	*T*	*F*	*F*

Use of formal languages will allow us to escape from the imprecision and ambiguities of natural languages. But this is not done without cost; our formal languages will have a sharply limited degree of expressiveness.

In order to describe a formal language we will generally give three pieces of information:

1. We will specify the set of symbols (the alphabet). In the present case of sentential logic some of the symbols are

$$(,), \rightarrow, \neg, \mathbf{A}_1, \mathbf{A}_2, \ldots .$$

2. We will specify the rules for forming the "grammatically correct" finite sequences of symbols. (Such sequences will be called *well-formed formulas* or *wffs*.) For example, in the present case

$$(\mathbf{A}_1 \rightarrow (\neg \mathbf{A}_2))$$

will be a wff, whereas

$$)) \rightarrow \mathbf{A}_3$$

will not.

3. We will also indicate the allowable translations between English and the formal language. The symbols $\mathbf{A}_1$, $\mathbf{A}_2$, ... can be translations of declarative English sentences.

Only in this third part is any meaning assigned to the wffs. This process of assigning meaning guides and motivates all we do. But it will also be observed that it would theoretically be possible to carry out various manipulations with wffs in complete ignorance of any possible meaning. A person aware of only the first two pieces of information listed above could perform some of the things we will do, but it would make no sense to him.

Before proceeding, let us look briefly at another class of formal languages of widespread interest today. These are the languages used by (or at least in connection with) digital computers.

There are many of these languages. In one of them a typical wff is

0110101101010001111000100000111 1010.

In another a typical wff is

STEP#ADDIMAX,A.

(Here # is a symbol called a blank; it is brought into the alphabet so that a wff will be a string of symbols.) A well-known language called Fortran has wffs such as

DO#3#I=IMIN,IMAX.

In all cases there is a given way of translating the wffs into English, and (for a restricted class of English sentences) a way of translating from English into the formal language. But the computer is unaware of the English language. An unthinking automaton, the computer juggles symbols and follows its program slavishly. We could approach formal languages that way too, but it would not be much fun.

§ 1.1 THE LANGUAGE OF SENTENTIAL LOGIC

We assume we are given an infinite sequence of distinct objects which we will call symbols, and to which we now give names (Table II). We further assume that no one of these symbols is a finite sequence of other symbols.

TABLE II

Symbol	Verbose name	Remarks
(	left parenthesis	punctuation
)	right parenthesis	punctuation
$\neg$	negation symbol	English: not
$\wedge$	conjunction symbol	English: and
$\vee$	disjunction symbol	English: or (inclusive)
$\rightarrow$	conditional symbol	English: if__, then__
$\leftrightarrow$	biconditional symbol	English: if and only if
$\mathbf{A}_1$	first sentence symbol	
$\mathbf{A}_2$	second sentence symbol	
. . .		
$\mathbf{A}_n$	*n*th sentence symbol	
. . .		

Several remarks are now in order:

1. The five symbols

$$\neg, \wedge, \vee, \rightarrow, \leftrightarrow$$

are called *sentential connective* symbols. Their use is suggested by the English translation given above. The sentential connective symbols, together with the parentheses, are the *logical symbols*. In translating to and from English, they always play the same role. The sentence symbols are the *parameters* (or *nonlogical symbols*). Their translation is not fixed; instead they will be open to a variety of interpretations, as we shall illustrate shortly.

2. We have included only countably many sentence symbols. Most of the things we say in this chapter would apply equally well if we had allowed an arbitrary set of sentence symbols. (The exceptions are primarily in Section 1.7.)

3. Some logicians prefer to call $\mathbf{A}_n$ the nth *proposition* symbol (and to speak of propositional logic instead of sentential logic). This stems from wanting the word "sentence" to refer to a particular utterance, and wanting a proposition to be that which a sentence asserts.

4. We call these objects "symbols," but we remain neutral as to what the exact ontological status of symbols might be. In the leftmost column of our list of symbols, names of the symbols are printed. For example, $\mathbf{A}_{243}$ is one symbol, namely the two hundred forty-third sentence symbol. (On the other hand, "$\mathbf{A}_{243}$" is a name of that symbol. The conditional symbol may or may not have the geometric property of being shaped like an arrow, although its name "$\rightarrow$" does.) The symbols themselves can be sets, numbers, marbles, or objects from a universe of linguistic objects. In the last case, it is conceivable that they are actually the same things as the names we use for them. Another possibility, which will be exploited in the next chapter, is that the sentence symbols are themselves formulas in another language.

5. We have assumed that no symbol is a finite sequence of other symbols. We mean by this that not only are the symbols listed distinct (e.g., $\mathbf{A}_3 \neq \leftrightarrow$), but no one of them is a finite sequence of two or more symbols. For example, we demand that $\mathbf{A}_3 \neq \langle \neg, \mathbf{A}_4, (\rangle$. (It is a fact of set theory that $\mathbf{A}_3 \neq \langle \neg, \mathbf{A}_3, (\rangle$.) The purpose of this assumption is to assure that finite sequences of symbols be uniquely decomposable. If

$$\langle a_1, \ldots, a_m \rangle = \langle b_1, \ldots, b_n \rangle$$

and each a_i and each b_j is a symbol, then $m = n$ and $a_i = b_i$. (See Chapter 0, Lemma 0A, and subsequent remarks.)

An *expression* is a finite sequence of symbols. We can specify an expression by concatenating the names of the symbols; thus $(\neg\mathbf{A}_1)$ is the sequence $\langle (, \neg, \mathbf{A}_1,) \rangle$. This notation is extended: If α and β are sequences of symbols, then $\alpha\beta$ is the sequence consisting first of the symbols of the sequence α followed by the symbols of the sequence β.

For example, if α and β are the expressions given by the equations

$$\alpha = (\neg\mathbf{A}_1),$$

$$\beta = \mathbf{A}_2,$$

then $(\alpha \rightarrow \beta)$ is the expression

$$((\neg \mathbf{A}_1) \rightarrow \mathbf{A}_2).$$

We should now look at a few examples of possible translations of English sentences into expressions of the formal language. Let $\mathbf{A}, \mathbf{B}, \ldots, \mathbf{Z}$ be the first twenty-six sentence symbols. (For example, $\mathbf{E} = \mathbf{A}_5$.)

1. English: The suspect must be released from custody. Translation: **R**.
English: The evidence obtained is admissible. Translation: **E**.
English: The evidence obtained is inadmissible. Translation: $(\neg\mathbf{E})$.
English: The evidence obtained is admissible, and the suspect need not be released from custody. Translation: $(\mathbf{E} \wedge (\neg\mathbf{R}))$.
English: Either the evidence obtained is admissible, or the suspect must be released from custody (or possibly both). Translation: $(\mathbf{E} \vee \mathbf{R})$.
English: Either the evidence obtained is admissible, or the suspect must be released from custody, but not both. Translation: $((\mathbf{E} \vee \mathbf{R}) \wedge (\neg(\mathbf{E} \wedge \mathbf{R})))$. We intend always to use the symbol $\vee$ to translate the word "or" in its inclusive meaning of "and/or."
English: The evidence obtained is inadmissible, but the suspect need not be released from custody. Translation: $((\neg\mathbf{E}) \wedge (\neg\mathbf{R}))$. On the other hand, the expression $((\neg\mathbf{E}) \vee (\neg\mathbf{R}))$ translates the English: Either the evidence obtained is inadmissible or the suspect need not be released from custody.

2. English: If wishes are horses, then beggars will ride. Translation: $(\mathbf{W} \rightarrow \mathbf{B})$.
English: Beggars will ride if and only if wishes are horses. Translation: $(\mathbf{B} \leftrightarrow \mathbf{W})$.

3. English: This commodity constitutes wealth if and only if it is transferable, limited in supply, and either productive of pleasure or preventive of pain. Translation: $(\mathbf{W} \leftrightarrow (\mathbf{T} \wedge (\mathbf{L} \wedge (\mathbf{P} \vee \mathbf{Q}))))$. Here **W** is the translation of "This commodity constitutes wealth." Of course in the preceding example we used **W** to translate a different sentence. We are not tied to any one translation.

One note of caution: Do not confuse a sentence in the English language (Roses are red) with a translation of that sentence in the formal language (e.g., **R**). These are different. The English sentence is presumably either true or false. But the formal expression is just a sequence of symbols. It may indeed be interpreted in one context as a true (or false) English sentence, but it can have other interpretations in other contexts.

Now some expressions cannot be obtained as translations of any English sentences but are mere nonsense, like

$$\rightarrow ((\mathbf{A}_3.$$

We want to define the well-formed formulas (wffs) to be the "grammatically correct" expressions; the nonsensical expressions must be excluded. The definition will have the following consequences:

(a) Every sentence symbol is a wff.

(b) If α and β are wffs, then so are $(\neg\alpha)$, $(\alpha \wedge \beta)$, $(\alpha \vee \beta)$, $(\alpha \rightarrow \beta)$, and $(\alpha \leftrightarrow \beta)$.

(c) No expression is a wff unless it is compelled to be one by (a) and (b).

There are two equivalent ways of making this third property (about compulsion) precise. The first way defines the set of wffs "from the top down." Let us say that a set S of expressions is *inductive* iff it has the properties

(a_S) Every sentence symbol is in S.

(b_S) If expressions α and β are in S, then so also are $(\neg\alpha)$, $(\alpha \wedge \beta)$, $(\alpha \vee \beta)$, $(\alpha \rightarrow \beta)$, and $(\alpha \leftrightarrow \beta)$.

An expression is a *well-formed formula* (*wff*) iff it is a member of every inductive set. Then it is not hard to see that (a) and (b) are satisfied. As for (c), we can say that the set of all wffs is as small as it can be, in the sense that it is a subset of any other inductive set.

The second (and equivalent) definition works "from the bottom up." An expression is a wff if and only if it can be built up from the sentence symbols by applying some finite number of times the *formula-building operations* (on expressions) defined by the equations

$$\begin{aligned}
\mathscr{E}_{\neg}(\alpha) &= (\neg\alpha),\\
\mathscr{E}_{\wedge}(\alpha, \beta) &= (\alpha \wedge \beta),\\
\mathscr{E}_{\vee}(\alpha, \beta) &= (\alpha \vee \beta),\\
\mathscr{E}_{\rightarrow}(\alpha, \beta) &= (\alpha \rightarrow \beta),\\
\mathscr{E}_{\leftrightarrow}(\alpha, \beta) &= (\alpha \leftrightarrow \beta).
\end{aligned}$$

For example,

$$((\mathbf{A}_1 \wedge \mathbf{A}_{10}) \rightarrow ((\neg\mathbf{A}_3) \vee (\mathbf{A}_8 \leftrightarrow \mathbf{A}_3)))$$

is a wff, as can be seen by contemplating its ancestral tree:

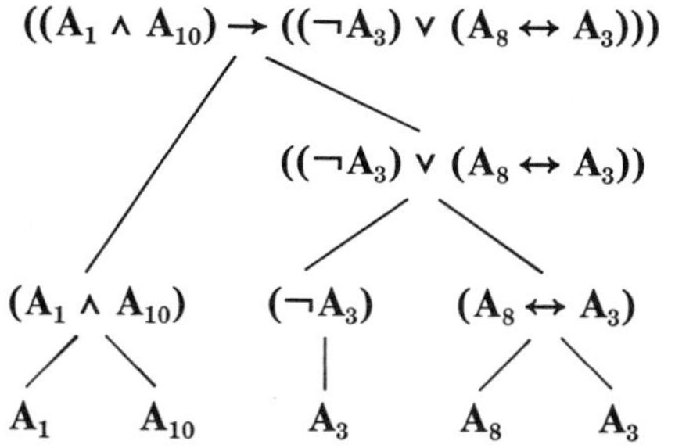

The tree illustrates the construction of the expression from four sentence symbols by five applications of formula-building operations.

These two definitions of wffs are discussed in a more general setting in Section 1.2. Here we only note that any inductive set of wffs is actually the set of all wffs; this fact will be called the *induction principle*. This principle will receive much use in the coming pages. In the following example we use it to show that certain expressions are *not* wffs.

EXAMPLE Any expression with more left parentheses than right parentheses is not a wff.

Proof The idea is that we start with sentence symbols (having zero left parentheses and zero right parentheses), and then apply formula-building operations which add parentheses only in matched pairs. We can rephrase this argument as follows: The set of "balanced" wffs (having equal numbers of left and right parentheses) contains all sentence symbols and is closed under the formula-building operations. Thus the set of balanced wffs is inductive; the induction principle then assures us that all wffs are balanced. ■

EXERCISES

1. Give three sentences in English together with translations into our formal language. The sentences should be chosen so as to have an interesting structure, and the translations should each contain fifteen or more symbols.

2. Show that there are no wffs of length 2, 3, or 6, but that any other length is possible.

3. Let α be a wff; let c be the number of places at which binary connective symbols ($\wedge$, $\vee$, $\rightarrow$, $\leftrightarrow$) occur in α; let s be the number of places at which sentence symbols occur in α. Show that $s = c + 1$.

§ 1.2 INDUCTION AND RECURSION

Induction

There is one special type of construction which occurs frequently both in logic and in other branches of mathematics. We may want to construct a certain subset of a set U by starting with some initial elements of U, and applying certain operations to them over and over again. The set we seek will be the smallest set containing the initial elements and closed under the operations. Its members will be those elements of U which can be built up from the initial elements by applying the operations a finite number of times.

In the special case of immediate interest to us, U is the set of expressions, the initial elements are the sentence symbols, and the operations are $\mathcal{E}_\neg$, $\mathcal{E}_\wedge$, etc. The set to be constructed is the set of wffs. But we will encounter other special cases later, and it will be helpful to view the situation abstractly here.

To simplify our discussion, we will consider an initial set $B \subseteq U$ and a class $\mathcal{F}$ of functions containing just two members f and g, where

$$f : U \times U \to U \quad \text{and} \quad g : U \to U.$$

Thus f is a binary operation on U and g is a unary operation. (Actually $\mathcal{F}$ need not be finite; it will be seen that our simplified discussion here is, in fact, applicable to a more general situation. $\mathcal{F}$ can be any set of relations on U, and in Chapter 2 this greater generality will be utilized. But the case discussed here is easier to visualize and is general enough to illustrate the ideas. For a less restricted version, see Exercise 3.)

If B contains points a and b, then the set C we wish to construct will contain, for example,

$$b, f(b, b), g(a), f(g(a), f(b, b)), g(f(g(a), f(b, b))).$$

Of course these might not all be distinct. The idea is that we are given certain bricks to work with, and certain types of mortar, and we want C to contain just the things we are able to build.

In defining C more formally, we have our choice of two definitions. We can define it "from the top down" as follows: Say that a subset S of U is *closed* under f and g iff whenever elements x and y belong to S, then so do $f(x, y)$ and $g(x)$. Say that S is *inductive* iff $B \subseteq S$ and S is closed under f and g. Let C^* be the intersection of all the inductive subsets of U; thus

$x \in C^*$ iff x belongs to every inductive subset of U. It is not hard to see (and the reader should check) that C^* is itself inductive. Furthermore, C^* is the smallest such set, being included in all the other inductive sets.

The second (and equivalent) definition works "from the bottom up." We want C_* to contain the things which can be reached from B by applying f and g a finite number of times. Temporarily define a *construction sequence* to be a finite sequence $\langle x_0, \ldots, x_n \rangle$ of elements of U such that for each $i \leq n$ we have at least one of

$$x_i \in B,$$

$$x_i = f(x_j, x_k) \qquad \text{for some} \quad j < i,\ k < i,$$

$$x_i = g(x_j) \qquad \text{for some} \quad j < i.$$

Then let C_* be the set of all points x such that some construction sequence ends with x.

Let C_n be the set of points x such that some construction sequence of length n ends with x. Then $C_1 = B$,

$$C_1 \subseteq C_2 \subseteq C_3 \subseteq \cdots,$$

and $C_* = \bigcup_n C_n$. For example, $g(f(a, f(b, b)))$ is in C_5 and hence in C_*, as can be seen by contemplating the tree shown:

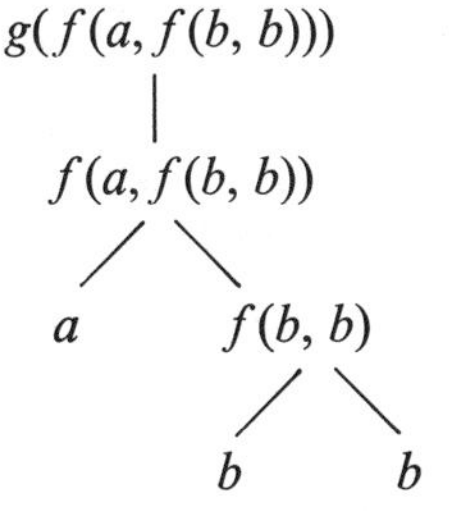

We obtain a construction sequence for $g(f(a, f(b, b)))$ by squashing this tree into a linear ordering.

Examples 1. The natural numbers. Let U be the set of all real numbers, and let $B = \{0\}$. Take one operation S, where $S(x) = x + 1$. Then

$$C_* = \{0, 1, 2, \ldots\}.$$

The set C_* of natural numbers contains exactly those numbers obtainable from 0 by applying the successor operation repeatedly.

2. The integers. Let U be the set of all real numbers; let $B = \{0\}$. This time take two operations, the successor operation S and the predecessor operation P:

$$S(x) = x + 1 \quad \text{and} \quad P(x) = x - 1.$$

Now C_* contains all the integers,

$$C_* = \{\ldots, -2, -1, 0, 1, 2, \ldots\}.$$

Notice that there is more than one way of obtaining 2 as a member of C_*. For 2 is $S(S(0))$, but it is also $S(P(S(S(0))))$.

3. The algebraic functions. Let U contain all functions whose domain and range are each sets of real numbers. Let B contain the identity function and all constant functions. Let $\mathscr{F}$ contain the operations (on functions) of addition, multiplication, division, and root extraction. Then C_* is the class of algebraic functions.

4. The well-formed formulas. Let U be the set of all expressions and let B be the set of sentence symbols. Let $\mathscr{F}$ contain the five formula-building operations on expressions: $\mathscr{E}_\neg$, $\mathscr{E}_\wedge$, $\mathscr{E}_\vee$, $\mathscr{E}_\rightarrow$, and $\mathscr{E}_\leftrightarrow$. Then C_* is the set of all wffs.

At this point we should verify that our two definitions are actually equivalent, i.e., that $C^* = C_*$.

To show that $C^* \subseteq C_*$ we need only check that C_* is inductive, i.e., that $B \subseteq C_*$ and C_* is closed under the functions. Clearly $B = C_1 \subseteq C_*$. If x and y are in C_*, then we can concatenate their construction sequences and append $f(x, y)$ to obtain a construction sequence placing $f(x, y)$ in C_*. Similarly, C_* is closed under g.

Finally, to show that $C_* \subseteq C^*$ we consider a point in C_* and a construction sequence $\langle x_0, \ldots, x_n \rangle$ for it. By ordinary induction on i, we can see that $x_i \in C^*$, $i \leq n$. First $x_0 \in B \subseteq C^*$. For the inductive step we use the fact that C^* is closed under the functions. Thus we conclude that

$$\bigcup_n C_n = C_* = C^* = \bigcap \{S : S \text{ is inductive}\}.$$

(A parenthetical remark: Suppose our present study is embedded in axiomatic set theory, where the natural numbers are usually defined from the top down. Then our definition of C_* (employing finiteness and hence natural numbers) is not really different from our definition of C^*. But we are not working within axiomatic set theory; we are working within intuitive mathematics. And the notion of natural number seems to be a solid, well-understood intuitive concept.)

Since $C^* = C_*$, we call the set simply C and refer to it as the set *generated from B by* the functions in $\mathscr{F}$. We will often want to prove theorems by using the following:

Induction Principle Assume that C is the set generated from B by the functions in $\mathscr{F}$. If S is a subset of C which includes B and is closed under the functions of $\mathscr{F}$, then $S = C$.

Proof S is inductive, so $C = C^* \subseteq S$. We are given the other inclusion. ∎

The special case now of interest to us is, of course, Example 4. Here C is the class of wffs generated from the set of sentence symbols by the formula-building operations. This special case has interesting features of its own. Both α and β are proper segments of $\mathscr{E}_\wedge(\alpha, \beta)$, i.e., of $(\alpha \wedge \beta)$. More generally, if we look at the family tree of a wff, we see that each constituent is a proper segment of the end product.

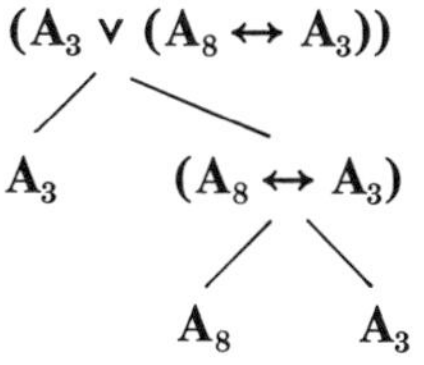

Suppose, for example, that we temporarily call an expression *special* if the only sentence symbols in it are among $\{\mathbf{A}_2, \mathbf{A}_3, \mathbf{A}_5\}$ and the only connective symbols in it are among $\{\neg, \rightarrow\}$. Then no special wff requires $\mathbf{A}_9$ or $\mathscr{E}_\wedge$ for its construction. In fact, every special wff belongs to the set C_s generated from $\{\mathbf{A}_2, \mathbf{A}_3, \mathbf{A}_5\}$ by $\mathscr{E}_\neg$ and $\mathscr{E}_\rightarrow$. (For we can use the induction principle to show that every wff either belongs to C_s or is not special.)

Recursion[1]

We return now to the more abstract case. There is a set U (such as the set of all expressions), a subset B of U (such as the set of sentence symbols),

[1] The reader not already familiar with recursion is advised to postpone reading this subsection until after reading Section 1.3, where a specific application is encountered.

and two functions f and g, where

$$f : U \times U \to U \quad \text{and} \quad g : U \to U.$$

C is the set generated from B by f and g.

The problem we now want to consider is that of defining a function on C recursively. That is, we suppose we are given

1. Rules for computing $\bar{h}(x)$ for $x \in B$.
2a. Rules for computing $\bar{h}(f(x, y))$, making use of $\bar{h}(x)$ and $\bar{h}(y)$.
2b. Rules for computing $\bar{h}(g(x))$, making use of $\bar{h}(x)$.

(For example, this is the situation discussed in Section 1.3, where $\bar{h}$ is the extension of a truth assignment for B.) It is not hard to see that there can be at most one function $\bar{h}$ on C meeting all the given requirements.

But it is possible that no such $\bar{h}$ exists; the rules may be contradictory. For example, let

$$\begin{aligned} U &= \text{the set of real numbers}, \\ B &= \{0\}, \\ f(x, y) &= x \cdot y, \\ g(x) &= x + 1. \end{aligned}$$

Then C is the set of natural numbers. Suppose we impose the following requirements on $\bar{h}$:

1. $\bar{h}(0) = 0$.
2a. $\bar{h}(f(x, y)) = f(\bar{h}(x), \bar{h}(y))$.
2b. $\bar{h}(g(x)) = \bar{h}(x) + 2$.

Then no such function $\bar{h}$ can exist. (Try computing $\bar{h}(1)$, noting that we have both $1 = g(0)$ and $1 = f(g(0), g(0))$.

A similar situation is encountered in algebra.[1] Suppose that you have a group G which is generated from B by the group multiplication and inverse operation. Then an arbitrary map of B into a group H is not necessarily extendible to a homomorphism of the entire group G into H. But if G happens to be a free group with set B of independent generators, then any such map is extendible to a homomorphism of the entire group.

[1] It is hoped that examples such as this will be useful to the reader with some algebraic experience. The other readers will be glad to know that these examples are merely illustrative and not essential to our development of the subject.

Say that C is *freely* generated from B by f and g iff in addition to the requirements for being generated we have

1. f_C and g_C are one-to-one, and
2. The range of f_C, the range of g_C, and the set B are pairwise disjoint.

(Here f_C and g_C are the restrictions of f and g to C.)

Recursion Theorem Assume that the subset C of U is freely generated from B by f and g, where

$$f : U \times U \to U,$$
$$g : U \to U.$$

Further assume that V is a set and F, G, and h functions such that

$$h : B \to V,$$
$$F : V \times V \to V,$$
$$G : V \to V.$$

Then there is a unique function

$$\bar{h} : C \to V$$

such that

(i) For x in B, $\bar{h}(x) = h(x)$.
(ii) For x, y in C,

$$\bar{h}(f(x, y)) = F(\bar{h}(x), \bar{h}(y)),$$
$$\bar{h}(g(x)) = G(\bar{h}(x)).$$

Viewed algebraically, the conclusion of this theorem says that any map h of B into V can be extended to a homomorphism $\bar{h}$ from C (with operations f and g) into V (with operations F and G).

If the content of the recursion theorem is not immediately apparent, try looking at it chromatically. You want to have a function $\bar{h}$ which paints each member of C some color. You have before you

1. h, telling you how to color the initial elements in B;
2. F, which tells you how to combine the color of x and y to obtain the color of $f(x, y)$ (i.e., it gives $\bar{h}(f(x, y))$ in terms of $\bar{h}(x)$ and $\bar{h}(y)$);
3. G, which similarly tells you how to convert the color of x into the color of $g(x)$.

The danger is that of a conflict in which, for example, F is saying "green" but G is saying "red" for the same point (unlucky enough to be equal both to $f(x, y)$ and $g(z)$ for some x, y, z). But if C is freely generated, then this danger is avoided.

EXAMPLES Consider again the examples of the preceding subsection.

1. $B = \{0\}$ with one operation, the successor operation S. Then C is the set N of natural numbers. Since the successor operation is one-to-one, C is freely generated from $\{0\}$ by S. Therefore, by the recursion theorem, for any set V, any $a \in V$, and any $F : V \rightarrow V$ there is a unique $\bar{h} : N \rightarrow V$ such that $\bar{h}(0) = a$ and $\bar{h}(S(x)) = F(\bar{h}(x))$ for each $x \in N$. For example, there is a unique $\bar{h} : N \rightarrow N$ such that $\bar{h}(0) = 0$ and $\bar{h}(S(x)) = 1 - \bar{h}(x)$. This function has the value 0 at even numbers and the value 1 at odd numbers.

2. The integers are generated from $\{0\}$ by the successor and predecessor operations but not freely generated.

3. Freeness fails also for the generation of the algebraic functions in the manner described.

4. The wffs are freely generated from the sentence symbols by the five formula-building operations. The purpose of Section 1.4 is to prove this fact. It follows, for example, that there is a unique function $\bar{h}$ defined on the set of wffs such that

$$\begin{aligned}\bar{h}(A) &= 1 \text{ for a sentence symbol } A,\\ \bar{h}((\neg\alpha)) &= 3 + \bar{h}(\alpha),\\ \bar{h}((\alpha \wedge \beta)) &= 3 + \bar{h}(\alpha) + \bar{h}(\beta),\end{aligned}$$

and similarly for $\vee$, $\rightarrow$, and $\leftrightarrow$. This function gives the length of each wff.

Proof of the recursion theorem The idea is to let $\bar{h}$ be the union of many approximating functions. Temporarily call a function v (which maps part of C into V) *acceptable* if it meets the conditions imposed on $\bar{h}$ by (i) and (ii). More precisely, v is acceptable iff the domain of v is a subset of C, the range a subset of V, and

(i′) If x belongs to B and to the domain of v, then $v(x) = h(x)$.

(ii′) If $f(x, y)$ belongs to the domain of v, then so do x and y, and $v(f(x, y)) = F(v(x), v(y))$. If $g(x)$ belongs to the domain of v, then so does x, and $v(g(x)) = G(v(x))$.

Let K be the collection of all acceptable functions, and let $\bar{h}$ be the union of K. Thus

$$\langle x, y\rangle \in \bar{h} \qquad \text{iff } v(x) = y \text{ for some } v \in K.$$

We claim that $\bar{h}$ meets our requirements. We will outline the procedure for checking this, leaving many details to the reader. (We feel that a detailed understanding of this set-theoretic proof, while nice, is not essential here. But some understanding of its outline should make the theorem itself more comprehensible.)

1. We claim that $\bar{h}$ is a function. Let

$$S = \{x \in C\text{: For at most one } y, \langle x, y\rangle \in \bar{h}\}.$$

It is easy to verify that S is inductive, by using (i′) and (ii′). Hence $S = C$ and $\bar{h}$ is a function.

2. We claim that $\bar{h} \in K$; i.e., that $\bar{h}$ is an acceptable function. This follows fairly easily from the definition of $\bar{h}$ and the fact that it is a function.

3. We claim that $\bar{h}$ is defined throughout C. It suffices to show that the domain of $\bar{h}$ is inductive. It is here that the assumption of freeness is used. For example, one case is the following: Suppose that x is in the domain of $\bar{h}$. Then $\bar{h}$; $\langle g(x), G(\bar{h}(x))\rangle$ is acceptable. (The freeness is required in showing that it is a function.) Consequently, $g(x)$ is in the domain of $\bar{h}$.

4. We claim that $\bar{h}$ is unique. For given two such functions, let S be the set on which they agree. Then S is inductive, and so equals C. ■

It is interesting to note that there is an alternative way of describing the proof of the recursion theorem, by presenting the desired function $\bar{h}$ as the set (of pairs) generated from a set by some operations. For let

$$\hat{U} = U \times V,$$
$$\hat{B} = h, \qquad \text{a subset of } \hat{U},$$
$$\hat{F}(\langle x, u\rangle, \langle y, v\rangle) = \langle f(x, y), F(u, v)\rangle,$$
$$\hat{G}(\langle x, u\rangle) = \langle g(x), G(u)\rangle.$$

Thus $\hat{F}$ is the binary operation on $\hat{U}$ obtained as the product of the operations f and F. Now let $\bar{h}$ be the subset of $\hat{U}$ generated from $\hat{B}$ by $\hat{F}$ and $\hat{G}$. Then it can be checked that $\bar{h}$ has the desired properties. The freeness must be used in showing that $\bar{h}$ is a function.

One final comment on induction and recursion: The induction principle we have stated is not the only one possible. It is entirely possible to give proofs by induction (and definitions by recursion) on the length of expressions, the number of places at which connective symbols occur, etc. Such methods are inherently less basic but may be necessary in some situations.

EXERCISES

1. Suppose that C is generated from a set $B = \{a, b\}$ by the binary operation f and unary operation g. List all the members of C_2. How many members might C_3 have? C_4?

2. Obviously $(\mathbf{A}_3 \rightarrow \wedge \mathbf{A}_4)$ is not a wff. But prove that it is not a wff.

3. We can generalize the discussion in this section by requiring of $\mathscr{F}$ only that it be a class of relations on U. C_* is defined as before, except that $\langle x_0, x_1, \ldots, x_n \rangle$ is now a construction sequence provided that for each $i \leq n$ we have either $x_i \in B$ or $\langle x_{j_1}, \ldots, x_{j_k}, x_i \rangle \in R$ for some $R \in \mathscr{F}$ and some $j_1, \ldots, j_k$ all less than i. Give the correct definition of C^* and show that $C^* = C_*$.

§ 1.3 TRUTH ASSIGNMENTS

We want to define what it means for one wff of our language to follow logically from other wffs. For example, $\mathbf{A}_1$ should follow from $(\mathbf{A}_1 \wedge \mathbf{A}_2)$. For no matter how the parameters $\mathbf{A}_1$ and $\mathbf{A}_2$ are translated back into English, if the translation of $(\mathbf{A}_1 \wedge \mathbf{A}_2)$ is true, then the translation of $\mathbf{A}_1$ must be true. But the notion of all possible translations back into English is unworkably vague. Luckily the spirit of this notion can be expressed in a simple and precise way.

Fix once and for all a set $\{T, F\}$ of *truth values* consisting of two distinct points:

T, called *truth*,

F, called *falsity*.

(It makes no difference what these points themselves are; they might as well be the numbers 1 and 0.) Then a *truth assignment* v for a set $\mathscr{S}$ of sentence symbols is a function

$$v : \mathscr{S} \rightarrow \{T, F\}$$

assigning either T or F to each symbol in $\mathscr{S}$. These truth assignments will be used in place of the translations into English mentioned in the preceding paragraph.

(At this point we have committed ourselves to *two-valued* logic. It is also possible to study three-valued logic, in which case one has a set of three possible truth values. And then, of course, it is a small additional step to allow 512 or $\aleph_0$ truth values; or to take as the set of truth values the unit interval [0, 1] or some other convenient space. A particularly interesting case is that for which the truth values form a complete Boolean algebra. But it is two-valued logic that has always had the greatest significance, and we will be content to confine ourselves to this case.)

Let $\bar{\mathscr{S}}$ be the set of wffs generated from $\mathscr{S}$ by the five formula-building operations. ($\bar{\mathscr{S}}$ can also be characterized as the set of wffs whose sentence symbols are all in $\mathscr{S}$; see the remarks at the end of the subsection on induction in Section 1.2.) We want an extension $\bar{v}$ of v,

$$\bar{v} : \bar{\mathscr{S}} \to \{T, F\},$$

which assigns the correct truth value to each wff in $\bar{\mathscr{S}}$. It should meet the following conditions:

0. For any $A \in \mathscr{S}$, $\bar{v}(A) = v(A)$. (Thus $\bar{v}$ is an extension of v.)

For any α, β in $\bar{\mathscr{S}}$:

1. $\bar{v}((\neg\alpha)) = \begin{cases} T \text{ if } \bar{v}(\alpha) = F, \\ F \text{ otherwise.} \end{cases}$

2. $\bar{v}((\alpha \wedge \beta)) = \begin{cases} T \text{ if } \bar{v}(\alpha) = T \text{ and } \bar{v}(\beta) = T, \\ F \text{ otherwise.} \end{cases}$

3. $\bar{v}((\alpha \vee \beta)) = \begin{cases} T \text{ if } \bar{v}(\alpha) = T \text{ or } \bar{v}(\beta) = T \text{ (or both)}, \\ F \text{ otherwise.} \end{cases}$

4. $\bar{v}((\alpha \to \beta)) = \begin{cases} F \text{ if } \bar{v}(\alpha) = T \text{ and } \bar{v}(\beta) = F, \\ T \text{ otherwise.} \end{cases}$

5. $\bar{v}((\alpha \leftrightarrow \beta)) = \begin{cases} T \text{ if } \bar{v}(\alpha) = \bar{v}(\beta), \\ F \text{ otherwise.} \end{cases}$

(Conditions 1–5 are given in tabular form in Table III. It is at this point that the intended meaning of, for example, the conjunction symbol enters into our formal proceedings. Note especially the intended meaning of $\to$. Whenever α is assigned F, then $(\alpha \to \beta)$ is considered "vacuously true" and is assigned the value T. For this and the other connectives, it is certainly possible to question how accurately we have reflected the common meaning in everyday speech of "if . . . , then," "or," etc. But our ultimate concern lies more with mathematical statements than with the subtle nuances of everyday speech.)

TABLE III

α	β	$\neg\alpha$	$\alpha \wedge \beta$	$\alpha \vee \beta$	$\alpha \rightarrow \beta$	$\alpha \leftrightarrow \beta$
T	T	F	T	T	T	T
T	F	F	F	T	F	F
F	T	T	F	T	T	F
F	F	T	F	F	T	T

As an example of the calculation of $\bar{v}$, let α be the wff

$$((\mathbf{A}_2 \rightarrow (\mathbf{A}_1 \rightarrow \mathbf{A}_6)) \leftrightarrow ((\mathbf{A}_2 \wedge \mathbf{A}_1) \rightarrow \mathbf{A}_6))$$

and let v be the truth assignment for $\{\mathbf{A}_1, \mathbf{A}_2, \mathbf{A}_6\}$ such that

$$v(\mathbf{A}_1) = T,$$
$$v(\mathbf{A}_2) = T,$$
$$v(\mathbf{A}_6) = F.$$

We want to compute $\bar{v}(\alpha)$. We can look at the tree which displays the construction of α:

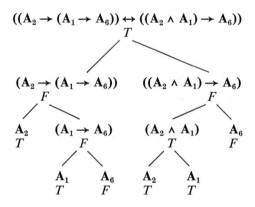

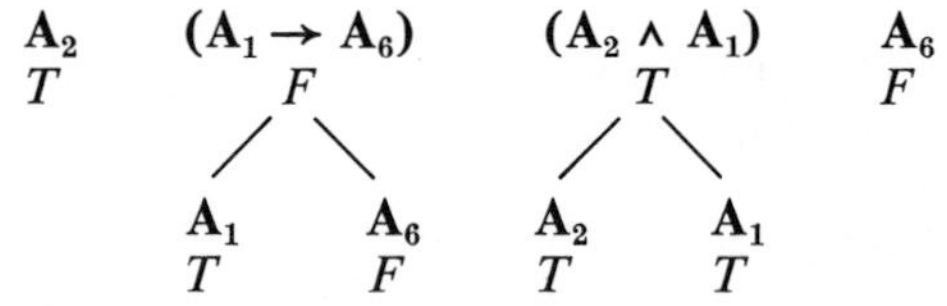

Working from the bottom up, we can assign to each vertex β of the tree the value $\bar{v}(\beta)$. So as a first step we compute

$$\bar{v}((\mathbf{A}_1 \rightarrow \mathbf{A}_6)) = F \quad \text{and} \quad \bar{v}((\mathbf{A}_2 \wedge \mathbf{A}_1)) = T.$$

Next we compute $\bar{v}((\mathbf{A}_2 \rightarrow (\mathbf{A}_1 \rightarrow \mathbf{A}_6))) = F$, and so forth. Finally, at the top of the tree we arrive at $\bar{v}(\alpha) = T$.

Actually this computation can be carried out with much less writing. First, the tree can be given in starker form:

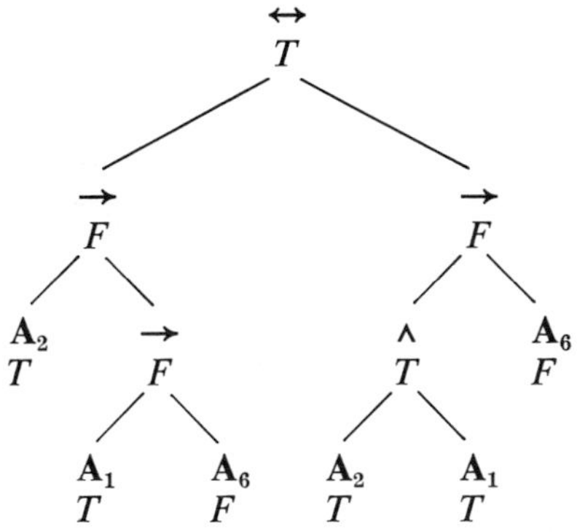

And even this can be compressed into a single line (with the parentheses restored):

$$\begin{array}{c}((\mathbf{A}_2 \to (\mathbf{A}_1 \to \mathbf{A}_6)) \leftrightarrow ((\mathbf{A}_2 \wedge \mathbf{A}_1) \to \mathbf{A}_6)). \\ T\ \ F\ \ T\ \ F\ F\quad T\quad T\ \ T\ T\quad F\ F\end{array}$$

Theorem 13A For any truth assignment v for a set $\mathscr{S}$ there is a unique function $\bar{v} : \bar{\mathscr{S}} \to \{T, F\}$ meeting the aforementioned conditions 0–5.

This theorem follows from the recursion theorem of Section 1.2 and the unique readability theorem of Section 1.4. But it should already seem extremely plausible, especially in light of the preceding example. In proving the existence of $\bar{v}$, the crucial issue will, in effect, be the uniqueness of the trees mentioned in the example.

We say that a truth assigment v *satisfies* φ iff $\bar{v}(\varphi) = T$. (Of course for this to happen, the domain of v must contain all sentence symbols in φ.) Now consider a set Σ of wffs (thought of as hypotheses) and another wff τ (thought of as a possible conclusion).

Definition Σ *tautologically implies* τ ($\Sigma \models \tau$) iff every truth assignment for the sentence symbols in Σ and τ which satisfies every member of Σ also satisfies τ.

This definition reflects our intuitive feeling that a conclusion follows from a set of hypotheses, if the assumption that the hypotheses are true guarantees that the conclusion is true.

Several special cases of the concept of tautological implication deserve mention. First, take the special case in which Σ is the empty set $\varnothing$. Observe that it is vacuously true that any truth assignment satisfies every member of $\varnothing$. (How could this fail? Only if there was some unsatisfied member of $\varnothing$, which is absurd.) Hence we are left with: $\varnothing \models \tau$ iff every truth assignment (for the sentence symbols in τ) satisfies τ. In this case we say that τ is a *tautology* ($\models \tau$). For example, the wff $\mathbf{((A_2 \rightarrow (A_1 \rightarrow A_6)) \leftrightarrow ((A_2 \wedge A_1) \rightarrow A_6))}$ was in a recent example found to be satisfied by one of the eight possible truth assignments for $\{\mathbf{A_1}, \mathbf{A_2}, \mathbf{A_6}\}$. In fact, it is satisfied by the other seven as well, and hence is a tautology.

Another special case is that in which no truth assignment satisfies every member of Σ. Then for any τ, it is vacuously true that $\Sigma \models \tau$. For example,

$$\{\mathbf{A}, (\neg \mathbf{A})\} \models \mathbf{B}.$$

There is no deep principle involved here; it is just a by-product of our definitions.

EXAMPLE $\{\mathbf{A}, (\mathbf{A} \rightarrow \mathbf{B})\} \models \mathbf{B}$. There are four truth assignments for $\{\mathbf{A}, \mathbf{B}\}$. It is easy to check that only one of these four satisfies both $\mathbf{A}$ and $(\mathbf{A} \rightarrow \mathbf{B})$, namely the v for which $v(\mathbf{A}) = v(\mathbf{B}) = T$. This v also satisfies $\mathbf{B}$.

If Σ is singleton $\{\sigma\}$, then we write "$\sigma \models \tau$" in place of "$\{\sigma\} \models \tau$." If both $\sigma \models \tau$ and $\tau \models \sigma$, then σ and τ are said to be *tautologically equivalent* ($\sigma \models\!\!\!\dashv \tau$). For example, in Section 1.0 we encountered the wffs $(\neg(\mathbf{C} \vee \mathbf{K}))$ and $((\neg \mathbf{C}) \wedge (\neg \mathbf{K}))$ as alternative translations of an English sentence. We can now assert that they are tautologically equivalent.

We can state here a nontrivial fact whose proof will be given later (in Section 1.7).

Compactness Theorem Let Σ be an infinite set of wffs such that for any finite subset Σ_0 of Σ, there is a truth assignment which satisfies every member of Σ_0. Then there is a truth assignment which satisfies every member of Σ.

This theorem can be restated more simply as: If every finite subset of Σ is satisfiable, then Σ itself is satisfiable. (The reader familiar with some general topology should try to discover why this is called "compactness theorem"; it does assert the compactness of a certain topological space. He should then prove the theorem for himself, using Tychonoff's theorem on product spaces.)

Truth tables

There is a systematic procedure, which we will now illustrate, for checking, given wffs $\sigma_1, \ldots, \sigma_k$, and τ, whether or not

$$\{\sigma_1, \ldots, \sigma_k\} \models \tau.$$

In particular (when $k = 0$), the procedure will, given a wff, decide whether or not it is a tautology.

As a first example, we can show that

$$(\neg(\mathbf{A} \wedge \mathbf{B})) \models ((\neg\mathbf{A}) \vee (\neg\mathbf{B})).$$

To do this, we consider all truth assignments for $\{\mathbf{A}, \mathbf{B}\}$. There are four such assignments; in general there are 2^n truth assignments for a set of n sentence symbols. The four can be listed in a table:

A	**B**
T	*T*
T	*F*
F	*T*
F	*F*

This table can then be expanded to include $(\neg(\mathbf{A} \wedge \mathbf{B}))$ and $((\neg\mathbf{A}) \vee (\neg\mathbf{B}))$. For each formula we compute the T's and F's the way described before, writing the truth value under the correct connective (Table IV). (The two leftmost columns of Table IV are actually unnecessary.) We can now see from this table that all those truth assignments satisfying $(\neg(\mathbf{A} \wedge \mathbf{B}))$, and there are three such, also satisfy $((\neg\mathbf{A}) \vee (\neg\mathbf{B}))$. In fact, the converse holds also, and thus

$$(\neg(\mathbf{A} \wedge \mathbf{B})) \models\!\!\dashv ((\neg\mathbf{A}) \vee (\neg\mathbf{B})).$$

TABLE IV

A	**B**	$(\neg(\mathbf{A} \wedge \mathbf{B}))$	$((\neg\mathbf{A}) \vee (\neg\mathbf{B}))$
T	*T*	*F* *T T T*	*F T* *F* *F T*
T	*F*	*T* *T F F*	*F T* *T* *T F*
F	*T*	*T* *F F T*	*T F* *T* *F T*
F	*F*	*T* *F F F*	*T F* *T* *T F*

To show that $(\neg(\mathbf{A} \wedge \mathbf{B})) \not\models ((\neg\mathbf{A}) \wedge (\neg\mathbf{B}))$ we can construct the table as before. But only one line of the table is needed to establish that there is indeed a truth assignment satisfying $(\neg(\mathbf{A} \wedge \mathbf{B}))$ which fails to satisfy $((\neg\mathbf{A}) \wedge (\neg\mathbf{B}))$.

The more generally applicable a procedure it is, the less efficient it is likely to be. For example, to show that

$$\models((\mathbf{A} \vee (\mathbf{B} \wedge \mathbf{C})) \leftrightarrow ((\mathbf{A} \vee \mathbf{B}) \wedge (\mathbf{A} \vee \mathbf{C}))),$$

we could apply the truth-table method. But this requires eight lines (for the eight possible truth assignments for $\{\mathbf{A}, \mathbf{B}, \mathbf{C}\}$). With a little cleverness the tedium can be reduced:

$$\begin{array}{l}
((\mathbf{A} \vee (\mathbf{B} \wedge \mathbf{C})) \leftrightarrow ((\mathbf{A} \vee \mathbf{B}) \wedge (\mathbf{A} \vee \mathbf{C}))). \\
\quad T\ T \qquad\qquad\quad T \quad T\ T \quad\ T\ T\ T \\
\quad F\ F\ F\ F \qquad\ T \quad F\ F\ F\ F\ \ F \\
\quad F\ T\ T\ T\ T \quad T \quad F\ T\ T\ T\ \ F\ T\ T
\end{array}$$

In the first line we assumed only that $v(\mathbf{A}) = T$. Since that is enough information to obtain T for the wff, we assume in all later lines that $v(\mathbf{A}) = F$. In the second line we assume that $v(\mathbf{B}) = F$; this again lets us obtain T for the wff. So we may limit ourselves to the case $v(\mathbf{B}) = T$. Since the expression is symmetric in **B** and **C**, we may further suppose $v(\mathbf{C}) = T$. This leaves only the third line, whereupon we are done.

As an example of the nonuse of a sixteen-line table, consider the following tautology:

$$\begin{array}{l}
((((\mathbf{P} \wedge \mathbf{Q}) \rightarrow \mathbf{R}) \rightarrow \mathbf{S}) \rightarrow ((\mathbf{P} \rightarrow \mathbf{R}) \rightarrow \mathbf{S})). \\
\qquad\qquad\qquad\qquad\qquad\quad T \qquad\qquad\ T\ T \\
\qquad F \qquad T \qquad F\ F\ T \\
\quad T\ T\ T\ \ R\ R\ \bar{R}\ F\ \ T\ \ \ T\ R\ R\ \bar{R}\ F
\end{array}$$

Here in the first line we dispose of the case where $v(\mathbf{S}) = T$. In the second line we dispose of the case where either $v(\mathbf{P}) = F$ or $v(\mathbf{Q}) = F$. The third line incorporates the two remaining possibilities; here R is the truth value assigned to **R** and $\bar{R}$ is the opposite value.

For the above example it is possible to see directly why it is a tautology. The stronger the hypothesis, the weaker the conditional. Thus

$$(\mathbf{P} \wedge \mathbf{Q}) \models \mathbf{P},$$

$$(\mathbf{P} \rightarrow \mathbf{R}) \models ((\mathbf{P} \wedge \mathbf{Q}) \rightarrow \mathbf{R}),$$

$$(((\mathbf{P} \wedge \mathbf{Q}) \rightarrow \mathbf{R}) \rightarrow \mathbf{S}) \models ((\mathbf{P} \rightarrow \mathbf{R}) \rightarrow \mathbf{S}).$$

The problem of developing effective procedures which reduce the tedium is important for theorem proving by machine. Some of the programs here may require testing wffs of sentential logic having thousands of sentence symbols. Truth tables are far too cumbersome for anything this large. The problem of developing highly efficient methods is a current area of research in computer science.

A SELECTED LIST OF TAUTOLOGIES

1. Associative and commutative laws for $\wedge, \vee, \leftrightarrow$.
2. Distributive laws:

$$((\mathbf{A} \wedge (\mathbf{B} \vee \mathbf{C})) \leftrightarrow ((\mathbf{A} \wedge \mathbf{B}) \vee (\mathbf{A} \wedge \mathbf{C}))).$$
$$((\mathbf{A} \vee (\mathbf{B} \wedge \mathbf{C})) \leftrightarrow ((\mathbf{A} \vee \mathbf{B}) \wedge (\mathbf{A} \vee \mathbf{C}))).$$

3. Negation:

$$((\neg(\neg \mathbf{A})) \leftrightarrow \mathbf{A}).$$
$$((\neg(\mathbf{A} \rightarrow \mathbf{B})) \leftrightarrow (\mathbf{A} \wedge (\neg \mathbf{B}))).$$
$$((\neg(\mathbf{A} \leftrightarrow \mathbf{B})) \leftrightarrow ((\mathbf{A} \wedge (\neg \mathbf{B})) \vee ((\neg \mathbf{A}) \wedge \mathbf{B}))).$$

De Morgan's laws:

$$((\neg(\mathbf{A} \wedge \mathbf{B})) \leftrightarrow ((\neg \mathbf{A}) \vee (\neg \mathbf{B}))).$$
$$((\neg(\mathbf{A} \vee \mathbf{B})) \leftrightarrow ((\neg \mathbf{A}) \wedge (\neg \mathbf{B}))).$$

4. Other

Excluded middle:	$(\mathbf{A} \vee (\neg \mathbf{A}))$.
Contradiction:	$(\neg(\mathbf{A} \wedge (\neg \mathbf{A})))$.
Contraposition:	$((\mathbf{A} \rightarrow \mathbf{B}) \leftrightarrow ((\neg \mathbf{B}) \rightarrow (\neg \mathbf{A})))$.
Exportation:	$(((\mathbf{A} \wedge \mathbf{B}) \rightarrow \mathbf{C}) \leftrightarrow (\mathbf{A} \rightarrow (\mathbf{B} \rightarrow \mathbf{C})))$.

EXERCISES

1. Show that neither of the following two formulas tautologically implies the other:

$$(\mathbf{A} \leftrightarrow (\mathbf{B} \leftrightarrow \mathbf{C})),$$
$$((\mathbf{A} \wedge (\mathbf{B} \wedge \mathbf{C})) \vee ((\neg \mathbf{A}) \wedge ((\neg \mathbf{B}) \wedge (\neg \mathbf{C})))).$$

2. Is $(((\mathbf{P} \rightarrow \mathbf{Q}) \rightarrow \mathbf{P}) \rightarrow \mathbf{P})$ a tautology?

3. Show that

(a) $\Sigma ; \alpha \models \beta$ iff $\Sigma \models (\alpha \rightarrow \beta)$.
(b) $\alpha \models\!\!\!\!\!\dashv \beta$ iff $\models (\alpha \leftrightarrow \beta)$.
(Recall that $\Sigma ; \alpha = \Sigma \cup \{\alpha\}$.)

4. Prove or refute each of the following assertions:

(a) If either $\Sigma \models \alpha$ or $\Sigma \models \beta$, then $\Sigma \models (\alpha \vee \beta)$.
(b) If $\Sigma \models (\alpha \vee \beta)$, then either $\Sigma \models \alpha$ or $\Sigma \models \beta$.

5. (a) Show that if v_1 and v_2 are truth assignments which agree on all the sentence symbols in the wff α, then $\bar{v}_1(\alpha) = \bar{v}_2(\alpha)$.

(b) Let $\mathcal{S}$ be a set of sentence symbols which includes those in Σ and τ (and possibly more). Show that $\Sigma \models \tau$ iff every truth assignment for $\mathcal{S}$ which satisfies every member of Σ also satisfies τ.

6. You are in a land inhabited by people who either always tell the truth or always tell falsehoods. You come to a fork in the road and you need to know which fork leads to the capital. There is a local resident there, but he has time only to reply to one yes-or-no question. What one question should you ask so as to learn which fork to take?

7. (Substitution) Consider a sequence $\alpha_1, \alpha_2, \ldots$ of wffs. For a wff φ, let φ^* be the result of replacing the sentence symbol $\mathbf{A}_n$ by α_n, for $n = 1, 2, \ldots$.

(a) Let v be a truth assignment for the set of all sentence symbols; define u to be the truth assignment for which $u(\mathbf{A}_n) = \bar{v}(\alpha_n)$. Show that $\bar{u}(\varphi) = \bar{v}(\varphi^*)$.

(b) Show that if φ is a tautology, then so is φ^*.

8. (Duality) Let α be a wff whose only connective symbols are $\wedge$, $\vee$, and $\neg$. Let α^* be the result of interchanging $\wedge$ and $\vee$ and replacing each sentence symbol by its negation. Show that α^* is tautologically equivalent to $(\neg\alpha)$.

9. Say that a set Σ_1 of wffs is *equivalent* to a set Σ_2 of wffs iff for any wff α, $\Sigma_1 \models \alpha$ iff $\Sigma_2 \models \alpha$. A set Σ is *independent* iff no member of Σ is tautologically implied by the remaining members in Σ. Show that

(a) A finite set of wffs has an independent equivalent subset.
(b) A infinite set need not have an independent equivalent subset.
*(c) Let $\Sigma = \{\sigma_0, \sigma_1, \ldots\}$; show that there is an independent equivalent set Σ'.

10. Show that a truth assignment v satisfies the wff

$$(\cdots(\mathbf{A}_1 \leftrightarrow \mathbf{A}_2) \leftrightarrow \cdots \leftrightarrow \mathbf{A}_n)$$

iff $v(\mathbf{A}_i) = F$ for an even number of i's, $1 \leq i \leq n$.

11. There are three suspects for a murder: Adams, Brown, and Clark. Adams says "I didn't do it. The victim was an old acquaintance of Brown's. But Clark hated him." Brown states "I didn't do it. I didn't even know the guy. Besides I was out of town all that week." Clark says "I didn't do it. I saw both Adams and Brown downtown with the victim that day; one of them must have done it." Assume that the two innocent men are telling the truth, but that the guilty man might not be. Who did it?

§ 1.4 UNIQUE READABILITY

The purpose of this section is to prove that we have used enough parentheses to eliminate any ambiguity in analyzing wffs. (The existence of the extension $\bar{v}$ of a truth assignment v will hinge on this lack of ambiguity.[1])

It is instructive to consider the result of not having parentheses at all. The resulting ambiguity is illustrated by the wff

$$\mathbf{A}_1 \vee \mathbf{A}_2 \wedge \mathbf{A}_3,$$

which can be formed in two ways, corresponding to $((\mathbf{A}_1 \vee \mathbf{A}_2) \wedge \mathbf{A}_3)$ and to $(\mathbf{A}_1 \vee (\mathbf{A}_2 \wedge \mathbf{A}_3))$. If $v(\mathbf{A}_1) = T$ and $v(\mathbf{A}_3) = F$, then there is an unresolvable conflict which arises in trying to compute $\bar{v}(\mathbf{A}_1 \vee \mathbf{A}_2 \wedge \mathbf{A}_3)$.

We must show that with our parentheses this type of ambiguity does not arise but that on the contrary each wff is formed in a unique way. There is one sense in which this fact is unimportant: If it failed, we would simply change notation until it was true. For example, instead of building formulas by means of concatenation, we could have used ordered pairs and triples: $\langle \neg, \alpha \rangle$, $\langle \alpha, \wedge, \beta \rangle$, etc. (This is, in fact, a tidy, but untraditional, method.) The unique readability theorem would then be immediate. But we do not have to resort to this device, and we will now prove that we do not.

Lemma 14A Every wff has the same number of left as right parentheses.

Proof This was done as an example at the end of Section 1.1. ∎

[1] If the reader has already accepted the existence of $\bar{v}$, then he may omit almost all of this section. The final subsection, on omitting parentheses, should still be read.

Lemma 14B Any proper initial segment of a wff contains an excess of left parentheses. Thus no proper initial segment of a wff can itself be a wff.

Proof We show that the set S of wffs possessing the desired property (that proper initial segments are left-heavy) is inductive. A wff consisting of a sentence symbol alone has no proper initial segments and hence is in S vacuously. To verify that S is closed under $\mathscr{E}_\wedge$, consider α and β in S. The proper initial segments are

1. (.
2. $(\alpha_0$, where α_0 is a proper initial segment of α.
3. $(\alpha$.
4. $(\alpha \wedge$.
5. $(\alpha \wedge \beta_0$, where β_0 is a proper initial segment of β.
6. $(\alpha \wedge \beta$.

By applying the inductive hypothesis that α and β are in S (in cases 2 and 5), we obtain the desired conclusion. ■

Unique Readability Theorem The five formula-building operations, when restricted to the set of wffs,

(a) have ranges which are disjoint from each other and from the set of sentence symbols, and
(b) are one-to-one.

In the language of Section 1.2, this asserts that the set of wffs is *freely* generated from the set of sentence symbols by the five operations.

Proof To show that the restriction of $\mathscr{E}_\wedge$ is one-to-one, suppose that

$$(\alpha \wedge \beta) = (\gamma \wedge \delta),$$

where α, β, γ, and δ are wffs. Delete the first symbol of each sequence, obtaining

$$\alpha \wedge \beta) = \gamma \wedge \delta).$$

Then we must have $\alpha = \gamma$, lest one be a proper initial segment of the other (in contradiction with the preceding lemma). And then it follows at once that $\beta = \delta$. The same argument applies to $\mathscr{E}_\vee$, $\mathscr{E}_\rightarrow$, and $\mathscr{E}_\leftrightarrow$; for $\mathscr{E}_\neg$ a simpler argument suffices.

A similar line of reasoning tells us that the operations have disjoint ranges. For example, if

$$(\alpha \wedge \beta) = (\gamma \rightarrow \delta),$$

where α, β, γ, and δ are wffs, then as in the above paragraph we have $\alpha = \gamma$. But that implies that $\wedge = \rightarrow$, contradicting the fact that our symbols are distinct. Hence $\mathscr{E}_\wedge$ and $\mathscr{E}_\rightarrow$ (when restricted to wffs) have disjoint ranges. Similarly for any two binary connectives.

The remaining cases are simple. If $(\neg\alpha) = (\beta \wedge \gamma)$, then β begins with $\neg$, which no wff does. No sentence symbol is a sequence of symbols beginning with (. ■

Now let us return to the question of extending a truth assignment v to $\bar{v}$. First consider the special case where v is a truth assignment for the set of all sentence symbols. Then by applying the unique readability theorem and the recursion theorem (of Section 1.2) we conclude that there is a unique extension $\bar{v}$ to the set of all wffs with the desired properties.

Next take the general case where v is a truth assignment for a set $\mathscr{S}$ of sentence symbols. The set $\bar{\mathscr{S}}$ generated from $\mathscr{S}$ by the five formula-building operations is freely generated, as a consequence of the unique readability theorem. So by the recursion theorem there is a unique extension $\bar{v}$ of v to that set, having the desired properties.

An algorithm

Our proof of the unique readability theorem can be converted from a proof-by-contradiction into an algorithm which, given a wff, will produce its unique family tree. The algorithm has the further advantage that if it is given an expression which is not a wff, it will detect that fact.

Assume that we are given an expression. Initially it is the only vertex in the tree (and so is the minimum one), but as the procedure progresses the tree will grow downward from the given expression.

1. If all minimal vertices have sentence symbols, then the procedure is completed. Otherwise, select a minimal vertex which has an expression which is not a sentence symbol.

2. The first symbol must[1] be (. If the second symbol is the negation symbol, skip to step 4.

3. Scan the expression from the left until first reaching (α, where α is an expression having a balance between left and right parentheses.[2] Then α is the first constituent. The next symbol must[1] be $\wedge$, $\vee$, $\rightarrow$, or $\leftrightarrow$ and is the principal connective. The remainder of the expression, β), must[1]

[1] If not, then the original expression was not a wff.

[2] If the end of the expression is reached before finding such an α, then the original expression is not a wff.

consist of an expression β and a right parenthesis. The second constituent is β. This completes the decomposition of selected expression; return to step 1.

4. If the second symbol is the negation symbol, then that is the principal connective. The remainder of the expression, $\beta)$, must[1] consist of an expression β and a right parenthesis. β is the constituent. This completes the decomposition of the selected expression; return to step 1.

Now for some comments about the algorithm. First we claim that given any expression, the procedure halts after a finite number of steps. This is because any vertex contains a shorter expression than the one above it, so the depth of the tree is bounded by the length of the given expression.

Second, we should remark on the uniqueness of the procedure. For example, in step 3 we arrive at an expression α. We could not use less than α for a constituent, for it would not have a balance between left and right parentheses. We could not use more than α, for that would have the proper initial segment α that was balanced. Thus α is forced upon us. And then the choice of the principal connective is inevitable.

It is clear that if our algorithm is given a wff, it will not use the footnotes requiring the expression to be rejected. Conversely, suppose the expression given is such that the procedure does not reject it. Then, by working our way up the resulting tree, we discover inductively that every vertex has a wff, including the top vertex (which has the given expression).

We can also use the tree to see how $\bar{v}(\alpha)$ is obtained. For any wff α there is a unique tree constructing it. By working our way up this tree, we can unambiguously arrive at a value for $\bar{v}(\alpha)$.

Polish notation

It is possible to avoid both ambiguity and parentheses. This can be done by a very simple device. Instead of, for example, $(\alpha \land \beta)$ we use $\land\alpha\beta$. Let the set of P-wffs be the set generated from the sentence symbols by the five operations

$$\mathscr{D}_{\neg}(\alpha) = \neg\alpha, \qquad \mathscr{D}_{\lor}(\alpha, \beta) = \lor\alpha\beta,$$
$$\mathscr{D}_{\land}(\alpha, \beta) = \land\alpha\beta, \qquad \mathscr{D}_{\rightarrow}(\alpha, \beta) = \rightarrow\alpha\beta,$$
$$\mathscr{D}_{\leftrightarrow}(\alpha, \beta) = \leftrightarrow\alpha\beta.$$

For example, one P-wff is

$$\rightarrow\land\mathbf{A}\mathbf{D}\lor\neg\mathbf{B}\leftrightarrow\mathbf{C}\mathbf{B}.$$

[1] If not, then the original expression was not a wff.

Here the need for an algorithm to analyze the structure is quite apparent. Even for the short example above, it requires some thought to see how it was built up. We will give a unique readability theorem for such expressions in Section 2.3.

This way of writing formulas (but with N, K, A, C, and E in place of $\neg$, $\wedge$, $\vee$, $\rightarrow$, and $\leftrightarrow$, respectively) was introduced by the Polish logician Łukasiewicz. The notation is well suited to automatic processing. Computer compiler programs often begin by converting the formulas given them into Polish notation.

Omitting parentheses

Hereafter when naming wffs, we will not feel compelled to mention explicitly every parenthesis. To establish a more compact notation, we now adopt the following conventions:

1. The outermost parentheses need not be explicitly mentioned. For example, when we write "$\mathbf{A} \wedge \mathbf{B}$" we are referring to $(\mathbf{A} \wedge \mathbf{B})$.

2. The negation symbol applies to as little as possible. For example, $\neg\mathbf{A} \wedge \mathbf{B}$ is $(\neg\mathbf{A}) \wedge \mathbf{B}$, i.e., $((\neg\mathbf{A}) \wedge \mathbf{B})$. It is not the same as $(\neg(\mathbf{A} \wedge \mathbf{B}))$.

3. The conjunction and disjunction symbols apply to as little as possible, given that convention 2 is to be observed. For example,

$$\mathbf{A} \wedge \mathbf{B} \rightarrow \neg\mathbf{C} \vee \mathbf{D} \text{ is } ((\mathbf{A} \wedge \mathbf{B}) \rightarrow ((\neg\mathbf{C}) \vee \mathbf{D})).$$

4. Where one connective symbol is used repeatedly, grouping is to the right:

$$\alpha \wedge \beta \wedge \gamma \text{ is } \alpha \wedge (\beta \wedge \gamma),$$
$$\alpha \rightarrow \beta \rightarrow \gamma \text{ is } \alpha \rightarrow (\beta \rightarrow \gamma).$$

It must be admitted that these conventions violate what was said on page 18 about naming expressions. We can get away with this only because we no longer have any interest in naming expressions which are not wffs.

EXERCISES

1. Rewrite the tautologies in the "selected list" at the end of Section 1.3, but using the conventions of the present section to minimize the number of parentheses.

2. Give an example of wffs α and β and expressions γ and δ such that $(\alpha \wedge \beta) = (\gamma \wedge \delta)$ but $\alpha \neq \gamma$.

3. Suppose that we modify our definition of wff by omitting all right parentheses. Thus instead of

$$((\mathbf{A} \wedge (\neg \mathbf{B})) \rightarrow (\mathbf{C} \vee \mathbf{D}))$$

we use

$$((\mathbf{A} \wedge (\neg \mathbf{B} \rightarrow (\mathbf{C} \vee \mathbf{D}.$$

Show that we still have unique readability. *Suggestion*: These expressions have the same number of parentheses as connective symbols.

4. The English language has a tendency to use two-part connectives: "both . . . and . . . ," "either . . . or . . . ," "if . . . , then . . . ." How does this affect unique readability in English?

5. We have given an algorithm for analyzing a wff by constructing its tree from the top down. There are also ways of constructing the tree from the bottom up. This can be done by looking through the formula for innermost pairs of parentheses. Give a complete description of an algorithm of this sort.

§ 1.5 SENTENTIAL CONNECTIVES

We have thus far employed five sentential connective symbols. Even in the absence of a general definition of "connective," it is clear that the five familiar ones are not the only ones possible. Would we gain anything by adding more connectives to the language? Would we lose anything by omitting some we already have?

In this section we make these questions precise and give some answers. First consider an informal example. We could expand the language by adding a three-place sentential connective symbol $\#$, called the majority symbol. We allow now as a wff the expression $(\#\alpha\beta\gamma)$ whenever α, β, and γ are wffs. In other words, we add a sixth formula-building operator to our list:

$$\mathscr{E}_{\#}(\alpha, \beta, \gamma) = (\#\alpha\beta\gamma).$$

Then we must give the interpretation of this symbol. That is, we must say how to compute $\bar{v}((\#\alpha\beta\gamma))$, given the values $\bar{v}(\alpha)$, $\bar{v}(\beta)$, and $\bar{v}(\gamma)$. We choose to define

$$\bar{v}((\#\alpha\beta\gamma)) \text{ is to agree with the majority of } \bar{v}(\alpha), \bar{v}(\beta), \bar{v}(\gamma).$$

We claim that this extension has gained us nothing, in the following precise sense: For any wff in the extended language, there is a tautologically equivalent wff in the original language. (On the other hand, the wff in the original language may be much longer than the wff in the extended language.) We will prove this (in a more general situation) below; here we just note that it relies on the fact that $(\#\alpha\beta\gamma)$ is tautologically equivalent to

$$(\alpha \wedge \beta) \vee (\alpha \wedge \gamma) \vee (\beta \wedge \gamma).$$

(We note parenthetically that our insistence that $\bar{v}((\#\alpha\beta\gamma))$ be calculable from $\langle \bar{v}(\alpha), \bar{v}(\beta), \bar{v}(\gamma)\rangle$ plays a definite role here. In everyday speech, there are unary operators like "it is possible that" or "I believe that." We can apply one of these operators to a sentence, producing a new sentence whose truth or falsity cannot be determined solely on the basis of the truth or falsity of the original one.)

In generalizing the foregoing example, the formal language will be more of a hindrance than a help. We can restate everything using only functions. Say that a k-place *Boolean function* is a function from $\{T, F\}^k$ into $\{T, F\}$. (A *Boolean function* is then anything which is a k-place Boolean function for some k. We stretch this slightly by permitting T and F themselves to 0-place Boolean functions.) Some sample Boolean functions are defined by the equations (where $X \in \{T, F\}$)

$$I_i^n(X_1, \ldots, X_n) = X_i,$$
$$N(T) = F, \qquad N(F) = T,$$
$$K(T, T) = T, \qquad K(F, X) = K(X, F) = F,$$
$$A(F, F) = F, \qquad A(T, X) = A(X, T) = T,$$
$$C(T, F) = F, \qquad C(F, X) = C(X, T) = T,$$
$$E(X, X) = T, \qquad E(T, F) = E(F, T) = F.$$

From a wff α we can extract a Boolean function. For example, if α is the wff $\mathbf{A}_1 \wedge \mathbf{A}_2$, then we can make a table, Table V. The 2^2 lines of the table correspond to the 2^2 truth assignments for $\{\mathbf{A}_1, \mathbf{A}_2\}$. For each of the 2^2 pairs $\vec{X}$, we set $B_\alpha(\vec{X})$ equal to the truth value α receives when its sentence symbols are given the values indicated by $\vec{X}$.

TABLE V

$\mathbf{A}_1$	$\mathbf{A}_2$	$\mathbf{A}_1 \wedge \mathbf{A}_2$	
F	F	F	$B_\alpha(F, F) = F$
F	T	F	$B_\alpha(F, T) = F$
T	F	F	$B_\alpha(T, F) = F$
T	T	T	$B_\alpha(T, T) = T$

In general, suppose that α is a wff whose sentence symbols are at most $\mathbf{A}_1, \ldots, \mathbf{A}_n$. We define an n-place Boolean function B_α^n (or just B_α if n seems unnecessary), the Boolean function *realized* by α, by

$$B_\alpha^n(X_1, \ldots, X_n) = \text{the truth value given to } \alpha \text{ when } \mathbf{A}_1, \ldots, \mathbf{A}_n \text{ are given the values } X_1, \ldots, X_n.$$

Or, in other words, $B_\alpha^n(X_1, \ldots, X_n) = \bar{v}(\alpha)$, where v is the truth assignment for $\{\mathbf{A}_1, \ldots, \mathbf{A}_n\}$ for which $v(\mathbf{A}_i) = X_i$. Thus B_α^n comes from looking at $\bar{v}(\alpha)$ as a function of v, with α fixed.

For example, the Boolean functions listed previously are obtainable in this way:

$$\begin{aligned} I_i^n &= B_{\mathbf{A}_i}^n, \\ N &= B_{\neg\mathbf{A}_1}^1, \\ K &= B_{\mathbf{A}_1 \wedge \mathbf{A}_2}^2, \\ A &= B_{\mathbf{A}_1 \vee \mathbf{A}_2}^2, \\ C &= B_{\mathbf{A}_1 \to \mathbf{A}_2}^2, \\ E &= B_{\mathbf{A}_1 \leftrightarrow \mathbf{A}_2}^2. \end{aligned}$$

From these functions we can compose others. For example,

$$B_{\neg\mathbf{A}_1 \vee \neg\mathbf{A}_2}^2(X_1, X_2) = A(N(I_1^2(X_1, X_2)), N(I_2^2(X_1, X_2))).$$

(The right-hand side of this equation can be compared with the result of putting $\neg\mathbf{A}_1 \vee \neg\mathbf{A}_2$ into Polish notation.) We will shortly come to the question whether every Boolean function is obtainable in this fashion.

As the theorem below states, in shifting attention from wffs to the Boolean functions they realize, we have in effect identified tautologically equivalent wffs. Impose an ordering on $\{T, F\}$ by defining $F < T$. (If $T = 1$ and $F = 0$, then this is the natural order.)

Theorem 15A Let α and β be wffs whose sentence symbols are among $\mathbf{A}_1, \ldots, \mathbf{A}_n$. Then

(a) $\alpha \models \beta$ iff for all $\vec{X} \in \{T, F\}^n$, $B_\alpha(\vec{X}) \leq B_\beta(\vec{X})$.

(b) $\alpha \models\!\!\!\dashv \beta$ iff $B_\alpha = B_\beta$.

(c) $\models \alpha$ iff ran $B_\alpha = \{T\}$.

Proof of (a) $\alpha \models \beta$ iff for all 2^n truth assignments v for $\mathbf{A}_1, \ldots, \mathbf{A}_n$, whenever $\bar{v}(\alpha) = T$, then also $\bar{v}(\beta) = T$. (This is true even if the sentence symbols in α and β do not include all of $\mathbf{A}_1, \ldots, \mathbf{A}_n$; cf. Exercise 5 of

Section 1.3.) Thus

$$
\begin{aligned}
\alpha \models \beta \quad & \text{iff for all } 2^n \text{ assignments, } v,\ \bar{v}(\alpha) = T \Rightarrow \bar{v}(\beta) = T,\\
& \text{iff for all } 2^n\ n\text{-tuples } \vec{X},\ B^n_\alpha(\vec{X}) = T \Rightarrow B^n_\beta(\vec{X}) = T,\\
& \text{iff for all } 2^n\ n\text{-tuples } \vec{X},\ B^n_\alpha(\vec{X}) \leq B^n_\beta(\vec{X}),
\end{aligned}
$$

where $F < T$. ■

In addition to identifying tautologically equivalent wffs, we have freed ourselves from the formal language. We are now at liberty to consider any Boolean function, whether it is realized by a wff or not. But this freedom is only apparent:

Theorem 15B Let G be an n-place Boolean function, $n \geq 1$. We can find a wff α such that $G = B^n_\alpha$, i.e., such that α realizes the function G.

Proof Case I: ran $G = \{F\}$. Let $\alpha = \mathbf{A}_1 \wedge \neg\mathbf{A}_1$.

Case II: Otherwise there are k points at which G has the value T, $k > 0$. List these:

$$
\begin{aligned}
\vec{X}_1 &= \langle X_{11}, X_{12}, \ldots, X_{1n} \rangle,\\
\vec{X}_2 &= \langle X_{21}, X_{22}, \ldots, X_{2n} \rangle,\\
&\cdots\\
\vec{X}_k &= \langle X_{k1}, X_{k2}, \ldots, X_{kn} \rangle.
\end{aligned}
$$

Let

$$
\begin{aligned}
\beta_{ij} &= \begin{cases} \mathbf{A}_j & \text{if } X_{ij} = T,\\ (\neg\mathbf{A}_j) & \text{if } X_{ij} = F, \end{cases}\\
\gamma_i &= \beta_{i1} \wedge \cdots \wedge \beta_{in},\\
\alpha &= \gamma_1 \vee \gamma_2 \vee \cdots \vee \gamma_k.
\end{aligned}
$$

We claim that $G = B^n_\alpha$.

At this point it might be helpful to consider a concrete example. Let G be the three-place Boolean function as follows:

$$
\begin{aligned}
G(F, F, F) &= F,\\
G(F, F, T) &= T,\\
G(F, T, F) &= T,\\
G(F, T, T) &= F,\\
G(T, F, F) &= T,\\
G(T, F, T) &= F,\\
G(T, T, F) &= F,\\
G(T, T, T) &= T.
\end{aligned}
$$

Then the list of triples at which G assumes the value T has four members:

$$\begin{array}{ll} FFT & \neg \mathbf{A}_1 \wedge \neg \mathbf{A}_2 \wedge \mathbf{A}_3, \\ FTF & \neg \mathbf{A}_1 \wedge \mathbf{A}_2 \wedge \neg \mathbf{A}_3, \\ TFF & \mathbf{A}_1 \wedge \neg \mathbf{A}_2 \wedge \neg \mathbf{A}_3, \\ TTT & \mathbf{A}_1 \wedge \mathbf{A}_2 \wedge \mathbf{A}_3. \end{array}$$

To the right of each triple above is written the corresponding conjunction γ_i. Then α is the formula

$$(\neg \mathbf{A}_1 \wedge \neg \mathbf{A}_2 \wedge \mathbf{A}_3) \vee (\neg \mathbf{A}_1 \wedge \mathbf{A}_2 \wedge \neg \mathbf{A}_3) \vee (\mathbf{A}_1 \wedge \neg \mathbf{A}_2 \wedge \neg \mathbf{A}_3) \vee (\mathbf{A}_1 \wedge \mathbf{A}_2 \wedge \mathbf{A}_3).$$

Notice how α lists explicitly the triples at which G assumes the value T.

To return to the proof of the theorem, note first that $B_\alpha^n(\vec{X}_i) = T$ for $1 \le i \le k$. (For the truth assignment corresponding to $\vec{X}_i$ satisfies γ_i and hence satisfies α.) On the other hand, only one truth assignment for $\{\mathbf{A}_1, \ldots, \mathbf{A}_n\}$ can satisfy γ_i, whence only k such truth assignments can satisfy α. Hence $B_\alpha^n(\vec{Y}) = F$ for the $2^n - k$ other n-tuples $\vec{Y}$. Thus in all cases, $B_\alpha^n(\vec{Y}) = G(\vec{Y})$. ■

From this theorem we know that every Boolean function is realizable. Of course the α which realizes G is not unique; any tautologically equivalent wff will also realize the same function. It is sometimes of interest to choose α to be as short as possible. (In the example done above, the wff

$$\mathbf{A}_1 \leftrightarrow \mathbf{A}_2 \leftrightarrow \mathbf{A}_3$$

also realizes G.)

As a corollary to the above theorem, we may conclude that we have enough (in fact, more than enough) sentential connectives. For suppose that we expand the language by adding some exotic new sentential connectives (such as the majority connective discussed at the beginning of this section). Any wff φ of this expanded language realizes a Boolean function B_φ^n. By the above theorem we have a wff α of the original language such that $B_\varphi^n = B_\alpha^n$. Hence φ and α are tautologically equivalent, by Theorem 15A.

In fact, the proof shows that α can be of a rather special form. For one thing, the only sentential connective symbols in α are $\wedge$, $\vee$, and $\neg$. Furthermore, α is in so-called *disjunctive normal form*. That is,

$$\alpha = \gamma_1 \vee \cdots \vee \gamma_k,$$

where

$$\gamma_i = \beta_{i1} \wedge \cdots \wedge \beta_{in_i}$$

and each β_{ij} is a sentence symbol or a negation of a sentence symbol. (The advantages of wffs in disjunctive normal form stem from the fact that they explicitly list the truth assignments satisfying the formula.) Thus we have

Corollary 15C For any wff φ, we can find a tautologically equivalent wff α in disjunctive normal form.

Because every function $G : \{T, F\}^n \to \{T, F\}$ for $n \geq 1$ can be realized by a wff using only the connective symbols in $\{\wedge, \vee, \neg\}$, we say that the set $\{\wedge, \vee, \neg\}$ is *complete*. (Actually the completeness is more a property of the Boolean functions K, A, and N which correspond to these symbols. But the above terminology is convenient.) Once we have a complete set of connectives, we know that any wff is tautologically equivalent to one all of whose connectives are in that set. (For given any wff φ, we can make α using those connectives and realizing B_φ. Then $\alpha \models\!\!\!\dashv \varphi$.) The completeness of $\{\wedge, \vee, \neg\}$ can be improved upon:

Theorem 15D Both $\{\neg, \wedge\}$ and $\{\neg, \vee\}$ are complete.

Proof We must show that any Boolean function G can be realized by a wff using only, in the first case, $\{\neg, \wedge\}$. We begin with a wff α using $\{\neg, \wedge, \vee\}$, which realizes G. It suffices to find a tautologically equivalent α' which uses only $\{\neg, \wedge\}$. For this we use De Morgan's law:

$$\beta \vee \gamma \models\!\!\!\dashv \neg(\neg\beta \wedge \neg\gamma).$$

By applying this repeatedly, we can completely eliminate $\vee$ from α.

(More formally, we can prove by induction on α that there is a tautologically equivalent α' in which only the connectives $\wedge$, $\neg$ occur. Two cases in the inductive step are

Case $\neg$: If α is $(\neg\beta)$, then let α' be $(\neg\beta')$.

Case $\vee$: If α is $(\beta \vee \gamma)$, then let α' be $\neg(\neg\beta' \wedge \neg\gamma')$. Since β' and γ' are tautologically equivalent to β and γ, respectively,

$$\begin{aligned} \alpha' &= \neg(\neg\beta' \wedge \neg\gamma') \\ &\models\!\!\!\dashv \neg(\neg\beta \wedge \neg\gamma) \\ &\models\!\!\!\dashv \beta \vee \gamma \\ &= \alpha. \end{aligned}$$

In future proofs that a set of connectives is complete, this induction will be omitted. Instead we will just give, for example, the method of simulating $\vee$ by using $\neg$ and $\wedge$.) ■

Showing that a certain set of connectives is *not* complete is usually more difficult than showing that one is complete. The basic method is first to show (usually by induction) that for any wff α using only those connectives, the function B_α^n has some peculiarity, and second to show that some Boolean function lacks that peculiarity.

EXAMPLE $\{\wedge, \rightarrow\}$ is not complete.

Proof The idea is that with these connectives, if the sentence symbols are assigned T, then the entire formula is assigned T. In particular, there is nothing tautologically equivalent to $\neg\mathbf{A}$.

In more detail, we can show by induction that for any wff α using only these connectives and having $\mathbf{A}$ as its only sentence symbol, we have $\mathbf{A} \models \alpha$. (In terms of functions, this says that $X \leq B_\alpha^1(X)$.) ■

For each n there are 2^{2^n} n-place Boolean functions. Hence if we identify a connective with its Boolean function (e.g., $\wedge$ with the function K mentioned before), we have 2^{2^n} n-ary connectives. We will now catalog these for $n \leq 2$.

0-ary connectives

There are two 0-place Boolean functions, T and F. For the corresponding connective symbols we take $\top$ and $\bot$. Now an n-ary connective symbol combines with n wffs to produce a new wff. When $n = 0$, we have that $\bot$ is a wff all by itself. It differs from the sentence symbols in that $\bar{v}(\bot) = F$ for every v; i.e., $\bot$ is a logical symbol always assigned the value F. Similarly, $\top$ is a wff, and $\bar{v}(\top) = T$ for every v. Then, for example, $\mathbf{A} \rightarrow \bot$ is a wff, tautologically equivalent to $\neg\mathbf{A}$, as can be seen from a two-line truth table.

Unary connectives

There are four unary connectives but only one of any interest. The interesting case is negation. The other three one-place Boolean functions are the identity function and the two constant functions.

Binary connectives

There are sixteen binary connectives, but only the last ten listed in Table VI are "really binary."

TABLE VI

Symbol	Equivalent	Remarks
	$\top$	two-place constant, essentially 0-ary
	$\bot$	two-place constant, essentially 0-ary
	A	projection, essentially unary
	B	projection, essentially unary
	$\neg$**A**	negation, essentially unary
	$\neg$**B**	negation, essentially unary
$\wedge$	**A** $\wedge$ **B**	and; if $T = 1$ and $F = 0$, then this gives multiplication in the field $\{0, 1\}$
$\vee$	**A** $\vee$ **B**	or
$\rightarrow$	**A** $\rightarrow$ **B**	conditional
$\leftrightarrow$	**A** $\leftrightarrow$ **B**	biconditional
$\leftarrow$	**B** $\leftarrow$ **A**	reversed conditional
$+$	(**A** $\vee$ **B**) $\wedge$ $\neg$(**A** $\wedge$ **B**)	exclusive or, "**A** or **B** and not both"; if $T = 1$ and $F = 0$, then this gives the usual addition (modulo 2) in the field $\{0, 1\}$
$\downarrow$	$\neg$(**A** $\vee$ **B**)	nor, "neither **A** nor **B**"
$\mid$	$\neg$(**A** $\wedge$ **B**)	nand, "not both **A** and **B**"; the symbol is called the Sheffer stroke
$<$	($\neg$**A**) $\wedge$ **B**	the usual ordering, where $F < T$
$>$	**A** $\wedge$ ($\neg$**B**)	the usual ordering, where $F < T$

Ternary connectives

There are 256 ternary connectives; 2 are essentially 0-ary, 6 ($= 2 \cdot \binom{3}{1}$) are essentially unary, and 30 ($= 10 \cdot \binom{3}{2}$) are essentially binary. This leaves 218 which are really ternary. We have thus far mentioned only the majority connective $\#$. There is, similarly, the minority connective. In Exercise 7 we encounter $+^3$, ternary addition modulo 2. $+^3\alpha\beta\gamma$ is assigned the value T iff an odd number of α, β, and γ are assigned T. This formula is equivalent both to $\alpha + \beta + \gamma$ and to $\alpha \leftrightarrow \beta \leftrightarrow \gamma$. Another ternary connective arises in Exercise 8.

EXAMPLE $\{\mid\}$ and $\{\downarrow\}$ are complete.

Proof, for $\mid$

$$\neg\alpha \models\!\!\!\dashv \alpha \mid \alpha$$
$$\alpha \vee \beta \models\!\!\!\dashv (\neg\alpha) \mid (\neg\beta).$$

Since $\{\neg, \vee\}$ is complete and $\neg$, $\vee$ can be simulated using only $\mid$, $\{\mid\}$ is complete. ■

EXAMPLE $\{\neg, \rightarrow\}$ is complete. In fact, of the ten connectives which are really binary, eight have the property of forming, when added to $\neg$, a complete set. The two exceptions are $+$ and $\leftrightarrow$; see Exercise 6.

EXAMPLE $\{\bot, \rightarrow\}$ is complete.

EXERCISES

1. Let G be the following three-place Boolean function:

$$\begin{aligned} G(F, F, F) &= T, & G(T, F, F) &= T, \\ G(F, F, T) &= T, & G(T, F, T) &= F, \\ G(F, T, F) &= T, & G(T, T, F) &= F, \\ G(F, T, T) &= F, & G(T, T, T) &= F. \end{aligned}$$

Find a wff, using at most the connectives $\vee$, $\wedge$, and $\neg$, which realizes G. Then find such a wff in which connective symbols occur at not more than five places.

2. Show that $|$ and $\downarrow$ are the only binary connectives which are complete by themselves.

3. Show that $\{\neg, \#\}$ is not complete.

4. Let M be the ternary minority connective. (Thus $\bar{v}(\mathsf{M}\alpha\beta\gamma)$ always disagrees with the majority of $\bar{v}(\alpha)$, $\bar{v}(\beta)$, and $\bar{v}(\gamma)$.) Show that

(a) $\{\mathsf{M}, \bot\}$ is complete.
(b) $\{\mathsf{M}\}$ is not complete.

5. Show that $\{\top, \bot, \neg, \leftrightarrow, +\}$ is not complete. *Suggestion*: Show that any wff α using these connectives and the sentence symbols **A** and **B** has an even number of T's among the four possible values of $\bar{v}(\alpha)$.

6. Show that $\{\wedge, \leftrightarrow, +\}$ is complete but that no proper subset is complete.

7. Let $+^3$ be the ternary connective such that $+^3\alpha\beta\gamma$ is equivalent to $\alpha + \beta + \gamma$.

(a) Show that $\{\top, \bot, \wedge, +^3)$ is complete.
(b) Show that no proper subset is complete.

8. Let 1 be the ternary connective such that $1\alpha\beta\gamma$ is assigned the value T iff exactly one of the formulas α, β, γ is assigned the value T. Show that

there are no binary connectives $\circ$ and $\triangle$ such that $1\alpha\beta\gamma$ is equivalent to $(\alpha \circ \beta) \triangle \gamma$.

9. Add the 0-place connectives $\top$, $\bot$ to our language. For any wff φ and sentence symbol A, let $\varphi^A_\top$ be the wff obtained from φ by replacing A by $\top$. Similarly for $\varphi^A_\bot$ Then let $\varphi^A_* = (\varphi^A_\top \vee \varphi^A_\bot)$. Show that

(a) $\varphi \models \varphi^A_*$.

(b) If $\varphi \models \psi$ and A does not appear in ψ, then $\varphi^A_* \models \psi$.

(c) (Interpolation theorem) If $\alpha \models \beta$, then there is some γ all of whose sentence symbols occur both in α and in β and such that $\alpha \models \gamma \models \beta$.

§ 1.6 SWITCHING CIRCUITS[1]

Consider an electrical device (traditionally a black box) having n inputs and one output (Fig. 1). Assume that to each input we apply a signal having one of two values and that the output has one of two values. The two possible values we call T and F. (We could also define the F value as 0 potential and choose the unit of potential so that the T value has potential 1.) Further

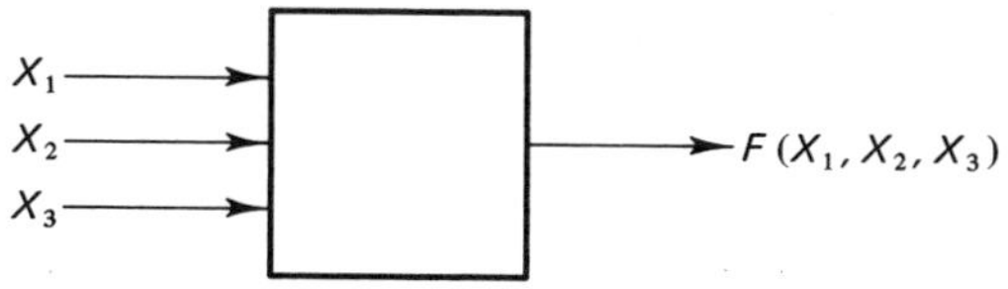

Figure 1. Electrical device with three inputs.

assume that the device has no memory; i.e., the present output level depends only on the present inputs (and not on past history). Then the performance of the device is described by a Boolean function:

$$F(X_1, \ldots, X_n) = \text{the output level given the input signals } X_1, \ldots, X_n.$$

Devices meeting all these assumptions constitute an essential part of digital-computer circuitry. There is, for example, the two-input AND gate, for which the output is the minimum of the inputs (where $F < T$). This device realizes the Boolean function K of the preceding section. It is convenient to attach the labels $\mathbf{A}_1$ and $\mathbf{A}_2$ to the inputs and to label the output $\mathbf{A}_1 \wedge \mathbf{A}_2$.

[1] This section, which discusses an application of the ideas of previous sections, may be omitted without loss of continuity.

Similar devices can be made for other sentential connectives. For a two-input OR gate (Fig. 2) the output voltage is the maximum of the input voltages. Corresponding to the negation connective there is the NOT device (or inverter), whose output voltage is the opposite of the input voltage.

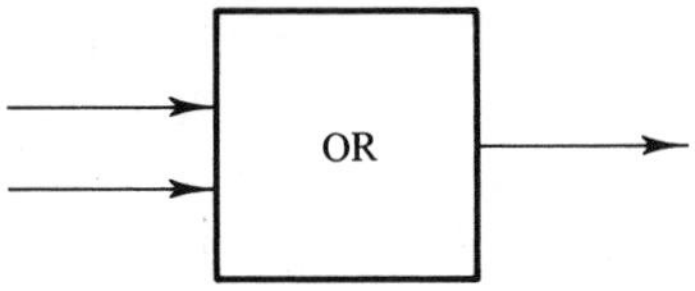

Figure 2. OR gate.

A circuit can be constructed from various devices of this sort. And it is again natural to use wffs of our formal language to label the voltages at different points (Fig. 3). Conversely, given the wff thus attached to the output, we can approximately reconstruct the circuit, which looks very much like the tree of the wff's formation.

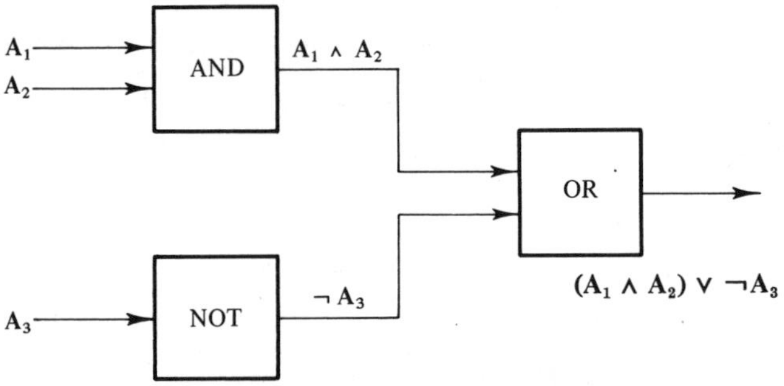

Figure 3. Circuit with wffs as labels.

For example, the circuit for

$$((\mathbf{A} \wedge \mathbf{B}) \wedge \mathbf{D}) \vee ((\mathbf{A} \wedge \mathbf{B}) \wedge \neg \mathbf{C})$$

would probably be as shown in Fig. 4. Duplication of the circuit for $\mathbf{A} \wedge \mathbf{B}$ would not usually be desirable.

Tautologically equivalent wffs yield circuits having ultimately the same performance, although possibly at different cost and (if the devices are not quite instantaneous in operation) different speed. Define the *delay* of a circuit as the maximum number of boxes through which the signal can pass in going from an input to the output. The corresponding notion for formulas is conveniently defined by recursion.

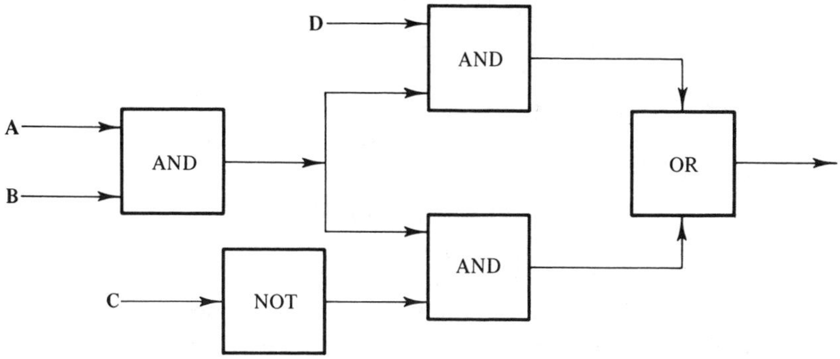

Figure 4. Circuit for **((A ∧ B) ∧ D) ∨ ((A ∧ B) ∧ ¬C).**

1. The delay of a sentence symbol is 0.
2. The delay of $\neg\alpha$ is one greater than the delay of α.
3. The delay of $\alpha \wedge \beta$ is one greater than the maximum of delay of α and the delay of β.

And similarly for any other connective.

For example, the circuit of $(\mathbf{A}_1 \wedge \mathbf{A}_2) \vee \neg\mathbf{A}_3$ uses three devices and has a delay of 2. The tautologically equivalent formula $\neg(\mathbf{A}_3 \wedge (\neg\mathbf{A}_1 \vee \neg\mathbf{A}_2))$ gives a circuit having five devices and a delay of 4. The problem facing many a computer engineer is: Given a circuit (or its wff), find an equivalent circuit (or a tautologically equivalent wff) for which the cost as a minimum, subject to constraints such as a maximum allowable delay. For this problem he has some catalog of available devices; for example, he might have available

NOT, two-input AND, three-input OR.

(It is desirable that the available devices correspond to a complete set of connectives.) The catalog of devices determines a formal language, having a connective symbol for each device.

Example 1 Inputs: **A, B, C**. Output: To agree with the majority of **A, B,** and **C**. Devices available: two-input OR, two-input AND. One solution is

$$((\mathbf{A} \wedge \mathbf{B}) \vee (\mathbf{A} \wedge \mathbf{C})) \vee (\mathbf{B} \wedge \mathbf{C}),$$

which uses five devices and has a delay of 3. But a better solution is

$$(\mathbf{A} \wedge (\mathbf{B} \vee \mathbf{C})) \vee (\mathbf{B} \wedge \mathbf{C}),$$

which uses four devices and has the same delay. Furthermore, there is no solution using only three devices; cf. Exercise 1.

EXAMPLE 2 Inputs: **A** and **B**. Output: *T* if the inputs agree, *F* otherwise; i.e., the circuit is to test for equality. Device available: two-input NOR. One solution is

$$((\mathbf{A} \downarrow \mathbf{A}) \downarrow \mathbf{B}) \downarrow ((\mathbf{B} \downarrow \mathbf{B}) \downarrow \mathbf{A}).$$

This uses five devices; is there a better solution? A deeper question is: Is there a systematic procedure for finding a minimal solution? These are questions which we merely raise here. In recent years a great deal of work has gone into investigating questions of this type.

EXAMPLE 3 (Relay switching) Inputs: **A**, ¬**A**, **B**, ¬**B**, ... Devices: OR (any number of inputs), AND (any number of inputs). Cost: Devices are free, but each use of an input costs one unit. To test for equality of **A** and **B** we could use

$$(\mathbf{A} \wedge \mathbf{B}) \vee (\neg\mathbf{A} \wedge \neg\mathbf{B}).$$

The wiring diagram for the circuit is shown in Fig. 5. The circuit will pass current iff **A** and **B** are assigned the same value. (This formula, equivalent to **A** ↔ **B**, has the property that its truth value changes whenever the value of one argument changes. For this reason, the circuit is used, with double-throw switches, in wiring hallway lights.)

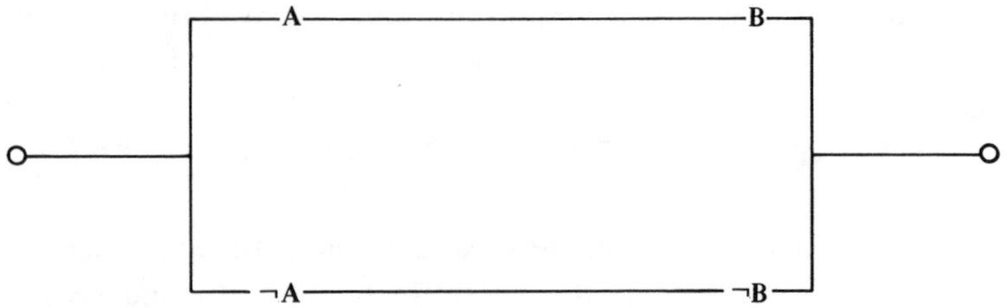

Figure 5. Wiring diagram for $(\mathbf{A} \wedge \mathbf{B}) \vee (\neg\mathbf{A} \wedge \neg\mathbf{B})$.

But there is one respect in which relay circuits do not fit the description given at the beginning of this section. Relays are bilateral devices; they will pass current in either direction. This feature makes "bridge" circuits possible (Fig. 6). The methods described here do not apply to such circuits.

EXAMPLE 4 There are four inputs, and the circuit is to realize the Boolean function *G*, where *G* is to have the value *T* at ⟨*F*, *F*, *F*, *T*⟩, ⟨*F*, *F*, *T*, *F*⟩ ⟨*F*,

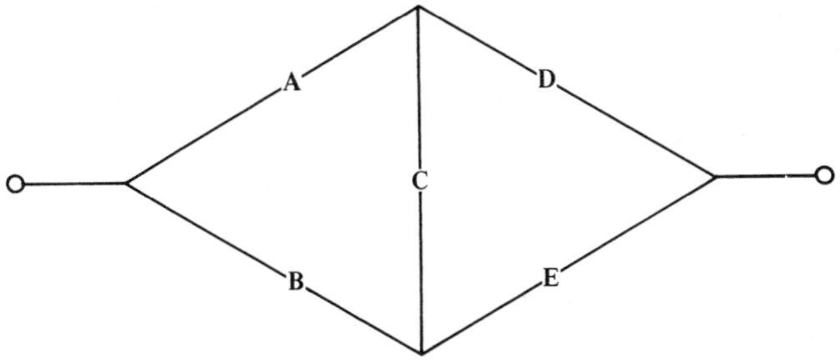

Figure 6. Bridge circuit.

$F, T, T\rangle$, $\langle F, T, F, F\rangle$, $\langle F, T, F, T\rangle$, $\langle F, T, T, F\rangle$, $\langle F, T, T, T\rangle$, and $\langle T, F, F, T\rangle$. G is to have the value F at $\langle T, F, F, F\rangle$, $\langle T, F, T, F\rangle$, $\langle T, T, F, F\rangle$, $\langle T, T, T, F\rangle$, and $\langle T, T, T, T\rangle$. At the remaining three points, $\langle F, F, F, F\rangle$, $\langle T, F, T, T\rangle$, and $\langle T, T, F, T\rangle$, we don't care about the value of G. (The application of the circuit is such that these three combinations never occur.)

We know that G can be realized by using, say, $\{\wedge, \vee, \neg\}$. But we want to do this in an efficient way. The first step is to represent the data in a more comprehensible form. We can do this by means of Fig. 7. Since $G(F, F, F, T) = T$, we have placed a T in the square with coordinates $\langle \neg\mathbf{A}, \neg\mathbf{B}, \neg\mathbf{C}, \mathbf{D}\rangle$. Similarly, there is an F in the square with coordinates

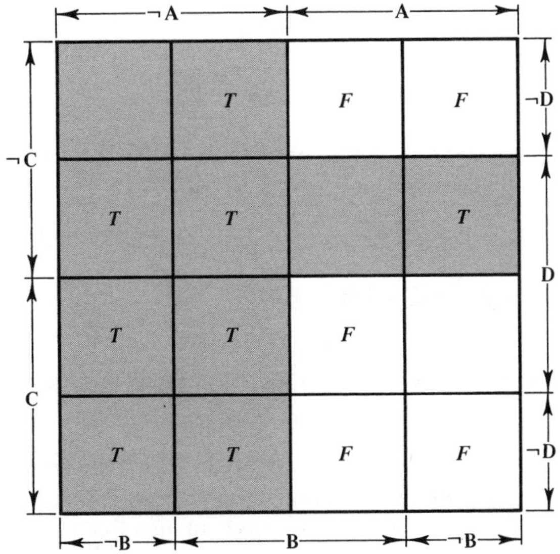

Figure 7. Diagram for Example 4.

$\langle \mathbf{A}, \mathbf{B}, \neg\mathbf{C}, \neg\mathbf{D} \rangle$ because $G(T, T, F, F) = F$. The three squares we do not care about are left blank.

Now we look for a simple geometrical pattern. The shaded area includes all T's and no F's. It corresponds to the formula

$$(\neg\mathbf{A}) \vee (\neg\mathbf{C} \wedge \mathbf{D}),$$

which is reasonably simple and meets all our requirements. Note that the input **B** is not needed at all.

EXERCISES

1. In Example 1 of this section, verify that there is no solution using only three devices.

2. Define a *literal* to be a wff which is either a sentence symbol or the negation of a sentence symbol. An *implicant* of φ is a conjunction α of literals (using distinct sentence symbols) such that $\alpha \models \varphi$. We showed in Section 1.5 (cf. Corollary 15C) that any satisfiable wff φ is tautologically equivalent to a disjunction $\alpha_1 \vee \cdots \vee \alpha_n$, where each α_i is an implicant of φ. An implicant α of φ is *prime* iff it ceases to be an implicant upon the deletion of any of its literals. Any disjunction of implicants equivalent to φ clearly must, if it is to be of minimum length, consist only of prime implicants.

(a) Find all prime implicants of

$$(\mathbf{A} \rightarrow \mathbf{B}) \wedge (\neg\mathbf{A} \rightarrow \mathbf{C}).$$

(b) Which disjunctions of prime implicants enjoy the property of being tautologically equivalent to the formula in part (a)?

3. Repeat (a) and (b) of Exercise 2, but for the formula

$$(\mathbf{A} \vee \neg\mathbf{B}) \wedge (\neg\mathbf{C} \vee \mathbf{D}) \rightarrow \mathbf{B} \wedge ((\mathbf{A} \wedge \mathbf{C}) \vee (\neg\mathbf{C} \wedge \mathbf{D})).$$

§ 1.7 COMPACTNESS AND EFFECTIVENESS

Compactness

We now give a proof of the compactness theorem mentioned earlier (Section 1.3). Call a set Σ of wffs *satisfiable* iff there is a truth assignment which satisfies every member of Σ.

Compactness Theorem A set of wffs is satisfiable iff every finite subset is satisfiable.

Let us temporarily say that Σ is *finitely satisfiable* iff every finite subset of Σ is satisfiable. Then the compactness theorem asserts that this notion coincides with satisfiability. Notice that if Σ is satisfiable, then automatically it is finitely satisfiable. Also if Σ is finite, then the converse is trivial. (Every set is a subset of itself.) The nontrivial part is to show that if an infinite set is finitely satisfiable, then it is satisfiable.

Proof of the compactness theorem The proof consists of two distinct parts. In the first part we take our given finitely satisfiable set Σ and extend it to a maximal such set Δ. In the second part we utilize Δ to make a truth assignment which satisfies Σ.

For the first part let $\alpha_1, \alpha_2, \ldots$ be a fixed enumeration of the wffs. (This is possible since the set of sentence symbols, and hence the set of expressions, is countable; see Theorem 0B.) Define by recursion (on the natural numbers)

$$\Delta_0 = \Sigma,$$

$$\Delta_{n+1} = \begin{cases} \Delta_n ; \alpha_{n+1} & \text{if this is finitely satisfiable,} \\ \Delta_n ; \neg\alpha_{n+1} & \text{otherwise.} \end{cases}$$

(Recall that $\Delta_n ; \alpha_{n+1} = \Delta_n \cup \{\alpha_{n+1}\}$.) Then each Δ_n is finitely satisfiable; see Exercise 1. Let $\Delta = \bigcup_n \Delta_n$, the limit of the Δ_n's.

It is clear that (1) $\Sigma \subseteq \Delta$ and that (2) for any wff α either $\alpha \in \Delta$ or $(\neg\alpha) \in \Delta$. Furthermore, (3) Δ is finitely satisfiable. For any finite subset is already a finite subset of some Δ_n and hence is satisfiable.

This concludes the first part of the proof; we now have a set Δ having properties (1)–(3). There is in general not a unique such set, but there is at least one. (An alternative proof of the existence of such a Δ—and one that we can use even if there are uncountably many sentence symbols—employs Zorn's lemma. The reader familiar with uses of Zorn's lemma should perceive its applicability here.)

For the second part of the proof we define a truth assignment v for the set of all sentence symbols:

$$v(A) = T \qquad \text{iff } A \in \Delta$$

for any sentence symbol A. Then for any wff φ, we claim that

$$v \text{ satisfies } \varphi \qquad \text{iff } \varphi \in \Delta.$$

This is proved by induction on φ; see Exercise 2. Since $\Sigma \subseteq \Delta$, v must then satisfy every member of Σ. ■

Corollary 17A If $\Sigma \models \tau$, then there is a finite $\Sigma_0 \subseteq \Sigma$ such that $\Sigma_0 \models \tau$.

Proof We use the basic fact that $\Sigma \models \tau$ iff $\Sigma ; \neg\tau$ is unsatisfiable.

$$\begin{aligned} & \Sigma_0 \not\models \tau \quad \text{for every finite } \Sigma_0 \subseteq \Sigma \\ \Rightarrow\ & \Sigma_0 ; \neg\tau \text{ is satisfiable for every finite } \Sigma_0 \subseteq \Sigma \\ \Rightarrow\ & \Sigma ; \neg\tau \text{ is finitely satisfiable} \\ \Rightarrow\ & \Sigma ; \neg\tau \text{ is satisfiable} \\ \Rightarrow\ & \Sigma \not\models \tau. \end{aligned}$$ ■

In fact, the above corollary is equivalent to the compactness theorem; see Exercise 3.

Effectiveness

Although the method of truth tables is rather cumbersome to use, the existence of the method yields interesting theoretical conclusions. Suppose we ask of a set Σ of wffs whether or not there is an *effective* procedure which, given a wff τ, will decide whether or not $\Sigma \models \tau$. By an *effective* procedure we mean one meeting the following conditions:

1. There must be exact instructions, finitely long, explaining how to execute the procedure. These instructions should demand no cleverness on the part of the person (or machine) following them. The idea is that your secretary (who knows no mathematics) or your computing machine (which does not think at all) should be able to execute the procedure by mechanically following the instructions.
2. The procedure must avoid random devices (such as the flipping of a coin), or any such device which can, in practice, only be approximated.
3. In the case of a decision procedure, as asked for above, the procedure must be such that, given a wff τ, after a finite number of steps the procedure produces a "yes" or "no" answer.

On the other hand, we place no bound in advance on the number of steps required. Nor do we place any advance bound on the amount of scratch paper that might be required. These will depend on, among other things, the input τ. But for any one τ, the procedure is to require only a finite number of steps to produce the answer, and so only a finite amount of scratch paper will be consumed.

Of course the above description can hardly be considered a precise definition of the word "effective." And, in fact, that word will be used only in an informal intuitive way throughout this book. (In Chapter 3 we will meet a precise counterpart, "recursive.") But as long as we restrict ourselves to positive assertions that there *does* exist an effective procedure of a certain sort, the intuitive approach suffices. We simply display the procedure, show that it works, and people will agree that it is effective. (But this relies on the *empirical* fact that procedures which appear effective to one mathematician also appear so to others.) If we wanted a negative result, that there did *not* exist an effective procedure of a certain sort, then this intuitive viewpoint would be inadequate. (In Chapter 3 we do want to obtain just such negative results.) Because the notion of effectiveness is intuitive, definitions and theorems involving it will be marked with a star. For example:

⋆**Theorem 17B** There is an effective procedure which, given any expression ε, will decide whether or not it is a wff.

Proof See the algorithm in Section 1.4 and the footnotes thereto. ■

⋆**Definition** A set Σ of expressions is *decidable* iff there exists an effective procedure which, given an expression α, will decide whether or not $\alpha \in \Sigma$.

For example, any finite set is decidable. Some infinite sets are decidable but not all. For there are $2^{\aleph_0}$ sets of expressions but only countably many effective procedures. This is because the procedure is completely determined by its (finite) instructions. There are only $\aleph_0$ finite sequences of letters.

⋆**Theorem 17C** There is an effective procedure which, given a finite set $\Sigma\,;\tau$ of wffs, will decide whether or not $\Sigma \models \tau$.

Proof The truth-table procedure (Section 1.3) meets the requirement. ■

In this theorem we specified that $\Sigma\,;\tau$ was finite, since one cannot be "given" in any direct and effective way all of an infinite object.

⋆**Corollary 17D** For a finite set Σ, the set of tautological consequences of Σ is decidable. In particular, the set of tautologies is decidable.

If Σ is an infinite decidable set, then in general its set of tautological consequences may not be decidable. (See Chapter 3.) But we can obtain a weaker result, which is in a sense half of decidability.

Say that a set A of expressions is *effectively enumerable* iff there is an effective procedure which lists, in some order, the members of A. If A is infinite, then the procedure can never finish. But for any specified member of A, it must eventually (i.e., in a finite length of time) appear on the list.

Primarily to give the reader more of a feeling for this notion, we now state two elementary results about it.

⋆Theorem 17E A set A of expressions is effectively enumerable iff there is an effective procedure which, given any expression ε, produces the answer "yes" iff $\varepsilon \in A$.

(If $\varepsilon \notin A$, the procedure might produce the answer "no"; it might go on forever without producing any answer, but it must not produce the answer "yes.")

Proof If A is effectively enumerable, then given any ε we can examine the listing of A as our procedure churns it out. If and when ε appears, we say "yes." (Thus if $\varepsilon \notin A$, no answer is ever given. It is this that keeps A from being decidable. When ε has failed to occur among the first 10^{10} enumerated members of A, there is in general no way of knowing whether $\varepsilon \notin A$ (in which case one should give up looking) or whether ε will occur in the very next step.)

Conversely, suppose that we have the procedure described in the theorem. We want to create a listing of A. The idea is to enumerate all expressions, and to apply our given procedure to each. But we must budget our time sensibly. It is easy enough to enumerate effectively all expressions:

$$\varepsilon_1, \varepsilon_2, \varepsilon_3, \ldots .$$

Then proceed according to the following scheme:

1. Spend one minute testing ε_1 for membership in A (using the given procedure).
2. Spend two minutes testing ε_1, then two minutes testing ε_2.
3. Spend three minutes testing ε_1, three minutes testing ε_2, and three minutes testing ε_3.

And so forth. Of course whenever our procedure produces a "yes" answer, we put the accepted expression on the output list. Thus any member of A will eventually appear on the list. (It will appear infinitely many times, unless we modify the above instructions to check for duplication.) ∎

⋆Theorem 17F A set of expressions is decidable iff both it and its complement (relative to the set of all expressions) are effectively enumerable.

Proof Exercise 4.

Observe that if sets A and B are effectively enumerable, then so are $A \cup B$ and $A \cap B$. The class of decidable sets is also closed under union and intersection, and it is in addition closed under complementation.

Now for a more substantive result:

★Theorem 17G If Σ is a decidable set of wffs, then the set of tautological consequences of Σ is effectively enumerable.

Proof Actually it is enough for Σ to be effectively enumerated; consider an enumeration

$$\sigma_1, \sigma_2, \sigma_3, \ldots .$$

Given any wff τ, we can test (by truth tables) successively whether or not

$$\varnothing \models \tau,$$
$$\{\sigma_1\} \models \tau,$$
$$\{\sigma_1, \sigma_2\} \models \tau,$$
$$\{\sigma_1, \sigma_2, \sigma_3\} \models \tau,$$

and so forth. If any of these conditions is met, then we answer "yes." Otherwise, we keep trying.

This does produce an affirmative answer whenever $\Sigma \models \tau$, by the corollary to the compactness theorem. ■

EXERCISES

1. Assume that every finite subset of Σ is satisfiable. Show that the same is true of at least one of the sets $\Sigma ; \alpha$ and $\Sigma ; \neg\alpha$. (This is part of the proof of the compactness theorem.)

2. Let Δ be a set of wffs such that (i) every finite subset of Δ is satisfiable, and (ii) for every wff α, either $\alpha \in \Delta$ or $(\neg\alpha) \in \Delta$. Define the truth assignment v:

$$v(A) = \begin{cases} T & \text{if } A \in \Delta, \\ F & \text{if } A \notin \Delta \end{cases}$$

for each sentence symbol A. Show that for every wff φ, $\bar{v}(\varphi) = T$ iff $\varphi \in \Delta$. (This is part of the proof of the compactness theorem.)

3. Show that from the corollary to the compactness theorem we can prove the compactness theorem itself (far more easily than we can starting from scratch).

4. Prove Theorem 17F.

*5. The notions of decidability and effective enumerability can be applied not only to sets of expressions but also to sets of integers or to sets of pairs of expressions or integers. Show that a set A of expressions is effectively enumerable iff there is a decidable set B of pairs $\langle \alpha, n \rangle$ (consisting of an expression α and an integer n) such that $A = \text{dom } B$.

6. Let Σ be an effectively enumerable set of wffs. Assume that for each wff τ, either $\Sigma \models \tau$ or $\Sigma \models \neg\tau$. Show that the set of tautological consequences of Σ is decidable.

CHAPTER TWO

First-Order Logic

§ 2.0 PRELIMINARY REMARKS

In the preceding chapter we presented the first of our mathematical models of deductive thought. It was a simple model, indeed too simple. It is easy to think of examples of intuitively correct deductions which cannot be adequately mirrored in the model of sentential logic.

Suppose we begin with a collection of hypotheses (in English) and a possible conclusion. By translating everything to the language of sentential logic we obtain a set Σ of hypotheses and a possible conclusion τ. Now if $\Sigma \models \tau$, then we feel that the original English-language deduction was valid. But if $\Sigma \not\models \tau$, then we are unsure. It may well be that the model of sentential logic was just too crude to mirror the subtlety of the original deduction.

In this chapter we present a system of logic of much greater ability. In fact, when the "working mathematician" finds a proof, he almost invariably means a proof that can be mirrored in the system of this chapter.

First, we want to give an informal description of the features our first-order languages might have (or at least might be able to simulate). We begin with a special case, the first-order language for number theory. For

TABLE VII

Formal expression	Intended translation
$\mathbf{0}$	"zero." Here $\mathbf{0}$ is a constant symbol, intended to name the number 0.
$\mathbf{S}t$	"the successor of t." Here $\mathbf{S}$ is a one-place function symbol. t is to be an expression which names some number a. Then $\mathbf{S}t$ names $S(a)$, the successor of a. For example, $\mathbf{S0}$ is intended to name the number 1.
$< \mathbf{v}_1 \mathbf{v}_2$	"$\mathbf{v}_1$ is less than $\mathbf{v}_2$." Here $<$ is a two-place predicate symbol. At the end of Section 2.1 we will adopt conventions letting us abbreviate the expression in the more usual style: $\mathbf{v}_1 < \mathbf{v}_2$.
$\forall$	"for every natural number." The symbol $\forall$ is the universal quantifier symbol. More generally, with each translation of the language into English there will be associated a certain set A (the so-called universe); $\forall$ will then become "for every member of the universe A."
$\forall \mathbf{v}_1 < \mathbf{0} \mathbf{v}_1$	"For every natural number $\mathbf{v}_1$, zero is less than $\mathbf{v}_1$." Or more euphoniously, "Every natural number is larger than $\mathbf{0}$." This formal sentence is false in the intended translation, since zero is not larger than itself.

this language there is a certain intended way of translating to and from English (Table VII).

One abbreviation is mentioned in Table VII. There will be more (Table VIII).

TABLE VIII

Abbreviated expression	Intended translation
$x \approx y$	"x equals y." In unabbreviated form this will become $\approx xy$.
$\exists v$	"there exists a natural number v such that." Or more generally, "there exists a member of the universe such that."
$\exists \mathbf{v}_1 \forall \mathbf{v}_2\, \mathbf{v}_1 \approx \mathbf{v}_2$	"There is exactly one natural number." Again this formal sentence is false in the intended translation.
$\forall \mathbf{v}_1 (\mathbf{0} < \mathbf{v}_1 \vee \mathbf{0} \approx \mathbf{v}_1)$	"Every natural number is greater than or equal to zero."

Actually we will not be quite as generous as the tables might suggest. There are two economy measures that we can take to obtain simplification without any essential loss of linguistic expressiveness.

First, we choose as our sentential connective symbols just $\neg$ and $\rightarrow$. We know from Section 1.5 that this is a complete set, so there is no real reason to use more.

Second, we forego the luxury of an existential quantifier, $\exists x$. In its place we use: $\neg\forall x\neg$. This is justified, since an English sentence,

There is something rotten in the state of Denmark,

is equivalent to

It is not the case that for every x, x is not rotten in the state of Denmark.

Thus the formula $\exists \mathbf{v}_1 \forall \mathbf{v}_2\, \mathbf{v}_1 \approx \mathbf{v}_2$ becomes, in unabbreviated from,

$$(\neg\forall \mathbf{v}_1(\neg\forall \mathbf{v}_2 \approx \mathbf{v}_1\mathbf{v}_2)).$$

For an example in an ad hoc language, we might translate "Socrates is a man" as Hs, where H is a one-place predicate symbol intended to translate "is a man" and s is a constant symbol intended to name Socrates. Similarly, to translate "Socrates is mortal" we take Ms. Then "All men are mortal" is translated as: $\forall \mathbf{v}_1(H\mathbf{v}_1 \rightarrow M\mathbf{v}_1)$.

The reader will possibly recognize the symbols $\forall$ and $\exists$ from previous mathematical contexts. Indeed, some mathematicians, when writing on the blackboard during their lectures, already use a nearly formalized language with only vestigial traces of English. That our first-order languages resemble theirs is no accident. We want to be able to take one step back and study not, e.g., sets or groups, but the sentences of set theory or group theory. (The term *metamathematics* is sometimes used; the word itself suggests the procedure of stepping back and examining what the mathematician is doing.) The objects you, the logician, now study are the sentences which you, the set theoretician, previously used in the study of sets. This requires formalizing the language of set theory. And we want our formal languages to incorporate the features used in, for example, set theory.

§ 2.1 FIRST-ORDER LANGUAGES

We assume henceforth that we have been given infinitely many distinct objects (which we call symbols), arranged as follows:

A. Logical symbols

0. Parentheses (,).
1. Sentential connective symbols: $\rightarrow$, $\neg$.
2. Variables (one for each positive integer n):

$$\mathbf{v}_1, \mathbf{v}_2, \ldots .$$

3. Equality symbol (optional): $\approx$.

B. Parameters

0. Quantifier symbol: $\forall$.
1. Predicate symbols: For each positive integer n, some set (possibly empty) of symbols, called n-place predicate symbols.
2. Constant symbols: Some set (possibly empty) of symbols.
3. Function symbols: For each positive integer n, some set (possibly empty) of symbols, called n-place function symbols.

In A.3 we allow for the possibility of the equality symbol's being present, but we do not assume its presence. Some languages will have it and others will not. The equality symbol is a two-place predicate symbol but is distinguished from the other two-place predicate symbols by being a logical symbol rather than a parameter. (This status will affect its behavior under translations into English.) We do assume that some n-place predicate symbol is present for some n.

In B.2, the constant symbols are also called 0-place function symbols. This will often allow a uniform treatment of the symbols in B.2 and B.3.

As before, we assume that the symbols are distinct and that no symbol is a finite sequence of other symbols.

In order to specify which language we have before us (as distinct from other first-order languages), we must (i) say whether or not the equality symbol is present, and (ii) say what the parameters are.

We now list some examples of what this language might be:

1. *Pure predicate language*

Equality: No.
n-place predicate symbols: $A_1^n, A_2^n, \ldots .$
Constant symbols: $a_1, a_2, \ldots .$
n-place function symbols ($n > 0$): None.

2. *Language of set theory*

Equality: Yes (usually).

Predicate parameters: One two-place predicate symbol $\in$.
Function symbols: None (or occasionally a constant symbol $\varnothing$).

3. *Language of elementary number theory* (as in Chapter 3)

Equality: Yes.
Predicate parameters: One two-place predicate symbol $<$.
Constant symbols: The symbol $\mathbf{0}$.
One-place function symbols: $\mathbf{S}$ (for successor).
Two-place function symbols: $+$ (for addition), $\cdot$ (for multiplication), and $\mathbf{E}$ (for exponentiation).

In examples 2 and 3 there are certain intended translations of the parameters. We will presently give a number of examples of sentences that can be translated into these languages and a few examples of sentences that cannot be so translated.

It is important to notice that our notion of language includes the language for set theory. For it is generally agreed that, by and large, mathematics can be embedded into set theory. By this is meant that

(a) statements in mathematics (like the fundamental theorem of calculus) can be expressed in the language of set theory; and

(b) the theorems of mathematics follow logically from the axioms of set theory.

Our model of first-order logic is fully adequate to mirror this procedure.

EXAMPLES in the language of set theory Here it is intended that $\forall$ should mean "for all sets" and $\in$ should mean "is a member of."

1. "There is no set of which every set is a member." We will translate this into the language of set theory using several steps. The intermediate sentences are neither in English nor in the formal language but are in a mixed language.

$\neg$[There is a set of which every set is a member]
$\neg\exists \mathbf{v}_1$[Every set is a member of $\mathbf{v}_1$]
$\neg\exists \mathbf{v}_1 \forall \mathbf{v}_2\ \mathbf{v}_2 \in \mathbf{v}_1$

Although it is tempting to stop here, we must now replace $\mathbf{v}_2 \in \mathbf{v}_1$ by $\in\mathbf{v}_2\mathbf{v}_1$, since predicate symbols will always go at the left in such contexts. Furthermore, $\exists \mathbf{v}_1$ must be replaced by $\neg\forall \mathbf{v}_1 \neg$, as mentioned earlier. And we must use the correct number of parentheses. The finished product is

$$(\neg(\neg\forall \mathbf{v}_1(\neg\forall \mathbf{v}_2 \in\mathbf{v}_2\mathbf{v}_1))).$$

2. Pair-set axiom: "For any two sets, there is a set whose members are exactly the two given sets." Again we carry out the translation in stages.

$\forall \mathbf{v}_1 \forall \mathbf{v}_2$[There is a set whose members are exactly $\mathbf{v}_1$ and $\mathbf{v}_2$],
$\forall \mathbf{v}_1 \forall \mathbf{v}_2\, \exists \mathbf{v}_3$[The members of $\mathbf{v}_3$ are exactly $\mathbf{v}_1$ and $\mathbf{v}_2$],
$\forall \mathbf{v}_1 \forall \mathbf{v}_2\, \exists \mathbf{v}_3 \forall \mathbf{v}_4 (\mathbf{v}_4 \in \mathbf{v}_3 \leftrightarrow \mathbf{v}_4 \approx \mathbf{v}_1 \vee \mathbf{v}_4 \approx \mathbf{v}_2)$.

Now we replace $\exists \mathbf{v}_3$ by $\neg\forall \mathbf{v}_3 \neg$, $\mathbf{v}_4 \in \mathbf{v}_3$ by $\in \mathbf{v}_4 \mathbf{v}_3$, and $\mathbf{v}_4 \approx \mathbf{v}_i$ by $\approx \mathbf{v}_4 \mathbf{v}_i$. In addition, we must eliminate $\leftrightarrow$ and $\vee$ in favor of our chosen connectives $\rightarrow$ and $\neg$. Thus

$$\alpha \vee \beta \text{ becomes } \neg\alpha \rightarrow \beta;$$

$$\alpha \leftrightarrow \beta \text{ becomes } \neg((\alpha \rightarrow \beta) \rightarrow \neg(\beta \rightarrow \alpha)).$$

The finished product is

$$\forall \mathbf{v}_1 \forall \mathbf{v}_2 (\neg\forall \mathbf{v}_3 (\neg\forall \mathbf{v}_4 (\neg((\in \mathbf{v}_4 \mathbf{v}_3 \rightarrow ((\neg \approx \mathbf{v}_4 \mathbf{v}_1) \rightarrow$$
$$\approx \mathbf{v}_4 \mathbf{v}_2)) \rightarrow (\neg(((\neg\approx \mathbf{v}_4 \mathbf{v}_1) \rightarrow \approx \mathbf{v}_4 \mathbf{v}_2) \rightarrow \in \mathbf{v}_4 \mathbf{v}_3)))))).$$

The finished product is not as pleasant to read as the version that preceded it. As we have no interest in deliberately making life unpleasant for ourselves, we will eventually adopt conventions allowing us to avoid seeing the finished product at all. But for the moment it should be regarded as an interesting, even if unattractive, novelty.

EXAMPLES in the language of elementary number theory Here it is intended that $\forall$ should mean "for all natural numbers" and that $<$, **0**, **S**, $+$, $\cdot$, and **E** should have the obvious meanings.

1. As a name for the natural number 2 we have the term **SS0**, since 2 is the successor of the successor of zero. Similarly, for 4 we have the term **SSSS0**. For the phrase "2 + 2" it is tempting to use **SS0** + **SS0**. But we will adopt the policy of always putting the function symbol at the left (i.e., we will use Polish notation for function symbols). Thus corresponding to the English phrase "2 + 2" we have the term + **SS0 SS0**. The English sentence "Two plus two is four" is translated as

$$\approx + \mathbf{SS0\ SS0\ SSSS0}.$$

(The spaces are inserted to help the reader, but they do not constitute an official feature of the language.)

2. "Any nonzero natural number is the successor of some number." We will perform the translation in three steps.

$$\forall \mathbf{v}_1[\text{If } \mathbf{v}_1 \text{ is nonzero, then } \mathbf{v}_1 \text{ is the successor of some number.}]$$

$$\forall \mathbf{v}_1(\mathbf{v}_1 \not\approx \mathbf{0} \rightarrow \exists \mathbf{v}_2\, \mathbf{v}_1 \approx \mathbf{S}\mathbf{v}_2).$$

$$\forall \mathbf{v}_1((\neg \approx \mathbf{v}_1 \mathbf{0}) \rightarrow (\neg \forall \mathbf{v}_2(\neg \approx \mathbf{v}_1 \mathbf{S}\mathbf{v}_2))).$$

3. "Any nonempty set of natural numbers has a least element." This cannot be translated into our language, because we cannot express "any... set." This requires either something like the (first-order) language for set theory or a second-order language for number theory. We could, however, translate, "The set of primes has a least element." (The first step is to convert this sentence into, "There is a smallest prime." We leave the other steps to the reader; hints can be found in the next section.)

Examples in ad hoc languages

1. "All apples are bad."

$$\forall \mathbf{v}_1(A\mathbf{v}_1 \rightarrow B\mathbf{v}_1).$$

2. "Some apples are bad."

Intermediate step: $\exists \mathbf{v}_1(A\mathbf{v}_1 \wedge B\mathbf{v}_1)$.

Finished product: $(\neg \forall \mathbf{v}_1(\neg(\neg(A\mathbf{v}_1 \rightarrow (\neg B\mathbf{v}_1))))).$

These two examples illustrate patterns which arise continually. An English sentence which asserts that everything in a certain category has some property, is translated

$$\forall v(\_\_\_\_ \rightarrow \_\_\_\_).$$

A sentence which asserts that there is some object or objects in the category and having the property, is translated

$$\exists v(\_\_\_\_ \wedge \_\_\_\_).$$

The reader should be cautioned against confusing the two patterns. For example,

$$\forall \mathbf{v}_1(A\mathbf{v}_1 \wedge B\mathbf{v}_1)$$

translates "Everything is an apple and is bad," which is a much stronger assertion than the sentence in the first example. Similarly, $\exists \mathbf{v}_1(A\mathbf{v}_1 \rightarrow B\mathbf{v}_1)$ translates "There is something which is bad, if it is an apple." This is a much weaker assertion than the sentence in the second example. It is true (vacuously) even if all apples are good, provided only that the world has something which is not an apple.

3. Bobby's father can beat up the father of any other kid on the block. Establish a language in which $\forall$ is intended to mean "for all people," Kx is to mean "x is a kid on the block," b is to mean "Bobby," Bxy is to mean "x can beat up y," and fx is to mean "the father of x." Then a translation is

$$\forall v_1(Kv_1 \rightarrow ((\neg \approx v_1 b) \rightarrow Bfbfv_1)).$$

Formulas

An expression is any finite sequence of symbols. Of course most expressions are nonsensical, but there are certain interesting expressions: the terms and the wffs.

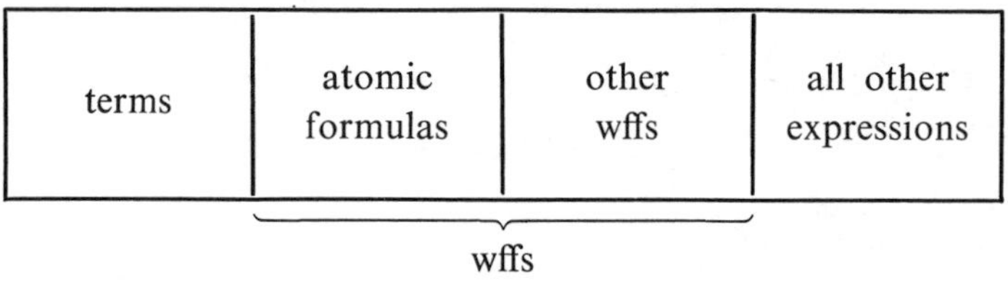
terms
atomic formulas
other wffs
all other expressions
wffs

The terms are the nouns and pronouns of our language; they are the expressions which can be interpreted as naming an object. The atomic formulas will be those wffs which have neither connective nor quantifier symbols.

The terms are defined to be those expressions which can be built up from the constant symbols and the variables by prefixing the function symbols. To restate this in the terminology of Section 1.2, we define for each n-place function symbol f, an n-place term-building operation $\mathscr{F}_f$ on expressions:

$$\mathscr{F}_f(\varepsilon_1, \ldots, \varepsilon_n) = f\varepsilon_1 \cdots \varepsilon_n.$$

Definition The set of *terms* is the set of expressions generated from the constant symbols and variables by the $\mathscr{F}_f$ operations.

If there are no function symbols, then the terms are just the constant symbols and the variables. In this case we do not need an inductive definition.

Notice that we use Polish notation for terms by placing the function symbol at the left. The terms do not contain parentheses or commas. We will later prove a unique readability result, showing that the set of terms is freely generated.

The terms are the expressions which are translated as names of objects (noun phrases), in contrast to the wffs which are translated as assertions about objects.

Some examples of terms in the language of number theory are

$$+ \, \mathbf{v}_2 \, \mathbf{S0},$$
$$\mathbf{SSSS0},$$
$$+ \, \mathbf{E} \, \mathbf{v}_1 \, \mathbf{SS0} \, \mathbf{E} \, \mathbf{v}_2 \, \mathbf{SS0}.$$

The atomic formulas will play a role roughly analogous to that played by the sentence symbols in sentential logic. An *atomic formula* is an expression of the form

$$Pt_1 \cdots t_n,$$

where P is an n-place predicate symbol and $t_1, \ldots, t_n$ are terms.

For example, $\approx \mathbf{v}_1 \mathbf{v}_2$ is an atomic formula, since $\approx$ is a two-place predicate symbol and each variable is a term. In the language of set theory we have the atomic formula $\in \mathbf{v}_5 \mathbf{v}_3$.

Notice that the atomic formulas are not defined inductively. Instead we have simply said explicitly just which expressions are atomic formulas.

The *well-formed formulas* are those expressions which can be built up from the atomic formulas by use of the connective symbols and the quantifier symbol. We can restate this in the terminology of Section 1.2 by first defining some formula-building operations on expressions:

$$\mathscr{E}_{\neg}(\gamma) = (\neg\gamma),$$
$$\mathscr{E}_{\rightarrow}(\gamma, \delta) = (\gamma \rightarrow \delta),$$
$$\mathscr{Q}_i(\gamma) = \forall \mathbf{v}_i \, \gamma.$$

Definition The set of *well-formed formulas* (*wffs*, or just *formulas*) is the set of expressions generated from the atomic formulas by the operations $\mathscr{E}_{\neg}$, $\mathscr{E}_{\rightarrow}$, and $\mathscr{Q}_i (i = 1, 2, \ldots)$.

For example, $\neg \mathbf{v}_3$ is not a wff. (Why?) On the other hand,

$$\forall \mathbf{v}_1((\neg \forall \mathbf{v}_3(\neg \in \mathbf{v}_3 \mathbf{v}_1)) \rightarrow (\neg \forall \mathbf{v}_2(\in \mathbf{v}_2 \mathbf{v}_1 \rightarrow$$
$$(\neg \forall \mathbf{v}_4(\in \mathbf{v}_4 \mathbf{v}_2 \rightarrow (\neg \in \mathbf{v}_4 \mathbf{v}_1))))))$$

is a wff, as is demonstrated by the following tree:

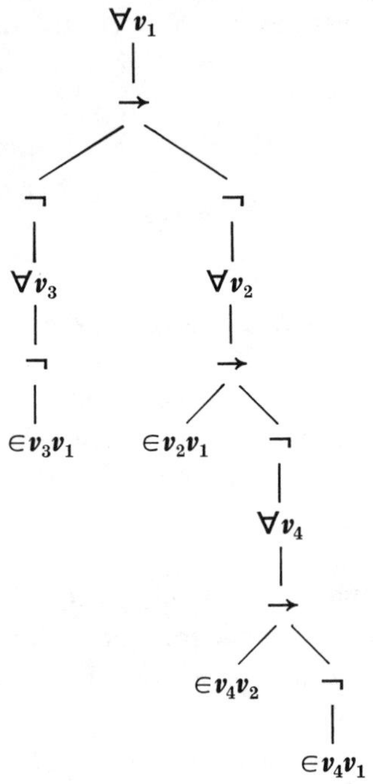

But it requires some study to discover that this wff is the axiom of regularity for set theory.

Free variables

Two examples of wffs are $\forall \boldsymbol{v}_2 \in \boldsymbol{v}_2\boldsymbol{v}_1$ and $(\neg\forall \boldsymbol{v}_1(\neg\forall \boldsymbol{v}_2 \in \boldsymbol{v}_2\boldsymbol{v}_1))$. But there is an important difference between the two examples. The second might be a translation from English of

There is a set such that every set is a member of it.

The first example, however, can only be a translation of an incomplete sentence, such as

Every set is a member of $\_\_\__1$.

We are unable to complete the sentence without knowing what to do with $\boldsymbol{v}_1$. In cases of this sort, we will say that $\boldsymbol{v}_1$ *occurs free* in the wff $\forall \boldsymbol{v}_2 \in \boldsymbol{v}_2\boldsymbol{v}_1$. In contrast, no variable occurs free in $(\neg\forall \boldsymbol{v}_1(\neg\forall \boldsymbol{v}_2 \in \boldsymbol{v}_2\boldsymbol{v}_1))$. But of course we need a precise definition which does not refer to possible translations to English but refers only to the symbols themselves.

Consider any variable x. We define, for each wff α, what it means for x to *occur free* in α. This we do by recursion:

1. For atomic α, x occurs free in α iff x occurs in (is a symbol of) α.
2. x occurs free in $(\neg\alpha)$ iff x occurs free in α.
3. x occurs free in $(\alpha \rightarrow \beta)$ iff x occurs free in α or in β.
4. x occurs free in $\forall \mathbf{v}_i\, \alpha$ iff x occurs free in α and $x \neq \mathbf{v}_i$.

This definition makes tacit use of the recursion theorem. We can restate the situation in terms of functions. We begin with the function h defined on atomic formulas:

$$h(\alpha) = \begin{cases} 1 & \text{if } x \text{ is in the atomic formula } \alpha, \\ 0 & \text{otherwise.} \end{cases}$$

And we want to extend h to a function $\bar{h}$ defined on all wffs in such a way that

$$\bar{h}(\mathscr{E}_{\neg}(\alpha)) = \bar{h}(\alpha),$$

$$\bar{h}(\mathscr{E}_{\rightarrow}(\alpha, \beta)) = \max\{\bar{h}(\alpha), \bar{h}(\beta)\},$$

$$\bar{h}(\mathscr{Q}_i(\alpha)) = \begin{cases} \bar{h}(\alpha) & \text{if } x \neq \mathbf{v}_i, \\ 0 & \text{if } x = \mathbf{v}_i. \end{cases}$$

Then we say that x occurs free in α iff $\bar{h}(\alpha) = 1$. The existence of a unique such $\bar{h}$ (and hence the meaningfulness of our definition) follows from the recursion theorem and from the fact (proved in Section 2.3) that the wffs are freely generated.

If no variable occurs free in the wff α, then α is a *sentence*. (The sentences are intuitively the wffs translatable into English without blanks, once we are told how to interpret the parameters.)

For example, $\forall \mathbf{v}_2(A\mathbf{v}_2 \rightarrow B\mathbf{v}_2)$ and $\forall \mathbf{v}_3(P\mathbf{v}_3 \rightarrow \forall \mathbf{v}_3\, Q\mathbf{v}_3)$ are sentences, but $\mathbf{v}_1$ occurs free in $(\forall \mathbf{v}_1\, A\mathbf{v}_1 \rightarrow B\mathbf{v}_1)$. The sentences are usually the most interesting wffs. The others lead a second-class existence; they are used primarily as building blocks for sentences.

In translating a sentence from English, the choice of particular variables is unimportant. We earlier translated "All apples are bad" as $\forall \mathbf{v}_1(A\mathbf{v}_1 \rightarrow B\mathbf{v}_1)$. We could equally well have used

$$\forall \mathbf{v}_{27}(A\mathbf{v}_{27} \rightarrow B\mathbf{v}_{27}).$$

The variable is, in effect, used as a pronoun, just as in English we might say,

"For any object whatsoever, if *it* is an apple, then *it* is bad." Since the choice of particular variables is unimportant, we will often not even specify the choice. Instead we will write, for example, $\forall x(Ax \rightarrow Bx)$, where it is understood that x is some variable. (The unimportance of the choice of variable will eventually become a theorem.)

Similar usages of variables occur elsewhere in mathematics. In

$$\sum_{i=1}^{7} a_{ij}$$

i is a "dummy" variable but j occurs free.

On notation

We can specify a wff (or indeed any expression) by writing a line which displays explicity every symbol. For example,

$$\forall \mathbf{v}_1((\neg \approx \mathbf{v}_1 \mathbf{0}) \rightarrow (\neg \forall \mathbf{v}_2(\neg \approx \mathbf{v}_1 \mathbf{S} \mathbf{v}_2))).$$

But this way of writing things, while splendidly complete, may not be readily comprehensible. The incomprehensibility can be blamed (in part) on the simplifications we wanted in the language (such as the lack of an existential quantifier symbol). We naturally want to have our cake and eat it too, so we now will agree on methods of specifying wffs in more indirect but more readable ways. These conventions will let us write a line such as

$$\forall \mathbf{v}_1(\mathbf{v}_1 \not\approx \mathbf{0} \rightarrow \exists \mathbf{v}_2\, \mathbf{v}_1 \approx \mathbf{S}\mathbf{v}_2)$$

to name the same wff as is named by the other line above.

Note well that we are *not* changing our definition of what a wff is. We are just conspiring to fix certain ways of naming wffs. In the (rare) cases where the exact sequence of symbols becomes important, we way have to drop these new conventions and revert to primitive notation.

We adopt then the following abbreviations and conventions. Here α and β are formulas, x is a variable, and u and t are terms.

$(\alpha \vee \beta)$ abbreviates $((\neg\alpha) \rightarrow \beta)$.
$(\alpha \wedge \beta)$ abbreviates $(\neg(\alpha \rightarrow (\neg\beta)))$.
$(\alpha \leftrightarrow \beta)$ abbreviates $((\alpha \rightarrow \beta) \wedge (\beta \rightarrow \alpha))$; i.e.,

$$(\neg((\alpha \rightarrow \beta) \rightarrow (\neg(\beta \rightarrow \alpha)))).$$

$\exists x\, \alpha$ abbreviates $(\neg \forall x(\neg\alpha))$.

$u \approx t$ abbreviates $\approx ut$ (and similarly for other two-place predicate symbols).

$u \not\approx t$ abbreviates $(\neg \approx ut)$; similarly $u \not< t$ abbreviates $(\neg < ut)$.

For parentheses we will use not only (and) but also [and], etc. And we omit mention of just as many as we possibly can. Toward that end we adopt the following conventions:

1. Outermost parentheses may be dropped. For example, $\forall x\, \alpha \leftrightarrow \beta$ is $(\forall x\, \alpha \rightarrow \beta)$.

2. $\neg$, $\forall$, and $\exists$ apply to as little as possible. For example,

$$\neg\alpha \wedge \beta \quad \text{is} \quad ((\neg\alpha) \wedge \beta), \quad \text{and not} \quad \neg(\alpha \wedge \beta);$$
$$\forall x\, \alpha \rightarrow \beta \quad \text{is} \quad (\forall x\, \alpha \rightarrow \beta), \quad \text{and not} \quad \forall x(\alpha \rightarrow \beta);$$
$$\exists x\, \alpha \wedge \beta \quad \text{is} \quad (\exists x\, \alpha \wedge \beta), \quad \text{and not} \quad \exists x(\alpha \wedge \beta).$$

In such cases we might even add gratuitous parentheses, as in $(\exists x\, \alpha) \wedge \beta$.

3. $\wedge$ and $\vee$ apply to as little as possible, subject to item 2. For example,

$$\neg\alpha \wedge \beta \rightarrow \gamma \quad \text{is} \quad ((\neg\alpha) \wedge \beta) \rightarrow \gamma.$$

4. When one connective is used repeatedly, the expression is grouped to the right. For example,

$$\alpha \rightarrow \beta \rightarrow \gamma \quad \text{is} \quad \alpha \rightarrow (\beta \rightarrow \gamma).$$

EXAMPLES of how we can eliminate abbreviations, rewriting the formula in an unabbreviated way that explicitly lists each symbol in order:

1. $\exists x(Ax \wedge Bx)$ is $(\neg\forall x(\neg(\neg(Ax \rightarrow (\neg Bx)))))$.

But $(\neg\forall x(Ax \rightarrow (\neg Bx)))$ would be an equivalent formula (in any reasonable notion of equivalence).

2. $\exists x\; Ax \rightarrow Bx$ is $((\neg\forall x(\neg Ax)) \rightarrow Bx)$.
$\exists x(Ax \rightarrow Bx)$ is $(\neg\forall x(\neg(Ax \rightarrow Bx)))$.

We will try to use the various alphabets in a systematic way. The system is listed below, but there will be occasional exceptions for special reasons.

Predicate symbols: Uppercase italic letters. Also $\in$, $<$.
Variables: $\mathbf{v}_i$, u, v, x, y, z.
Function symbols: f, g, h. Also $\mathbf{S}$, $+$, etc.
Constant symbols: a, b, $\ldots$. Also $\mathbf{0}$.
Terms: u, t.

Formulas: Lowercase Greek letters.

Sentences: σ, τ.

Sets of formulas: Uppercase Greek letters, plus certain italic letters which pretend to be Greek, viz., A (alpha) and T (tau).

Structures: Uppercase German letters.

EXERCISES

1. Assume that we have a language with the following parameters: $\forall$, intended to mean "for all things"; N, intended to mean "is a number"; I, intended to mean "is interesting"; $<$, intended to mean "is less than"; and $\mathbf{0}$, a constant symbol intended to denote zero. Translate into this language the English sentences listed below. If the English sentence is ambiguous, you will need more than one translation.

(a) Zero is less than any number.
(b) If any number is interesting, then zero is interesting.
(c) No number is less than zero.
(d) Any uninteresting number with the property that all smaller numbers are interesting certainly is interesting.
(e) There is no number such that all numbers are less than it.
(f) There is no number such that no number is less than it.

In 2–6, translate each English sentence into the first-order language specified. Make full use of the notational conventions and abbreviations to make the end result as readable as possible.

2. Neither a nor b is a member of every set. ($\forall$, for all sets; $\in$, is a member of; a, a; b, b.)

3. If horses are animals, then heads of horses are heads of animals. ($\forall$, for all things; E, is a horse; A, is an animal; hx, the head of x or (if x is headless) x itself.)

4. (a) You can fool some of the people all of the time. (b) You can fool all of the people some of the time. (c) You can't fool all of the people all of the time. ($\forall$, for all things; P, is a person; T, is a time; Fxy, you can fool x at y. One or more of the above may be ambiguous, in which case you will need more than one translation.)

5. (a) Adams can't do every job right. (b) Adams can't do any job right. ($\forall$, for all things; J, is a job; a, Adams; Dxy, x can do y right.)

6. Nobody likes everybody. ($\forall$, for all people; Lxy, x likes y.)

7. Give a precise definition of what it means for the variable x to occur free as the ith symbol in the wff α. (If x is the ith symbol of α but does not occur free there, then it is said to occur *bound* there.)

8. Rewrite each of the following wffs in a way which explicitly lists each symbol in the actual order:

(a) $\exists \mathbf{v}_1\, P\mathbf{v}_1 \wedge P\mathbf{v}_1$.
(b) $\forall \mathbf{v}_1\, A\mathbf{v}_1 \wedge B\mathbf{v}_1 \rightarrow \exists \mathbf{v}_2 \neg C\mathbf{v}_2 \vee D\mathbf{v}_2$.

In each case, say which variables occur free in the wff.

§ 2.2 TRUTH AND MODELS

In sentential logic we had truth assignments to tell us which sentence symbols were to be interpreted as being true and which as false. In first-order logic the analogous role is played by structures, which can be thought of as providing translations from the formal language into English. (Structures are sometimes called interpretations, but we prefer to reserve that word for another concept, to be encountered in Section 2.7.)

A structure for a first-order language will tell us

1. What collection of things the universal quantifier symbol ($\forall$) refers to, and
2. What the other parameters (the predicate and function symbols) denote.

Formally, a *structure* $\mathfrak{A}$ for our given first-order language is a function whose domain is the set of parameters and such that

1. $\mathfrak{A}$ assigns to the quantifier symbol $\forall$ a nonempty set $|\mathfrak{A}|$, called the *universe* of $\mathfrak{A}$.
2. $\mathfrak{A}$ assigns to each n-place predicate symbol P an n-ary relation $P^{\mathfrak{A}} \subseteq |\mathfrak{A}|^n$; i.e., $P^{\mathfrak{A}}$ is a set of n-tuples of members of the universe.
3. $\mathfrak{A}$ assigns to each constant symbol c a member $c^{\mathfrak{A}}$ of the universe $|\mathfrak{A}|$.
4. $\mathfrak{A}$ assigns to each n-place function symbol f an n-ary operation $f^{\mathfrak{A}}$ on $|\mathfrak{A}|$; i.e., $f^{\mathfrak{A}} : |\mathfrak{A}|^n \rightarrow |\mathfrak{A}|$.

The idea is that $\mathfrak{A}$ assigns meaning to the parameters. $\forall$ is to mean "for everything in $|\mathfrak{A}|$." The symbol c is to name the point $c^{\mathfrak{A}}$. The atomic formula $Pt_1 \cdots t_n$ is to mean that the n-tuple of points named by $t_1, \ldots, t_n$ is in the relation $P^{\mathfrak{A}}$. (We will shortly restate these conditions more carefully.)

Note that we require the universe $|\mathfrak{A}|$ to be nonempty. Notice also that $f^{\mathfrak{A}}$ must have all of $|\mathfrak{A}|^n$ for its domain; we have made no provision for partially-defined functions.

Example Consider the language for set theory, whose only parameter (other than $\forall$) is $\in$. Take the structure $\mathfrak{A}$ with

$$|\mathfrak{A}| = \text{the set of natural numbers,}$$
$$\in^{\mathfrak{A}} = \text{the set of pairs } \langle m, n \rangle \text{ such that } m < n.$$

(Thus we translate $\in$ as "less than.") In the presence of a structure we can translate sentences from the formal language into English and attempt to say whether these translations are true or false. The sentence of this first-order language

$$\exists x \, \forall y \, \neg y \in x$$

(or more formally, $(\neg \forall \boldsymbol{v}_1(\neg \forall \boldsymbol{v}_2(\neg \in \boldsymbol{v}_2 \boldsymbol{v}_1))))$, which under another translation asserts the existence of an empty set, is now translated under $\mathfrak{A}$ into

> There is a natural number such that no natural number is smaller,

which is true. Because of this we will say that $\exists x \, \forall y \, \neg y \in x$ is *true* in $\mathfrak{A}$, or that $\mathfrak{A}$ is a *model* of the sentence. On the other hand, $\mathfrak{A}$ is not a model of the pair-set axiom,

$$\forall x \, \forall y \, \exists z \, \forall t (t \in z \leftrightarrow t \approx x \vee t \approx y),$$

as the translation of this sentence under $\mathfrak{A}$ is false. For there is no natural number m such that for every n,

$$n < m \qquad \text{iff } n = 1.$$

(The reader familiar with axiomatic set theory can check that $\mathfrak{A}$ is a model of the extensionality axiom, the union axiom, and the axiom of regularity.)

In the above example it was intuitively pretty clear that certain sentences of the formal language were true in the structure and some were false. But we want a precise mathematical definition of "σ is true in $\mathfrak{A}$." This should be stated in set-theoretic terms, without employing translations into English or a supposed criterion for asserting that some English sentences are true while the others are false. (If you think you have such a criterion, try it on the sentence "This sentence is false.") In other words, we want to take our intuitive notion of "σ is true in $\mathfrak{A}$" and make it part of mathematics.

In order to define "σ is true in $\mathfrak{A}$,"

$$\models_{\mathfrak{A}} \sigma,$$

for sentences σ and structures $\mathfrak{A}$, we will find it desirable first to define a more general notion involving wffs. Let

φ be a wff of our language,
$\mathfrak{A}$ a structure for the language,
$s : V \to |\mathfrak{A}|$ a function from the set V of all variables into the universe $|\mathfrak{A}|$ of $\mathfrak{A}$.

Then we will define what it means for $\mathfrak{A}$ to *satisfy* φ *with* s,

$$\models_{\mathfrak{A}} \varphi\,[s].$$

The intuitive version is

$\models_{\mathfrak{A}} \varphi\,[s]$ if and only if the translation of φ determined by $\mathfrak{A}$, where the variable x is translated as $s(x)$ wherever it occurs free, is true.

The formal definition of satisfaction proceeds as follows:

I. *Terms.* We define the extension

$$\bar{s} : T \to |\mathfrak{A}|,$$

a function from the set T of all terms into the universe of $\mathfrak{A}$. The idea is that $\bar{s}(t)$ should be the member of the universe $|\mathfrak{A}|$ that is named by the term t. $\bar{s}$ is defined by recursion as follows:

1. For each variable x, $\bar{s}(x) = s(x)$.
2. For each constant symbol c, $\bar{s}(c) = c^{\mathfrak{A}}$.
3. If $t_1, \ldots, t_n$ are terms and f is an n-place function symbol, then

$$\bar{s}(ft_1 \cdots t_n) = f^{\mathfrak{A}}(\bar{s}(t_1), \ldots, \bar{s}(t_n)).$$

A commutative diagram, for $n = 1$, is

$$\begin{array}{ccc} T & \xrightarrow{\bar{s}} & |\mathfrak{A}| \\ {\scriptstyle \mathscr{F}_f}\downarrow & & \downarrow{\scriptstyle f^{\mathfrak{A}}} \\ T & \xrightarrow{\bar{s}} & |\mathfrak{A}| \end{array}$$

The existence of a unique such extension $\bar{s}$ of s follows from the recursion theorem (Section 1.2), by using the fact that the terms are freely generated (Section 2.3). Notice that $\bar{s}$ depends both on s and on $\mathfrak{A}$.

II. *Atomic formulas.* The atomic formulas were defined explicitly, not inductively. The definition of satisfaction of atomic formulas is therefore also explicit, and not recursive.

1. $\models_{\mathfrak{A}} \approx t_1 t_2\, [s]$ iff $\bar{s}(t_1) = \bar{s}(t_2)$.

(Thus $\approx$ means $=$. Note that $\approx$ is a logical symbol, not a parameter open to interpretation.)

2. For an n-place predicate parameter P,

$$\models_{\mathfrak{A}} Pt_1 \ldots t_n\, [s] \qquad \text{iff } \langle \bar{s}(t_1), \ldots, \bar{s}(t_n)\rangle \in P^{\mathfrak{A}}.$$

III. *Other wffs.* The wffs we defined inductively, and consequently here satisfaction is defined recursively.

1. For atomic formulas, the definition is above.
2. $\models_{\mathfrak{A}} \neg\varphi\, [s]$ iff $\not\models_{\mathfrak{A}} \varphi\, [s]$.
3. $\models_{\mathfrak{A}} (\varphi \rightarrow \psi)\, [s]$ iff either $\not\models_{\mathfrak{A}} \varphi\, [s]$ or $\models_{\mathfrak{A}} \psi\, [s]$ or both.

(In other words, if $\mathfrak{A}$ satisfies φ with s then $\mathfrak{A}$ satisfies ψ with s.)

4. $\models_{\mathfrak{A}} \forall x\, \varphi\, [s]$ iff for every $d \in |\mathfrak{A}|$, we have $\models_{\mathfrak{A}} \varphi\, [s(x \mid d)]$.

Here $s(x|d)$ is the function which is exactly like s except for one thing: At the variable x it assumes the value d. This can be expressed by the equation:

$$s(x|d)(y) = \begin{cases} s(y) & \text{if } y \neq x, \\ d & \text{if } y = x. \end{cases}$$

(Thus $\forall$ means "for all things in $|\mathfrak{A}|$.")

At this point the reader might want to reconsider the intuitive version of $\models_{\mathfrak{A}} \varphi\, [s]$ on page 81 and observe how it has been formalized. We should also remark that the definition of satisfaction is another application of the recursion theorem together with the fact that the wffs are freely generated. The definition can be restated in terms of functions to make it clearer how the recursion theorem of Section 1.2 applies:

(i) Consider one fixed $\mathfrak{A}$.

(ii) Define a function $\bar{h}$ (extending a function h defined on atomic formulas) such that for any wff φ, $\bar{h}(\varphi)$ is a set of functions from V into $|\mathfrak{A}|$.

(iii) Define

$$\models_{\mathfrak{A}} \varphi\, [s] \qquad \text{iff } s \in \bar{h}(\varphi).$$

We leave to the reader the exercise of writing down the explicit definition of h and the clauses which uniquely determine its extension $\bar{h}$. (See Exercise 7.) An elegant alternative is to have $\bar{h}(\varphi)$ be a set of functions on the set of those variables which occur free in φ.

Definition Let Γ be a set wffs, φ a wff. Then Γ *logically implies* φ, $\Gamma \models \varphi$, iff for every structure $\mathfrak{A}$ for the language and every function $s : V \to |\mathfrak{A}|$ such that $\mathfrak{A}$ satisfies every member of Γ with s, $\mathfrak{A}$ also satisfies φ with s.

We use the same symbol, "$\models$," that was used in Chaper 1 for tautological implication. But henceforth it will be used solely for logical implication. As before we will write "$\gamma \models \varphi$" in place of "$\{\gamma\} \models \varphi$." Say that φ and ψ are *logically equivalent* ($\varphi \models\!\!\dashv \psi$) iff $\varphi \models \psi$ and $\psi \models \varphi$.

The first-order analog of the tautologies are the valid formulas. A wff φ is *valid* iff $\varnothing \models \varphi$ (written just "$\models \varphi$"). Thus φ is valid iff for every $\mathfrak{A}$ and every $s : V \to |\mathfrak{A}|$, $\mathfrak{A}$ satisfies φ with s.

At this point we pause to verify that when we want to know whether or not a structure $\mathfrak{A}$ satisfies a wff φ with s, we don't really need all of the (infinite amount of) information s gives us. All that matter are the values of the functions s at the (finitely many) variables which occur free in s. In particular, if φ is a sentence, then s does not matter at all.

Theorem 22A Assume that s_1, s_2 are functions from V into $|\mathfrak{A}|$ which agree at all variables (if any) which occur free in the wff φ. Then

$$\models_{\mathfrak{A}} \varphi\,[s_1] \qquad \text{iff } \models_{\mathfrak{A}} \varphi\,[s_2].$$

Proof Because satisfaction was defined recursively, this proof uses induction. We consider the fixed structure $\mathfrak{A}$ and show by induction that every wff φ has the property that whenever two functions s_1, s_2 agree on the variables free in φ, then $\mathfrak{A}$ satisfies φ with s_1 iff it does so with s_2.

Case 1: $\varphi = Pt_1 \cdots t_n$ is atomic. Then any variable in φ occurs free. Thus s_1 and s_2 agree at all the variables in each t_i. It follows that $\bar{s}_1(t_i) = \bar{s}_2(t_i)$ for each i; a detailed proof would use induction on t_i. Consequently, $\mathfrak{A}$ satisfies $Pt_1 \cdots t_n$ with s_1 iff it does so with s_2.

Cases 2 and 3: φ has the form $\neg\alpha$ or $\alpha \to \beta$. These cases are immediate from the inductive hypothesis.

Case 4: $\varphi = \forall x\, \psi$. Then the variables free in φ are those free in ψ with the exception of x. Thus for any d in $|\mathfrak{A}|$, $s_1(x|d)$ and $s_2(x|d)$ agree at all variables free in ψ. By inductive hypothesis, then, $\mathfrak{A}$ satisfies ψ with $s_1(x|d)$ iff it does so with $s_2(x|d)$. From this and the definition of satisfaction we see that $\mathfrak{A}$ satisfies $\forall x\,\psi$ with s_1 iff it does so with s_2. ■

In effect, the above proof amounts to looking through the definition of satisfaction and seeing what information given by s was actually used. There is an analogous fact regarding structures: If $\mathfrak{A}$ and $\mathfrak{B}$ agree at all the parameters which occur in φ, then $\models_{\mathfrak{A}} \varphi\,[s]$ iff $\models_{\mathfrak{B}} \varphi\,[s]$.

Corollary 22B For a sentence σ, either

(a) $\mathfrak{A}$ satisfies σ with every function s from V into $|\mathfrak{A}|$, or
(b) $\mathfrak{A}$ does not satisfy σ with any such function.

If alternative (a) holds, then we say that σ is *true* in $\mathfrak{A}$ ($\models_{\mathfrak{A}} \sigma$) or that $\mathfrak{A}$ is a *model* of σ. And if alternative (b) holds, then of course σ is *false* in $\mathfrak{A}$. (They cannot both hold since $|\mathfrak{A}|$ is nonempty). $\mathfrak{A}$ is a *model* of a set Σ of sentences iff it is a model of every member of Σ.

Corollary 22C For a set $\Sigma ; \tau$ of sentences, $\Sigma \models \tau$ iff every model of Σ is also a model of τ.

The definition of logical implication is very much like the definition of tautological implication in Chapter 1. But there is an important difference of complexity. Suppose in sentential logic you want to know whether or not a wff α is a tautology. The definition requires that you consider finitely many truth assignments, each of which is a finite function. For each such truth assignment v, you must calculate $\bar{v}(\alpha)$, which can be effectively done in a finite length of time. (Consequently, the set of tautologies is decidable, as was observed before.)

In contrast to the finitary procedure for tautologies, suppose that you want to know whether or not a wff φ (of our first-order language) is valid. The definition requires that you consider every structure $\mathfrak{A}$. (In particular this requires using every nonempty set, of which there are a great many.) For each of these structures, you then must consider each function s from the set V of variables into $|\mathfrak{A}|$. And for each given $\mathfrak{A}$ and s, you must determine whether or not $\mathfrak{A}$ satisfies φ with s. When $|\mathfrak{A}|$ is infinite, this is a complicated notion in itself.

In view of these complications, it is not surprising that the set of valid formulas fails to be decidable (cf. Section 3.5). What is surprising is that the notion of validity turns out to be equivalent to another notion (deducibility) whose definition is much closer to being finitary. (See Section 2.4.)

Using that equivalence we will be able to show (under some reasonable assumptions) that the set of validities (i.e., the set of valid wffs) is effectively enumerable. The effective enumeration procedure yields a more concrete characterization of the set of valid formulas.

The notational conventions adopted earlier were done in a rational way:

1. $\models_{\mathfrak{A}} (\alpha \wedge \beta)\,[s]$ iff $\models_{\mathfrak{A}} \alpha\,[s]$ and $\models_{\mathfrak{A}} \beta\,[s]$; similarly for $\vee$ and $\leftrightarrow$.

2. $\models_{\mathfrak{A}} \exists x\, \alpha\,[s]$ iff there is some $d \in |\mathfrak{A}|$ such that $\models_{\mathfrak{A}} \alpha\,[s(x|d)]$.

The proof for the second of these is as follows:

$\models_{\mathfrak{A}} \exists x\, \alpha\,[s]$ iff $\models_{\mathfrak{A}} \neg\forall x \neg\alpha[s]$,
iff it is not the case that for all d in $|\mathfrak{A}|$, $\not\models_{\mathfrak{A}} \alpha\,[s(x|d)]$,
iff for some d in $|\mathfrak{A}|$, $\models_{\mathfrak{A}} \alpha\,[s(x|d)]$. ∎

Example Assume that our language has the parameters $\forall$, P (a two-place predicate symbol), f (a one-place function symbol), and c (a constant symbol). Let $\mathfrak{A}$ be the structure for his language defined as follows:

$$|\mathfrak{A}| = N, \text{ the set of all natural numbers},$$
$$P^{\mathfrak{A}} = \text{the set of pairs } \langle m, n\rangle \text{ such that } m \leq n,$$
$$f^{\mathfrak{A}} = \text{the successor function } S;\ f^{\mathfrak{A}}(n) = n + 1,$$
$$c^{\mathfrak{A}} = 0.$$

We can summarize this in one line, by suppressing the fact that $\mathfrak{A}$ is really a function and merely listing its components:

$$\mathfrak{A} = (N, \leq, S, 0).$$

This notation is unambiguous only when the context makes clear just which components go with which parameters.

Let $s : V \to N$ be the function for which $s(\boldsymbol{v}_i) = i - 1$; i.e., $s(\boldsymbol{v}_1) = 0$, $s(\boldsymbol{v}_2) = 1$, etc.

1. $\bar{s}(ff\boldsymbol{v}_3) = S(S(2)) = 4$.

2. $\bar{s}(ffc) = 2$; no use is made of s.

3. $\models_{\mathfrak{A}} Pcf\boldsymbol{v}_1\,[s]$. This is intuitively obvious, since when we translate back into English we get the true sentence "$0 \leq 1$." More formally, the reason is that

$$\langle \bar{s}(c), \bar{s}(f\boldsymbol{v}_1)\rangle = \langle 0, 1\rangle \in P^{\mathfrak{A}}.$$

4. $\models_{\mathfrak{A}} \forall \boldsymbol{v}_1\, Pc\boldsymbol{v}_1$. The translation into English is "0 is less than or equal to any natural number." Formally we must verify that for any n in $\mathbb{N}$,

$$\models_{\mathfrak{A}} Pc\boldsymbol{v}_1\, [s(\boldsymbol{v}_1|n)],$$

which reduces to

$$\langle 0, n\rangle \in P^{\mathfrak{A}}.$$

5. $\not\models_{\mathfrak{A}} \forall \boldsymbol{v}_1\, P\boldsymbol{v}_2\boldsymbol{v}_1\, [s]$ because there is a natural number m such that

$$\not\models_{\mathfrak{A}} P\boldsymbol{v}_2\boldsymbol{v}_1\, [s(\boldsymbol{v}_1|m)];$$

i.e.,

$$\langle s(\boldsymbol{v}_2), m\rangle \notin P^{\mathfrak{A}}.$$

In fact, since $s(\boldsymbol{v}_2) = 1$, we must take m to be 0.

The reader is to be cautioned against confusing, for example, the function *symbol* f with the *function* $f^{\mathfrak{A}}$.

EXAMPLES of logical implication. The reader is invited to convince himself of each of the following:

1. $\forall \boldsymbol{v}_1\, Q\boldsymbol{v}_1 \models Q\boldsymbol{v}_2$.
2. $Q\boldsymbol{v}_1 \not\models \forall \boldsymbol{v}_1\, Q\boldsymbol{v}_1$.
3. $\forall \boldsymbol{v}_1\, Q\boldsymbol{v}_1 \models \exists \boldsymbol{v}_2\, Q\boldsymbol{v}_2$.
4. $\exists x\, \forall y\, Pxy \models \forall y\, \exists x\, Pxy$.
5. $\forall y\, \exists x\, Pxy \not\models \exists x\, \forall y\, Pxy$.
6. $\models \exists x(Qx \to \forall x\, Qx)$.

EXAMPLE Suppose our given language has only the parameters $\forall$ and P, where P is a two-place predicate symbol. Then a structure $\mathfrak{A}$ is determined by the universe $|\mathfrak{A}|$ and the binary relation $P^{\mathfrak{A}}$. With some minor illegality we again write

$$\mathfrak{A} = (|\mathfrak{A}|, P^{\mathfrak{A}}).$$

Now consider the problem of characterizing the class of all models of the following sentences:

1. $\forall x \forall y\, x \approx y$. A structure (A, R) is a model of this iff A contains exactly one element. R can either be empty or can be the singleton $A \times A$.

2. $\forall x \forall y\, Pxy$. A structure (A, R) is a model of this iff $R = A \times A$. A can be any nonempty set.

3. $\forall x\, \forall y\, \neg Pxy$. A structure (A, R) is a model of this iff $R = \varnothing$.

4. $\forall x\, \exists y\, Pxy$. The condition for (A, R) to be a model of this is that dom R be A.

Definability of a class of structures

For a set Σ of sentences, let Mod Σ be the class of all models of Σ, i.e., the class of all structures for the language in which every member of Σ is true. For a single sentence τ we write simply "Mod τ" instead of "Mod $\{\tau\}$." (The reader familiar with axiomatic set theory will notice that Mod Σ, if nonempty, is a proper class; i.e., it is too large to be a set.)

A class $\mathcal{K}$ of structures for our language is an *elementary class* (EC) iff $\mathcal{K} = \text{Mod}\,\tau$ for some sentence τ. $\mathcal{K}$ is an *elementary class in the wider sense* (EC_Δ) iff $\mathcal{K} = \text{Mod}\,\Sigma$ for some set Σ of sentences. (The adjective "elementary" is synonymous with "first-order.")

EXAMPLES **1.** Assume that the language has equality and the parameters $\forall$ and P, where P is a two-place predicate symbol. As before, a structure (A, R) for the language consists of a nonempty set A together with a binary relation R on A. (A, R) is an *ordered set* iff R is transitive and satisfies the trichotomy condition (which states that for any a and b in A, exactly one of $\langle a, b\rangle \in R$, $a = b$, $\langle b, a\rangle \in R$ holds). Because these conditions can be translated into a sentence of the formal language, the class of nonempty ordered sets is an elementary class. It is, in fact, Mod τ, where τ is the conjunction of the three sentences

$$\forall x\,\forall y\,\forall z(xPy \to yPz \to xPz);$$
$$\forall x\,\forall y(xPy \vee x \approx y \vee yPx);$$
$$\forall x\,\forall y(xPy \to \neg yPx).$$

The next two examples assume that the reader has had some contact with algebra.

2. Assume that the language has $\approx$ and the parameters $\forall$ and $\circ$, where $\circ$ is a two-place function symbol. The class of all groups is an elementary class, being the class of all models of the conjunction of the group axioms:

$$\forall x\,\forall y\,\forall z\; x \circ (y \circ z) \approx (x \circ y) \circ z;$$
$$\forall x\,\forall y\;\exists z\; x \circ z \approx y;$$
$$\forall x\,\forall y\;\exists z\; z \circ x \approx y.$$

The class of all infinite groups is EC_Δ. To see this, let

$$\lambda_2 = \exists x\,\exists y\; x \not\approx y,$$
$$\lambda_3 = \exists x\,\exists y\,\exists z(x \not\approx y \wedge y \not\approx z \wedge x \not\approx z),$$
$$\ldots.$$

Thus λ_n translates, "There are at least n things." Then the group axioms together with $\{\lambda_2, \lambda_3, \ldots\}$ form a set Σ for which Mod Σ is exactly the class of infinite groups. We will eventually (in Section 2.6) be able to show that the class of infinite groups is not EC.

3. Assume that the language has equality and the parameters $\forall$, **0**, **1**, $+$, Fields can be regarded as structures for this language. The class of all fields is an elementary class. The class of fields of characteristic zero is EC_Δ. It is not EC, a fact which will follow from the compactness theorem for first-order logic (Section 2.6 again).

Definability within a structure

Consider a fixed structure $\mathfrak{A}$. Suppose that φ is a formula such that all variables occurring free in φ are included among $\mathbf{v}_1, \ldots, \mathbf{v}_k$. Then for elements $a_1, \ldots, a_k$ of $|\mathfrak{A}|$,

$$\models_{\mathfrak{A}} \varphi\, [\![a_1, \ldots, a_k]\!]$$

means that $\mathfrak{A}$ satisfies φ with some (and hence with any) function $s : V \to |\mathfrak{A}|$ for which $s(\mathbf{v}_i) = a_i$, $1 \leq i \leq k$.

With each such φ and structure $\mathfrak{A}$ we can associate the k-ary relation on $|\mathfrak{A}|$,

$$\{\langle a_1, \ldots, a_k\rangle : \models_{\mathfrak{A}} \varphi\, [\![a_1, \ldots, a_k]\!]\}.$$

Call this the k-ary relation φ *defines* in $\mathfrak{A}$. In general, a k-ary relation on $|\mathfrak{A}|$ is said to be *definable* in $\mathfrak{A}$ iff there is a formula (whose free variables are among $\mathbf{v}_1, \ldots, \mathbf{v}_k$) which defines it there.

EXAMPLE Assume that we have a part of the language for number theory, specifically that our language has the parameters $\forall$, **0**, **S**, $+$, and $\cdot$. Let $\mathfrak{N}$ be the intended structure:

$|\mathfrak{N}| = N$, the set of natural numbers.

$\mathbf{0}^{\mathfrak{N}} = 0$, the number 0.

$\mathbf{S}^{\mathfrak{N}}$, $+^{\mathfrak{N}}$, and $\cdot^{\mathfrak{N}}$ are S, $+$, and $\cdot$, the functions of successor, addition, and multiplication.

In one equation,

$$\mathfrak{N} = (N, 0, S, +, \cdot).$$

Some relations on N are definable in $\mathfrak{N}$ and some are not. We know that some are not definable because there are uncountably many relations on N but only $\aleph_0$ possible defining formulas. (There is, however, an inherent difficulty in giving a specific example. After all, if something is undefinable, then it is hard to say exactly what it is. Later we will get to see a specific example, the set of Gödel numbers of sentences true in $\mathfrak{N}$; see Section 3.5.)

1. The ordering relation $\{\langle m, n\rangle : m < n\}$ is defined in $\mathfrak{N}$ by the formula

$$\exists \boldsymbol{v}_3\, \boldsymbol{v}_1 + \mathbf{S}\boldsymbol{v}_3 \approx \boldsymbol{v}_2.$$

2. For any natural number n, $\{n\}$ is definable. For example, $\{2\}$ is defined by

$$\boldsymbol{v}_1 \approx \mathbf{SS0}.$$

Because of this we say that n is a *definable element* in $\mathfrak{N}$.

3. The set of primes is definable in $\mathfrak{N}$. We could use the formula

$$\mathbf{1} < \boldsymbol{v}_1 \wedge \forall \boldsymbol{v}_2 \forall \boldsymbol{v}_3 (\boldsymbol{v}_1 \approx \boldsymbol{v}_2 \cdot \boldsymbol{v}_3 \rightarrow \boldsymbol{v}_2 \approx \mathbf{1} \vee \boldsymbol{v}_3 \approx \mathbf{1})$$

if we had parameters $\mathbf{1}$ and $\mathbf{<}$ for 1 and $<$. But since $\{1\}$ and $<$ are definable in $\mathfrak{N}$, it is really quite unnecessary to add parameters for them; we can just use their definitions instead. Thus the set of primes is definable by

$$\exists \boldsymbol{v}_3\, \mathbf{S0} + \mathbf{S}\boldsymbol{v}_3 \approx \boldsymbol{v}_1 \wedge \forall \boldsymbol{v}_2 \forall \boldsymbol{v}_3 (\boldsymbol{v}_1 \approx \boldsymbol{v}_2 \cdot \boldsymbol{v}_3 \rightarrow \boldsymbol{v}_2 \approx \mathbf{S0} \vee \boldsymbol{v}_3 \approx \mathbf{S0}).$$

4. Exponentiation, $\{\langle m, n, p\rangle : p = m^n\}$ is also definable in $\mathfrak{N}$. This is much less obvious; we will give a proof later (in Section 3.7) using the Chinese remainder theorem.

In fact, we will argue later that any decidable relation on N is definable in $\mathfrak{N}$, as is any effectively enumerable relation and a great many others. To some extent the complexity of a definable relation can be measured by the complexity of the simplest defining formula. This idea will come up again at the end of Section 3.5.

Homomorphisms

Let $\mathfrak{A}$, $\mathfrak{B}$ be structures for the language. A *homomorphism* h of $\mathfrak{A}$ into $\mathfrak{B}$ is a function $h : |\mathfrak{A}| \rightarrow |\mathfrak{B}|$ such that

(a) For each n-place predicate symbol P and each n-tuple $\langle a_1, \ldots, a_n\rangle$ of elements of $|\mathfrak{A}|$,

$$\langle a_1, \ldots, a_n\rangle \in P^{\mathfrak{A}} \quad \text{iff } \langle h(a_1), \ldots, h(a_n)\rangle \in P^{\mathfrak{B}}.$$

(b) For each n-place function symbol f and each such n-tuple,

$$h(f^{\mathfrak{A}}(a_1, \ldots, a_n)) = f^{\mathfrak{B}}(h(a_1), \ldots, h(a_n)).$$

In the case of a constant symbol c this becomes

$$h(c^{\mathfrak{A}}) = c^{\mathfrak{B}}.$$

Conditions (a) and (b) are usually stated: "h preserves the relations and functions." (It must be admitted that some authors use a weakened version of condition (a); our homomorphisms are their "strong homomorphisms.")

If, in addition, h is one-to-one, it is then called an *isomorphism* of $\mathfrak{A}$ into $\mathfrak{B}$. If there is an isomorphism of $\mathfrak{A}$ onto $\mathfrak{B}$ (i.e., an isomorphism h for which $\operatorname{ran} h = |\mathfrak{B}|$), then $\mathfrak{A}$ and $\mathfrak{B}$ are said to be *isomorphic* ($\mathfrak{A} \cong \mathfrak{B}$).

The reader has quite possibly encountered this notion before in special cases such as structures which are groups or fields.

Example Assume that we have a language with the parameters $\forall$, $+$, and $\cdot$. Let $\mathfrak{A}$ be the structure $(\mathbb{N}, +, \cdot)$. We can define a function $h : \mathbb{N} \to \{e, o\}$ by

$$h(n) = \begin{cases} e & \text{if } n \text{ is even,} \\ o & \text{if } n \text{ is odd.} \end{cases}$$

Then h is a homomorphism of $\mathfrak{A}$ onto $\mathfrak{B}$, where $|\mathfrak{B}| = \{e, o\}$ and $+^{\mathfrak{B}}$, $\cdot^{\mathfrak{B}}$ are given by the following tables:

$+^{\mathfrak{B}}$	e	o
e	e	o
o	o	e

$\cdot^{\mathfrak{B}}$	e	o
e	e	e
o	e	o

It can then be verified that condition (b) of the definition is satisfied. For example, if a and b are both odd, then $h(a + b) = e$ and $h(a) +^{\mathfrak{B}} h(b) = o +^{\mathfrak{B}} o = e$.

Example Let $\mathbb{P}$ be the set of positive integers, let $<_P$ be the usual ordering relation on $\mathbb{P}$, and let $<_N$ be the usual ordering relation on $\mathbb{N}$. Then there is an isomorphism h from the structure $(\mathbb{P}, <_P)$ onto $(\mathbb{N}, <_N)$; we take $h(n) = n - 1$. Also the identity map $i : \mathbb{P} \to \mathbb{N}$ is an isomorphism of $(\mathbb{P}, <_P)$ *into* $(\mathbb{N}, <_N)$. Because of this last fact, we say that $(\mathbb{P}, <_P)$ is a *substructure* of $(\mathbb{N}, <_N)$.

More generally consider two structures $\mathfrak{A}$ and $\mathfrak{B}$ for the language such that $|\mathfrak{A}| \subseteq |\mathfrak{B}|$. It is clear from the definition of homomorphism that the identity map from $|\mathfrak{A}|$ into $|\mathfrak{B}|$ is an isomorphism of $\mathfrak{A}$ into $\mathfrak{B}$ iff

(a) $P^{\mathfrak{A}}$ is the restriction of $P^{\mathfrak{B}}$ to $|\mathfrak{A}|$, for each predicate symbol P;

(b) $f^{\mathfrak{A}}$ is the restriction of $f^{\mathfrak{B}}$ to $|\mathfrak{A}|$, for each function symbol f, and $c^{\mathfrak{A}} = c^{\mathfrak{B}}$ for each constant symbol c.

If these conditions are met, then $\mathfrak{A}$ is said to be a *substructure* of $\mathfrak{B}$, and $\mathfrak{B}$ is an *extension* of $\mathfrak{A}$.

These are basically algebraic notions, but the following theorem relates them to the logical notions of truth and satisfaction.

Homomorphism Theorem Let h be a homomorphism of $\mathfrak{A}$ into $\mathfrak{B}$, and let s map the set of variables into $|\mathfrak{A}|$.

(a) For any term t, $h(\bar{s}(t)) = \overline{h \circ s}(t)$, where $\bar{s}(t)$ is computed in $\mathfrak{A}$ and $\overline{h \circ s}(t)$ is computed in $\mathfrak{B}$.

(b) For any quantifier-free formula α not containing the equality symbol,

$$\models_{\mathfrak{A}} \alpha\,[s] \qquad \text{iff } \models_{\mathfrak{B}} \alpha\,[h \circ s].$$

(c) If h is one-to-one (i.e., is an isomorphism of $\mathfrak{A}$ into $\mathfrak{B}$), then in part (b) we may delete the restriction "not containing the equality symbol."

(d) If h is a homomorphism of $\mathfrak{A}$ *onto* $\mathfrak{B}$, then in (b) we may delete the restriction "quantifier-free."

Proof Part (a) uses induction on t; see Exercise 13.

(b) For an atomic formula such as Pt, we have

$$\begin{aligned} \models_{\mathfrak{A}} Pt\,[s] &\Leftrightarrow \bar{s}(t) \in P^{\mathfrak{A}} \\ &\Leftrightarrow h(\bar{s}(t)) \in P^{\mathfrak{B}} && \text{since } h \text{ is a homomorphism} \\ &\Leftrightarrow \overline{h \circ s}(t) \in P^{\mathfrak{B}} && \text{by (a)} \\ &\Leftrightarrow \models_{\mathfrak{B}} Pt\,[h \circ s]. \end{aligned}$$

An inductive argument is then required to handle the connective symbols $\neg$ and $\rightarrow$, but it is completely routine.

(c) In any case,

$$\begin{aligned} \models_{\mathfrak{A}} u \approx t\,[s] &\Leftrightarrow \bar{s}(u) = \bar{s}(t) \\ &\Rightarrow h(\bar{s}(u)) = h(\bar{s}(t)) \\ &\Leftrightarrow \overline{h \circ s}(u) = \overline{h \circ s}(t) && \text{by (a)} \\ &\Leftrightarrow \models_{\mathfrak{B}} u \approx t\,[h \circ s]. \end{aligned}$$

If h is one-to-one, the arrow in the second step can be reversed.

(d) We must extend the routine inductive argument of part (b) to include the quantifier step. That is, we must show that *if* φ has the property that for every s,

$$\models_{\mathfrak{A}} \varphi\,[s] \Leftrightarrow \models_{\mathfrak{B}} \varphi\,[h \circ s],$$

then $\forall x\,\varphi$ enjoys the same property. We have in any case (as a consequence of the inductive hypothesis on φ) the implication

$$\models_{\mathfrak{B}} \forall x\,\varphi\,[h \circ s] \Rightarrow \models_{\mathfrak{A}} \forall x\,\varphi\,[s].$$

This is intuitively very plausible; if φ is true of everything in the larger set $|\mathfrak{B}|$, then *a fortiori* it is true of everything in the smaller set $|\mathfrak{A}|$. The details are, for an element a of $|\mathfrak{A}|$,

$$\begin{aligned} \models_{\mathfrak{B}} \forall x\,\varphi\,[h \circ s] &\Rightarrow \models_{\mathfrak{B}} \varphi\,[(h \circ s)(x|h(a))] \\ &\Leftrightarrow \models_{\mathfrak{B}} \varphi\,[h \circ (s(x|a))], && \text{the functions being the same} \\ &\Leftrightarrow \models_{\mathfrak{A}} \varphi\,[s(x|a)] && \text{by the inductive hypothesis.} \end{aligned}$$

For the converse, suppose that $\not\models_{\mathfrak{B}} \forall x\,\varphi\,[h \circ s]$, so that $\models_{\mathfrak{B}} \neg\varphi\,[(h \circ s)(x|b)]$ for some element b in $|\mathfrak{B}|$. We need the implication

(∗) *If* for some b in $|\mathfrak{B}|$, $\models_{\mathfrak{B}} \neg\varphi\,[(h \circ s)(x|b)]$, *then* for some a in $|\mathfrak{A}|$, $\models_{\mathfrak{B}} \neg\varphi\,[(h \circ s)(x|h(a))]$.

For given (∗), we can proceed:

$$\begin{aligned} \models_{\mathfrak{B}} \neg\varphi\,[(h \circ s)(x|h(a))] &\Leftrightarrow \models_{\mathfrak{B}} \neg\varphi\,[h \circ (s(x|a))], && \text{the functions being the same} \\ &\Leftrightarrow \models_{\mathfrak{A}} \neg\varphi\,[s(x|a)] && \text{by the inductive hypothesis} \\ &\Rightarrow \not\models_{\mathfrak{A}} \forall x\,\varphi\,[s]. \end{aligned}$$

If h maps $|\mathfrak{A}|$ *onto* $|\mathfrak{B}|$, then (∗) is immediate; we take a such that $b = h(a)$. (But there might be other fortunate times when (∗) can be asserted even if h fails to have range $|\mathfrak{B}|$.) ■

Two structures $\mathfrak{A}$ and $\mathfrak{B}$ for the language are *elementarily equivalent* ($\mathfrak{A} \equiv \mathfrak{B}$) iff for any sentence σ,

$$\models_{\mathfrak{A}} \sigma \Leftrightarrow \models_{\mathfrak{B}} \sigma.$$

As a corollary to the above theorem we have: Isomorphic structures are elementarily equivalent. Actually more is true. Isomorphic structures are alike in every "structural" way; not only do they satisfy the same first-order sentences, they also satisfy the same second-order (and higher) sentences (i.e., they are secondarily equivalent and more).

There are elementarily equivalent structures which are not isomorphic. For example, it can be shown that the structure $(\mathbb{R}, <_R)$ consisting of the set of real numbers with its usual ordering relation is elementarily equivalent to the structure $(\mathbb{Q}, <_Q)$ consisting of the set of rational numbers with its ordering (see Section 2.6). But $\mathbb{Q}$ is a countable set whereas $\mathbb{R}$ is not, so these structures cannot be isomorphic. In Section 2.6 we will see how easy it is to make elementarily equivalent structures of differing cardinalities.

EXAMPLE, revisited We had an isomorphism h from $(\mathbb{P}, <_P)$ onto $(\mathbb{N}, <_N)$. So in particular, $(\mathbb{P}, <_P) \equiv (\mathbb{N}, <_N)$; these structures are indistinguishable by first-order sentences.

We furthermore noted that the identity map was an isomorphism of $(\mathbb{P}, <_P)$ into $(\mathbb{N}, <_N)$. Hence for a function $s : V \to \mathbb{P}$ and a quantifier-free φ,

$$\models_{(P,<_P)} \varphi\,[s] \Leftrightarrow \models_{(N,<_N)} \varphi\,[s].$$

This equivalence may fail if φ contains quantifiers. For example,

$$\models_{(P,<_P)} \forall v_2(v_1 \not\approx v_2 \to v_1 < v_2)\, [\![1]\!],$$

but

$$\not\models_{(N,<_N)} \forall v_2(v_1 \not\approx v_2 \to v_1 < v_2)\, [\![1]\!].$$

An *automorphism* of the structure $\mathfrak{A}$ is an isomorphism of $\mathfrak{A}$ onto $\mathfrak{A}$. The identity function on $|\mathfrak{A}|$ is trivially an automorphism of $\mathfrak{A}$. $\mathfrak{A}$ may or may not have nontrivial automorphisms. As a consequence of the homomorphism theorem, we can show that an automorphism must preserve the definable relations:

Corollary 22D Let h be an automorphism of the structure $\mathfrak{A}$, and let R be an n-ary relation on $|\mathfrak{A}|$ definable in $\mathfrak{A}$. Then for any $a_1, \ldots, a_n$ in $|\mathfrak{A}|$,

$$\langle a_1, \ldots, a_n\rangle \in R \Leftrightarrow \langle h(a_1), \ldots, h(a_n)\rangle \in R.$$

Proof Let φ be the formula which defines R in $\mathfrak{A}$. We need to know that

$$\models_{\mathfrak{A}} \varphi\, [\![a_1, \ldots, a_n]\!] \Leftrightarrow \models_{\mathfrak{A}} \varphi\, [\![h(a_1), \ldots, h(a_n)]\!].$$

But this is immediate from the homomorphism theorem. ■

This corollary is sometimes useful in showing that a given relation is *not* definable. Consider, for example, the structure $(\mathbb{R}, <)$ consisting of the set of real numbers with its usual ordering. An automorphism of this structure is simply a function h from $\mathbb{R}$ onto $\mathbb{R}$ which is strictly increasing:

$$a < b \Leftrightarrow h(a) < h(b).$$

One such automorphism is the function h for which $h(a) = a^3$. Since this function maps points outside of $\mathbb{N}$ into $\mathbb{N}$, the set $\mathbb{N}$ is not definable in this structure.

Another example is provided by elementary algebra books, which sometimes explain that the length of a vector in the plane cannot be defined in terms of vector addition and scalar multiplication. For the map which takes a vector $\boldsymbol{x}$ into the vector $2\boldsymbol{x}$ is an automorphism of the plane with respect to vector addition and scalar multiplication, but it is not length-preserving. From our viewpoint, the structure in question,

$$(E, +, f_r)_{r \in \mathbb{R}},$$

has for its universe the plane E, has the binary function $+$ of vector addition, and has (for each r in the set $\mathbb{R}$) the unary function f_r of scalar multiplication by r. (Thus the language in question has a one-place function symbol for each real number.) The doubling map described above is an automorphism of this structure. But it does not preserve the set of unit vectors,

$$\{\boldsymbol{x} : \boldsymbol{x} \in E \text{ and } \boldsymbol{x} \text{ has length } 1\}.$$

So this set cannot be definable in the structure. (Incidentally, the homomorphisms of vector spaces are normally called *linear transformations.*)

EXERCISES

1. Show that (a) $\Gamma ; \alpha \models \varphi$ iff $\Gamma \models (\alpha \rightarrow \varphi)$; and (b) $\varphi \models\mathrel{\mkern-3mu}\dashv \psi$ iff $\models (\varphi \leftrightarrow \psi)$.

2. Show that no one of the following sentences is logically implied by the other two. (This is done by giving a structure in which the sentence in question is false, while the other two are true.)

(a) $\forall x \forall y \forall z (Pxy \rightarrow Pyz \rightarrow Pxz)$. Recall that by our convention $\alpha \rightarrow \beta \rightarrow \gamma$ is $\alpha \rightarrow (\beta \rightarrow \gamma)$.

(b) $\forall x \forall y (Pxy \rightarrow Pyx \rightarrow x \approx y)$.

(c) $\forall x \exists y\, Pxy \rightarrow \exists y \forall x\, Pxy$.

3. Show that

$$\{\forall x(\alpha \to \beta), \forall x\, \alpha\} \models \forall x\, \beta.$$

4. Show that if x does not occur free in α, then $\alpha \models \forall x\, \alpha$.

5. Show that the formula $x \approx y \to Pzfx \to Pzfy$ (where f is a one-place function symbol and P is a two-place predicate symbol) is valid.

6. Show that a formula θ is valid iff $\forall x\, \theta$ is valid.

7. Restate the definition of "$\mathfrak{A}$ satisfies φ with s" in the way described on pages 82f. That is, define by recursion a function $\bar{h}$ such that $\mathfrak{A}$ satisfies φ with s iff $s \in \bar{h}(\varphi)$.

8. Assume that Σ is a set of sentences such that for any sentence τ, either $\Sigma \models \tau$ or $\Sigma \models \neg\tau$. Assume that $\mathfrak{A}$ is a model of Σ. Show that for any sentence τ, $\models_{\mathfrak{A}} \tau$ iff $\Sigma \models \tau$.

9. Assume that the language has equality and a two-place predicate symbol P. For each of the following conditions, find a sentence σ such that the structure $\mathfrak{A}$ $(= (|\mathfrak{A}|, P^{\mathfrak{A}}))$ is a model of σ iff the condition is met.

(a) $|\mathfrak{A}|$ has exactly two members.
(b) $P^{\mathfrak{A}}$ is a function from $|\mathfrak{A}|$ into $|\mathfrak{A}|$.
(c) $P^{\mathfrak{A}}$ is a permutation of $|\mathfrak{A}|$; i.e., $P^{\mathfrak{A}}$ is a one-to-one function with domain and range equal to $|\mathfrak{A}|$.

10. Show that

$$\models_{\mathfrak{A}} \forall \mathbf{v}_2\, Q\mathbf{v}_1\mathbf{v}_2\, [\![c^{\mathfrak{A}}]\!] \qquad \text{iff } \models_{\mathfrak{A}} \forall \mathbf{v}_2\, Qc\mathbf{v}_2.$$

Here Q is a two-place predicate symbol and c is a constant symbol.

11. For each of the following relations, give a formula which defines it in $(\mathbb{N}, +, \cdot)$. (The language is assumed to have equality and the parameters $\forall$, $+$, and $\cdot$.)

(a) $\{0\}$.
(b) $\{1\}$.
(c) $\{\langle m, n\rangle : n$ is the successor of m in $\mathbb{N}\}$.
(d) $\{\langle m, n\rangle : m < n$ in $\mathbb{N}\}$.

12. Let $\mathfrak{R}$ be the structure $(\mathbb{R}, +, \cdot)$. (The language is assumed to have equality and the parameters $\forall$, $+$, and $\cdot$. $\mathfrak{R}$ is the structure whose universe is the set $\mathbb{R}$ of real numbers and such that $+^{\mathfrak{R}}$ and $\cdot^{\mathfrak{R}}$ are the usual addition and multiplication operations.)

(a) Give a formula which defines in $\mathfrak{R}$ the set $[0, \infty)$.

(b) Give a formula which defines in $\mathfrak{R}$ the set $\{2\}$.

*(c) Show that any finite union of intervals, the endpoints of which are algebraic, is definable in $\mathfrak{R}$. (The converse is also true; these are the only definable sets in the structure. But we will not prove this fact.)

13. Prove part (a) of the homomorphism theorem.

14. What subsets of the real line $\mathbb{R}$ are definable in $(\mathbb{R}, <)$? What subsets of the plane $\mathbb{R} \times \mathbb{R}$ are definable in $(\mathbb{R}, <)$?

15. Show that the addition relation, $\{\langle m, n, p\rangle : p = m + n\}$, is not definable in $(\mathbb{N}, \cdot)$. *Suggestion*: Consider an automorphism of $(\mathbb{N}, \cdot)$ which switches two primes.

16. Let $\mathfrak{A}$ be a structure; let B be a set which includes $|\mathfrak{A}|$. Show that there is a structure $\mathfrak{B}$ whose universe is B and such that for any sentence σ not containing $\approx$, $\models_{\mathfrak{A}} \sigma$ iff $\models_{\mathfrak{B}} \sigma$. *Suggestion*: Choose some $a_0 \in |\mathfrak{A}|$. Let $h : B \to |\mathfrak{A}|$ map points in $|\mathfrak{A}|$ into themselves and map other points of B into a_0. Define $\mathfrak{B}$ so that h is a homomorphism of $\mathfrak{B}$ onto $\mathfrak{A}$.

17. (a) Consider a language with equality whose only parameter (aside from $\forall$) is a two-place predicate symbol P. Show that if $\mathfrak{A}$ is finite and $\mathfrak{A} \equiv \mathfrak{B}$, then $\mathfrak{A}$ is isomorphic to $\mathfrak{B}$.

*(b) Show that the result of part (a) holds regardless of what parameters the language contains.

18. A universal ($\forall_1$) formula is one of the form $\forall x_1 \cdots \forall x_n\, \theta$, where θ is quantifier-free. An existential ($\exists_1$) formula is of the dual form $\exists x_1 \cdots \exists x_n$ θ. Let $\mathfrak{A}$ be a substructure of $\mathfrak{B}$, and let $s : V \to |\mathfrak{A}|$.

(a) Show that if $\models_{\mathfrak{B}} \varphi\, [s]$ and φ is universal, then $\models_{\mathfrak{A}} \varphi\, [s]$. And if $\models_{\mathfrak{A}} \psi\, [s]$ and ψ is existential, then $\models_{\mathfrak{B}} \psi\, [s]$.

(b) Conclude that the sentence $\forall x\, Px$ is not logically equivalent to any existential sentence, nor $\exists x\, Px$ to any universal sentence.

19. An $\exists_2$ formula is one of the form $\exists x_1 \cdots \exists x_n\, \theta$, where θ is universal.

(a) Show that if an $\exists_2$ sentence not containing function symbols is true in $\mathfrak{A}$, then it is true in some finite substructure of $\mathfrak{A}$.

(b) Conclude that $\forall x\, \exists y\, Pxy$ is not logically equivalent to any $\exists_2$ sentence.

20. Assume the language has equality and a two-place predicate symbol P. Consider the two structures $(\mathbb{N}, <)$ and $(\mathbb{R}, <)$ for the language.

(a) Find a sentence true in one structure and false in the other.

*(b) Show that any $\exists_2$ sentence (as defined in the preceding exercise) true in $(\mathbb{R}, <)$ is also true in $(\mathbb{N}, <)$.

21. We could consider enriching the language by the addition of a new quantifier. The formula $\exists!x\,\alpha$ (read: there exists a unique x such that α) is to be satisfied in $\mathfrak{A}$ by s iff there is one and only one $a \in |\mathfrak{A}|$ such that $\models_{\mathfrak{A}} \alpha\,[s(x|a)]$. Assume that the language has the equality symbol and show that this apparent enrichment comes to naught, in the sense that we can find an ordinary formula α' logically equivalent to $\exists!x\,\alpha$.

22. Let $\mathfrak{A}$ be a structure and g a one-to-one function with dom $g = |\mathfrak{A}|$. Show that there is a unique structure $\mathfrak{B}$ such that g is an isomorphism of $\mathfrak{A}$ onto $\mathfrak{B}$.

23. Let h be an isomorphism of $\mathfrak{A}$ into $\mathfrak{B}$. Show that there is a structure $\mathfrak{C}$ isomorphic to $\mathfrak{B}$ such that $\mathfrak{A}$ is a substructure of $\mathfrak{C}$. *Suggestion*: Let g be a one-to-one function with domain $|\mathfrak{B}|$ such that $g(h(a)) = a$ for $a \in |\mathfrak{A}|$. Form $\mathfrak{C}$ such that g is an isomorphism of $\mathfrak{B}$ onto $\mathfrak{C}$.

24. Consider a fixed structure $\mathfrak{A}$. Expand the language by adding a new constant symbol c_a for each $a \in |\mathfrak{A}|$. Let $\mathfrak{A}^+$ be the structure for this expanded language which agrees with $\mathfrak{A}$ on the original parameters and which assigns to c_a the point a. A relation R on $|\mathfrak{A}|$ is said to be *definable from points* in $\mathfrak{A}$ iff R is definable in $\mathfrak{A}^+$. (This differs from ordinary definability only in that we now have parameters in the language for members of $|\mathfrak{A}|$.) Let $\mathfrak{R} = (\mathbb{R}, <, +, \cdot)$.

(a) Show that if A is a subset of $\mathbb{R}$ consisting of the union of finitely many intervals, then A is definable from points in $\mathfrak{R}$.

(b) Assume that $\mathfrak{A} \equiv \mathfrak{R}$. Show that any subset of $|\mathfrak{A}|$ which is nonempty, bounded (in the ordering $<^{\mathfrak{A}}$), and definable from points in $\mathfrak{A}$ has a least upper bound in $|\mathfrak{A}|$.

§ 2.3 UNIQUE READABILITY[1]

As in sentential logic, we need unique readability results in order to be able to apply the recursion theorem. If we had used Polish notation both for terms and for formulas, then we could give one proof that would yield

[1] This section may be omitted by a reader willing to accept the meaningfulness of our many definitions by recursion.

simultaneously the facts that the terms are freely generated and that the formulas are freely generated. But, in fact, we have used a mixed notation. Consequently, we will give first a unique readability proof for terms; actually that proof will be applicable to any use of Polish notation. And then we will give an extension of that proof to cover the formulas.

Recall that the set of terms is generated from the variables and constant symbols by operations corresponding to the function symbols. We now define a function K on the symbols involved such that for a symbol s, $K(s) = 1 - n$, where n is the number of terms which must follow s to obtain a term.

$K(x) = 1 - 0 = 1$ for a variable x;

$K(c) = 1 - 0 = 1$ for a constant symbol c;

$K(f) = 1 - n$ for an n-place function symbol f.

We then extend K to the set of expressions using these symbols by defining

$$K(s_1 s_2 \cdots s_n) = K(s_1) + K(s_2) + \cdots + K(s_n).$$

Since no symbol is a finite sequence of others, this definition is unambiguous.

Lemma 23A For any term t, $K(t) = 1$.

Proof Use induction on t. The inductive step, for an n-place function symbol f, is

$$K(ft_1 \cdots t_n) = (1 - n) + (1 + \cdots + 1) = 1. \quad \blacksquare$$

In fact, K was chosen to be the unique function on these symbols for which Lemma 23A holds. It follows from this lemma that if ε is a concatenation of m terms, then $K(\varepsilon) = m$.

By a *terminal* segment of a sequence $\langle s_1, \ldots, s_n\rangle$ of symbols we mean a sequence of the form $\langle s_k, s_{k+1}, \ldots, s_n\rangle$, where $1 \leq k \leq n$.

Lemma 23B Any terminal segment of a term is a concatenation of one or more terms.

Proof We use induction on the term. For a one-symbol term (i.e., a variable or a constant symbol) the conclusion follows trivially. For a term $ft_1 \cdots t_n$, any terminal segment (other than the term itself) must equal

$$t_k' t_{k+1} \cdots t_n,$$

where $k \leq n$ and t_k' is a terminal segment of t_k. By the inductive hypothesis

t_k' is a concatenation of, say, m terms, where $m \geq 1$. So altogether we have $m + (n - k)$ terms. ■

Corollary 23C No proper initial segment of a term is itself a term.

Proof Suppose a term t is divided into a proper initial segment t_1 and a terminal segment t_2. Then $1 = K(t) = K(t_1) + K(t_2)$, and by Lemma 23B, $K(t_2) \geq 1$. Hence $K(t_1) < 1$ and t_1 cannot be a term. ■

Unique Readability Theorem for Terms The set of terms is freely generated from the set of variables and constant symbols by the $\mathscr{F}_f$ operations.

Proof First, it is clear that if $f \neq g$, then ran $\mathscr{F}_f$ is disjoint from ran $\mathscr{F}_g$; this requires checking only the first symbol. Furthermore, both ranges are disjoint from the set of variables and constant symbols. It remains only to show that $\mathscr{F}_f$, when restricted to terms, is one-to-one. Suppose, for a two-place f, we have

$$ft_1t_2 = ft_3t_4.$$

By deleting the first symbol we are left with

$$t_1t_2 = t_3t_4.$$

If $t_1 \neq t_3$, then one would be a proper initial segment of the other, which is impossible for terms by the above corollary. So $t_1 = t_3$, and we are then left with $t_2 = t_4$. ■

To extend this argument to formulas, we now define K on the other symbols:

$$K(\,(\,) = -1;$$
$$K(\,)\,) = 1;$$
$$K(\forall) = -1;$$
$$K(\neg) = 0;$$
$$K(\rightarrow) = -1;$$
$$K(P) = 1 - n \text{ for an } n\text{-place predicate symbol } P.$$

The idea behind the definition is again that $K(s)$ should be $1 - n$, where n is number of things (right parentheses, terms, or formulas) required to go along with s. We extend K as usual to the set of all expressions:

$$K(s_1 \cdots s_n) = K(s_1) + \cdots + K(s_n).$$

Lemma 23D For any wff α, $K(\alpha) = 1$.

Proof Another straightforward induction. ■

Lemma 23E For any proper initial segment α' of a wff α, $K(\alpha') < 1$.

Proof Use induction on α. The details are left to Exercise 1. ■

Corollary 23F No proper initial segment of a formula is itself a formula.

Unique Readability Theorem for Formulas The set of wffs is freely generated from the set of atomic formulas by the operations $\mathscr{E}_\neg$, $\mathscr{E}_\rightarrow$, and $\mathscr{Q}_i (i = 1, 2, \ldots)$.

Proof The unary operations $\mathscr{E}_\neg$ and $\mathscr{Q}_i$ are obviously one-to-one. As in Section 1.4, we can show that the restriction of $\mathscr{E}_\rightarrow$ to wffs is one-to-one.

The disjointness half of the theorem follows from the ad hoc observations:

1. ran $\mathscr{E}_\neg$, ran $\mathscr{Q}_i$, ran $\mathscr{Q}_j$, and the set of atomic formulas are pairwise disjoint, for $i \neq j$. (Just look at the first two symbols.)

2. ran $\mathscr{E}_\rightarrow$, ran $\mathscr{Q}_i$, ran $\mathscr{Q}_j$, and the set of atomic formulas are similarly pairwise disjoint, for $i \neq j$.

3. For a wff β, $(\neg\alpha) \neq (\beta \rightarrow \gamma)$, because no wff begins with $\neg$. Hence ran $\mathscr{E}_\neg$ is disjoint from the range of the restriction of $\mathscr{E}_\rightarrow$ to wffs. ■

As in Section 1.4, we can obtain an effective algorithm for finding the unique tree displaying the pedigree of a term or formula. There we counted parentheses in initial segments; now we use the function K in the analogous role.

EXERCISES

1. Show that for a proper initial segment α' of a wff α, we have $K(\alpha') < 1$.

2. Let ε be an expression consisting of variables, constant symbols, and function symbols. Show that ε is a term iff $K(\varepsilon) = 1$ and for every terminal segment ε' of ε we have $K(\varepsilon') > 0$. *Suggestion*: Prove the stronger result that if $K(\varepsilon') > 0$ for every terminal segment ε' of ε, then ε is a concatenation of $K(\varepsilon)$ terms.

§ 2.4 A DEDUCTIVE CALCULUS

Suppose that $\Sigma \models \tau$. What methods of proof might be required to demonstrate that fact? Is there necessarily a proof at all?

Such questions lead immediately to considerations of what constitutes a proof. A proof is an argument you give to someone else and which completely convinces him of the correctness of your assertion (in this case, that $\Sigma \models \tau$).

Thus a proof should be finitely long, as you cannot give all of an infinite object to another person. If the set Σ of hypotheses is infinite, they cannot all be used. But the compactness theorem for first-order logic (which we will prove in Section 2.5 using the deductive calculus of this section) will ensure the existence of a finite $\Sigma_0 \subseteq \Sigma$ such that $\Sigma_0 \models \tau$.

Another essential feature of a proof (besides its being finite in length) is that it must be possible for someone else (if he is to be convinced by it) to check the proof to ascertain that it contains no fallacies. This check must be effective; it must be the sort of thing that can be carried out without brilliant flashes of insight on the part of the checker. In particular, the set of proofs from the empty set of hypotheses (i.e., proofs that $\models \tau$) must be decidable. This implies that the set of sentences provable without hypotheses must be effectively enumerable. For one could in principle enumerate provable sentences by generating all strings of symbols and sorting out the proofs from the nonproofs. When a proof is discovered, its last line is entered on the output list. (This issue will be examined more carefully at the end of Section 2.5.) But here again there is a theorem (the enumerability theorem, proved in Section 2.5) stating that, under reasonable conditions, the set of valid sentences is indeed effectively enumerable.

Thus the compactness theorem and the enumerability theorem are necessary conditions for proofs of logical implication always to exist. Conversely, we claim that these two theorems are sufficient for proofs to exist. For suppose that $\Sigma \models \tau$. By the compactness theorem, then, there is a finite set $\{\sigma_0, \ldots, \sigma_n\} \subseteq \Sigma$ which logically implies τ. Then $\sigma_0 \rightarrow \cdots \rightarrow \sigma_n \rightarrow \tau$ is valid (Exercise 1, Section 2.2). So to demonstrate conclusively that $\Sigma \models \tau$ one need only carry out a finite number of steps in the enumeration of the validities until $\sigma_0 \rightarrow \cdots \rightarrow \sigma_n \rightarrow \tau$ appears, and then verify that each $\sigma_i \in \Sigma$. (This should be compared with the complex procedure suggested by the original definition of logical implication, discussed on page 84.) The record of the enumeration procedure which produced $\sigma_0 \rightarrow \cdots \rightarrow \sigma_n \rightarrow \tau$ can then be regarded as a *proof* that $\Sigma \models \tau$. As a proof, it should be acceptable to anyone who accepts the correctness of your procedure for enumerating validities.

Against the foregoing general (and slightly vague) discussion, the outline of this section can be described as follows: We will introduce formal proofs but we will call them *deductions*, to avoid confusion with English language proofs. These will mirror (in our model of deductive thought) the proofs made by the working mathematician to convince his colleagues of certain truths. Then in Section 2.5 we will show that whenever $\Sigma \models \tau$, there is a deduction of τ from Σ (and only then). This will, as is suggested by the foregoing discussion, yield proofs of the compactness theorem and the enumerability theorem. And in the process we will get to see what methods of deduction are adequate to demonstrate that a given sentence, is, in fact, logically implied by certain others.

Formal deductions

We will shortly select an infinite set Λ of formulas to be called logical axioms. And we will have a rule of inference, which will enable us to obtain a new formula from certain others. Then for a set Γ of formulas, the *theorems* of Γ will be the formulas which can be obtained from $\Gamma \cup \Lambda$ by use of the rule of inference (some finite number of times). If φ is a theorem of Γ (written $\Gamma \vdash \varphi$), then a sequence of formulas which records (as explained below) how φ was obtained from $\Gamma \cup \Lambda$ with the rule of inference will be called a deduction of φ from Γ.

The choice of Λ and the choice of the rule (or rules) of inference are far from unique. In this section we are presenting one deductive calculus for first-order logic, chosen from the array of possible calculi. (For example, one can have $\Lambda = \varnothing$ by using many rules of inference. We will take the opposite extreme; our set Λ will be infinite but we will have only one rule of inference.)

Our one rule of inference is traditionally known as *modus ponens*. It is usually stated: From the formulas α and $\alpha \rightarrow \beta$ we may infer β:

$$\frac{\alpha, \ \alpha \rightarrow \beta}{\beta}.$$

Thus the theorems of a set Γ are the formulas obtainable from $\Gamma \cup \Lambda$ by use of modus ponens some finite number of times.

We can also look at modus ponens in the following way. Call a set Δ of formulas *closed under modus ponens* iff whenever two formulas α and $\alpha \rightarrow \beta$ are in Δ, then also β is in Δ. For a fixed set Γ, say that Δ is *inductive* iff $\Gamma \cup \Lambda \subseteq \Delta$ and Δ is closed under modus ponens. Then the set of theorems of Γ is simply the smallest inductive set. This differs from the situation of Section 1.2 only in that instead of closing the initial set under (everywhere-defined) functions, we are closing it under a relation which is only a "partially-defined" function. (Its domain consists not of all pairs of formulas, but only of pairs of the form $\langle\alpha, \alpha \rightarrow \beta\rangle$.) The ideas of Section 1.2 are, however, applicable to the present situation. Thus we may define φ to be a *theorem* of Γ (written $\Gamma \vdash \varphi$) iff φ belongs to the set generated from $\Gamma \cup \Lambda$ by modus ponens. (But the reader should be warned that the set of theorems of Γ is *not* freely generated. This reflects the fact that a theorem never has a unique deduction.)

For example, if the formulas β, γ, and $\gamma \rightarrow \beta \rightarrow \alpha$ are all in $\Gamma \cup \Lambda$, then $\Gamma \vdash \alpha$, as is evidenced by the tree

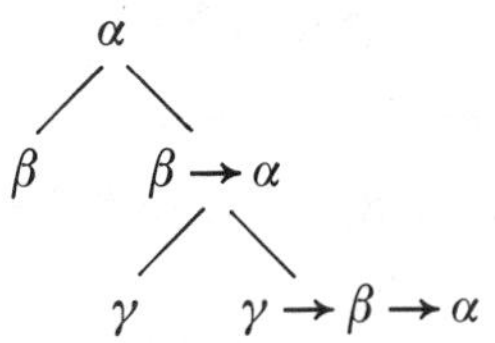

which displays how α was obtained. Although it is tempting (and in some ways more elegant) to define a deduction to be such a tree, it will be simpler to take deductions to be the linear sequences obtained by squashing such trees into straight lines.

Definition A *deduction of φ from Γ* is a sequence $\langle\alpha_0, \ldots, \alpha_n\rangle$ of formulas such that $\alpha_n = \varphi$ and for each $i \leq n$ either

(a) α_i is in $\Gamma \cup \Lambda$, or

(b) for some j and k less than i, α_i is obtained by modus ponens from α_j and α_k (i.e., $\alpha_k = \alpha_j \rightarrow \alpha_i$).

In Section 1.2 sequences of the sort just described (i.e., deductions) were called construction sequences. We showed there that C_* (the set of things having construction sequences) coincides with C^* (the intersection of all inductive sets). We now repeat that result, in the present case of generation from $\Gamma \cup \Lambda$ by modus ponens:

Theorem 24A There exists a deduction of α from Γ iff α is a theorem of Γ.

Proof If there is a deduction $\langle\alpha_0, \ldots, \alpha_n\rangle$, then (by induction on i) each α_i belongs to the set generated from $\Gamma \cup \Lambda$ by modus ponens. Conversely, the set of formulas for which deductions exist includes $\Gamma \cup \Lambda$ (whose members have one-line deductions) and is closed under modus ponens. (Why?) Consequently, this set includes all theorems of Γ. ■

This justifies defining φ to be *deducible* from Γ iff $\Gamma \vdash \varphi$.

Now at last we give the set Λ of logical axioms. These are arranged in six groups. Say that a wff φ is a *generalization* of ψ iff for some $n \geq 0$ and some variables $x_1, \ldots, x_n$,

$$\varphi = \forall x_1 \cdots \forall x_n \psi.$$

We include the case $n = 0$; any wff is a generalization of itself. The logical axioms are then all generalizations of wffs of the following forms, where x and y are variables and α and β are wffs:

1. Tautologies;
2. $\forall x \alpha \rightarrow \alpha_t^x$, where t is substitutable for x in α;
3. $\forall x(\alpha \rightarrow \beta) \rightarrow (\forall x \alpha \rightarrow \forall x \beta)$;
4. $\alpha \rightarrow \forall x \alpha$, where x does not occur free in α.

And if the language includes equality, then we add

5. $x \approx x$;
6. $x \approx y \rightarrow (\alpha \rightarrow \alpha')$, where α is atomic and α' is obtained from α by replacing x in zero or more (but not necessarily all) places by y.

For the most part groups 3–6 are self-explanatory; we will see various examples later. Groups 1 and 2 require explanation. But first we should admit that the above list of logical axioms may not appear very natural. Later it will be possible to see where each of the six groups originated.

Substitution

In axiom group 2 we find

$$\forall x \alpha \rightarrow \alpha_t^x.$$

Here α_t^x is the expression obtained from the formula α by replacing the variable x, wherever it occurs free in α, by the term t. This notion can also be (and for us officially is) defined by recursion:

1. For atomic α, α_t^x is the expression obtained from α by replacing the variable x by t. (This is elaborated upon in Exercise 1. Note that α_t^x is itself a formula.)
2. $(\neg\alpha)_t^x = (\neg\alpha_t^x)$.
3. $(\alpha \rightarrow \beta)_t^x = (\alpha_t^x \rightarrow \beta_t^x)$.
4. $(\forall y\, \alpha)_t^x = \begin{cases} \forall y\, \alpha \text{ if } x = y, \\ \forall y(\alpha_t^x) \text{ if } x \neq y. \end{cases}$

EXAMPLES

1. $\varphi^x_x = \varphi$.
2. $(Qx \to \forall x\, Px)^x_y = (Qy \to \forall x\, Px)$.
3. If α is $\neg\forall y\, x \approx y$, then $\forall x\, \alpha \to \alpha^x_z$ is

$$\forall x\, \neg\forall y\, x \approx y \to \neg\forall y\, z \approx y.$$

4. For α as in 3, $\forall x\, \alpha \to \alpha^x_y$ is

$$\forall x\, \neg\forall y\, x \approx y \to \neg\forall y\, y \approx y.$$

The last example above illustrates a hazard which must be guarded against. On the whole, $\forall x\, \alpha \to \alpha^x_t$ seems like a plausible enough axiom. ("If α is true of everything, then it should be true of t.") But in example 4, we have a sentence of the form $\forall x\, \alpha \to \alpha^x_t$, which is nearly always false. The antecedent, $\forall x\, \neg\forall y\, x \approx y$, is true in any structure whose universe contains two or more elements. But the succedent, $\neg\forall y\, y \approx y$, is false in any structure. So something has gone wrong.

The problem is that when y was substituted for x, it was immediately "captured" by the $\forall y$ quantifier. We must impose a restriction on axiom group 2 that will preclude this sort of quantifier capture. Informally, we can say that a term t is *not* substitutable for x in α if there is some variable y in t that is captured by a $\forall y$ quantifier in α^x_t. The real definition is given below by recursion. (Since the notion will be used later in proofs by induction, a recursive definition is actually the most usable variety.)

Let x be a variable, t a term. We define the phrase "t is *substitutable* for x in α" as follows:

1. For atomic α, t is always substitutable for x in α. (There are no quantifiers in α, so no capture could occur.)

2. t is substitutable for x in $(\neg\alpha)$ iff it is substitutable for x in α. t is substitutable for x in $(\alpha \to \beta)$ iff it is substitutable for x in both α and β.

3. t is substitutable for x in $\forall y\, \alpha$ iff either

(a) x does not occur free in $\forall y\, \alpha$, or
(b) y does not occur in t and t is substitutable for x in α.
(The point here is to be sure that nothing in t will be captured by the $\forall y$ prefix and that nothing has gone wrong inside α earlier.)

For example, x is always substitutable for itself in any formula. If t contains no variables which occur in α, then t is substitutable for x in α.

The reader is cautioned not to be confused about the choice of words. Even if t is not substitutable for x in α, still α_t^x is obtained from α by replacing x wherever it occurs free by t. Thus in forming α_t^x, we carry out the indicated substitution even if a prudent person would think it unwise to do so.

Axiom group 2 consists of all generalizations of formulas of the form

$$\forall x\, \alpha \to \alpha_t^x,$$

where the term t is substitutable for the variable x in the formula α. For example,

$$\forall \mathbf{v}_3(\forall \mathbf{v}_1(A\mathbf{v}_1 \to \forall \mathbf{v}_2\, A\mathbf{v}_2) \to (A\mathbf{v}_2 \to \forall \mathbf{v}_2\, A\mathbf{v}_2))$$

is in axiom group 2. Here x is $\mathbf{v}_1$, α is $A\mathbf{v}_1 \to \forall \mathbf{v}_2\, A\mathbf{v}_2$, and t is $\mathbf{v}_2$. On the other hand,

$$\forall \mathbf{v}_1\, \forall \mathbf{v}_2\, B\mathbf{v}_1\mathbf{v}_2 \to \forall \mathbf{v}_2\, B\mathbf{v}_2\mathbf{v}_2$$

is *not* in axiom group 2, since $\mathbf{v}_2$ is not substitutable for $\mathbf{v}_1$ in $\forall \mathbf{v}_2\, B\mathbf{v}_1\mathbf{v}_2$.

Tautologies

Axiom group 1 consists of generalizations of formulas to be called *tautologies*. These are the wffs obtainable from tautologies of sentential logic (having only the connectives $\neg$ and $\to$) by replacing each sentence symbol by a wff of the first-order language. For example,

$$\forall x[(\forall y\, \neg Py \to \neg Px) \to (Px \to \neg\forall y\, \neg Py)]$$

belongs to axiom group 1. It is a generalization of the formula in square brackets, which is obtained from a contraposition tautology

$$(\mathbf{A} \to \neg\mathbf{B}) \to (\mathbf{B} \to \neg\mathbf{A})$$

by replacing **A** by $\forall y\, \neg Py$ and **B** by Px.

There is another, more direct, way of looking at axiom group 1. Divide the wffs into two groups:

1. The *prime* formulas are the atomic formulas and those of the form $\forall x\, \alpha$.
2. The nonprime formulas are the others, i.e., those of the form $\neg\alpha$ or $\alpha \to \beta$.

Thus any formula is built up from prime formulas by the operations $\mathscr{E}_{\neg}$ and $\mathscr{E}_{\rightarrow}$. Now go back to sentential logic, but take the sentence symbols to be the prime formulas of our first-order language. Then any tautology of sentential logic (which uses only the connectives $\neg$, $\rightarrow$) is in axiom group 1. There is no need to *replace* sentence symbols here by first-order wffs; they already *are* first-order wffs. Conversely, anything in axiom group 1 is a generalization of a tautology of sentential logic. (The proof of this uses Exercise 7 of Section 1.3.)

EXAMPLE, revisited

$$(\forall y\, \neg Py \rightarrow \neg Px) \rightarrow (Px \rightarrow \neg \forall y\, \neg Py).$$

This has two sentence symbols (prime formulas), $\forall y\, \neg Py$ and Px. So its truth table has four lines:

$(\forall y\, \neg Py$	$\rightarrow \neg Px)$	$\rightarrow$	$(Px \rightarrow \neg \forall y\, \neg Py)$
T	$F\ F\ T$	$\mathbf{T}$	$T\ F\ F\ T$
T	$T\ T\ F$	$\mathbf{T}$	$F\ T\ F\ T$
F	$T\ F\ T$	$\mathbf{T}$	$T\ T\ T\ F$
F	$T\ T\ F$	$\mathbf{T}$	$F\ T\ T\ F$.

From the table we see that it is indeed a tautology.

On the other hand, neither $\forall x(Px \rightarrow Px)$ nor $\forall x\, Px \rightarrow Px$ is a tautology.

One remark: We have not assumed that our first-order language has only countably many formulas. So we are potentially employing an extension of Chapter 1 to the case of an uncountable set of sentence symbols.

A second remark: Now that first-order formulas are also wffs of sentential logic, we can apply notion from both Chapters 1 and 2 to them. If Γ tautologically implies φ, then it follows that Γ also logically impies φ. (See Exercise 3.) But the converse fails. For example, $\forall x\, Px$ logically implies Pc. But $\forall x\, Px$ does *not* tautologically imply Pc, as $\forall x\, Px$ and Pc are two different sentence symbols.

Theorem 24B $\Gamma \vdash \varphi$ iff $\Gamma \cup \Lambda$ tautologically implies φ.

Proof ($\Rightarrow$): This depends on the obvious fact that $\{\alpha, \alpha \rightarrow \beta\}$ tautologically implies β. Suppose that we have a truth assignment v which satisfies every member of $\Gamma \cup \Lambda$. By induction we can see that v satisfies any theorem of Γ. The inductive step uses exactly the abovementioned obvious fact.

($\Leftarrow$): Assume that $\Gamma \cup \Lambda$ tautologically implies φ. Then by the corollary to the compactness theorem (for sentential logic), there is a finite subset $\{\gamma_1, \ldots, \gamma_m, \lambda_1, \ldots, \lambda_n\}$ which tautologically implies φ. Consequently,

$$\gamma_1 \to \cdots \to \gamma_m \to \lambda_1 \to \cdots \to \lambda_n \to \varphi$$

is a tautology (cf. Exercise 3 of Section 1.3) and hence is in Λ. By applying modus ponens $m+n$ times to this tautology and to $\{\gamma_1, \ldots, \gamma_m, \lambda_1, \ldots, \lambda_n\}$, we obtain φ. ■

(The above proof uses sentential compactness for a possibly uncountable language.)

Deductions and metatheorems

We now have completed the description of the set Λ of logical axioms. The set of theorems of a set Γ is the set generated from $\Gamma \cup \Lambda$ by modus ponens. For example,

$$\vdash Px \to \exists y\, Py.$$

(Here $\Gamma = \varnothing$; we write "$\vdash \alpha$" in place of "$\varnothing \vdash \alpha$.") The formula $Px \to \exists y\, Py$ can be obtained by applying modus ponens (once) to two members of Λ, as displayed by the pedigree tree:

$$Px \to \neg\forall y \neg Py$$

$$\forall y\, \neg Py \to \neg Px$$
(in axiom group 2)

$$(\forall y\, \neg Py \to \neg Px) \to (Px \to \neg\forall y \neg Py)$$
(in axiom group 1)

We get a deduction of $Px \to \exists y\, Py$ (from $\varnothing$) by compressing this tree into a linear three-element sequence.

As a second example, we can obtain a generalization of the formula in the first example:

$$\vdash \forall x(Px \to \exists y\, Py).$$

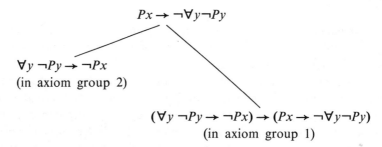

This fact is evidenced by the following tree, which displays the construction of $\forall x(Px \to \exists y\, Py)$ from Λ by modus ponens:

$$\forall x(Px \to \neg\forall y \neg Py)$$

$$\forall x(\forall y\, \neg Py \to \neg Px)$$
(in axiom group 2)

$$\forall x(\forall y\, \neg Py \to \neg Px) \to \forall x(Px \to \neg\forall y \neg Py)$$

(a member of axiom group 3)

$$\forall x[(\forall y\, \neg Py \to \neg Px) \to (Px \to \neg\forall y \neg Py)]$$
(in axiom group 1)

Again we can compress the tree into a deduction.

In these examples the pedigree trees may seem to have been pulled out of the air. But we will shortly develop techniques for generating such trees in a somewhat systematic manner. These techniques will rely heavily on the generalization theorem and the deduction theorem below.

Notice that we use the word "theorem" on two different levels. We say that α is a theorem of Γ if $\Gamma \vdash \alpha$. We also make numerous statements in English, each called a theorem, such as the one below. It seems unlikely that any confusion will arise. The English statements could have been labeled *metatheorems* to emphasize that they are results about deductions and theorems.

The generalization theorem reflects our informal feeling that if we can prove __x__ without any special assumptions about x, we then are entitled to say that "since x was arbitrary, we have $\forall x$__x__."

Generalization Theorem If $\Gamma \vdash \varphi$ and x does not occur free in any formula in Γ, then $\Gamma \vdash \forall x\, \varphi$.

Proof Consider a fixed set Γ and a variable x not free in Γ. We will show by induction that for any theorem φ of Γ, we have $\Gamma \vdash \forall x\, \varphi$. For this it suffices (by the induction principle) to show that the set

$$\{\varphi : \Gamma \vdash \forall x\, \varphi\}$$

includes $\Gamma \cup \Lambda$ and is closed under modus ponens. Notice that x *can* occur free in φ.

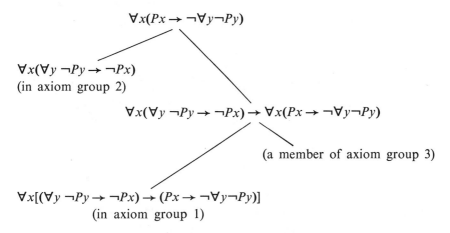

Case 1: φ is a logical axiom. Then $\forall x\,\varphi$ is also a logical axiom. And so $\Gamma \vdash \forall x\,\varphi$.

Case 2: $\varphi \in \Gamma$. Then x does not occur free in φ. Hence

$$\varphi \to \forall x\,\varphi$$

is in axiom group 4. Consequently, $\Gamma \vdash \forall x\,\varphi$, as is evidenced by

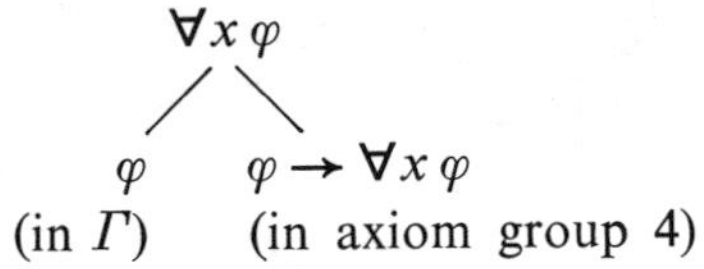

Case 3: φ is obtained by modus ponens from ψ and $\psi \to \varphi$. Then by inductive hypothesis we have $\Gamma \vdash \forall x\,\psi$ and $\Gamma \vdash \forall x(\psi \to \varphi)$. This is just the situation in which axiom group 3 is useful. That $\Gamma \vdash \forall x\,\varphi$ is evidenced by

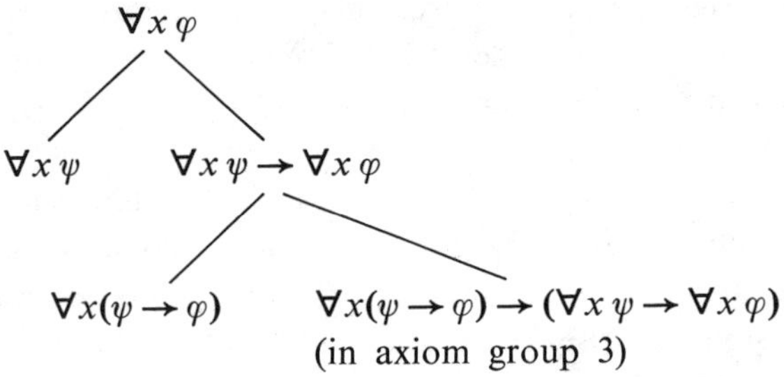

So by induction $\Gamma \vdash \forall x\,\varphi$ for every theorem φ of Γ. ■

(The sole reasons for having axiom groups 3 and 4 are indicated by the above proof.)

The restriction that x not occur free in Γ is essential. For example, $Px \not\models \forall x\, Px$, and so by the soundness theorem of Section 2.5, $Px \nvdash \forall x\, Px$. On the other hand, x will in general occur free in the formula φ. For example, at the beginning of this subsection we showed first that

$$\vdash (Px \to \exists y\, Py).$$

The second example there,

$$\vdash \forall x(Px \to \exists y\, Py),$$

was obtained from the first example as in case 3 of the above proof.

EXAMPLE $\forall x\, \forall y\, \alpha \vdash \forall y\, \forall x\, \alpha$.

The proof of the generalization theorem actually yields somewhat more than was stated. It shows how we can, given a deduction of φ from Γ, effectively transform it to obtain a deduction of $\forall x\, \varphi$ from Γ.

Lemma 24C (rule T) If $\Gamma \vdash \alpha_1, \ldots, \Gamma \vdash \alpha_n$ and $\{\alpha_1, \ldots, \alpha_n\}$ tautologically implies β, then $\Gamma \vdash \beta$.

Proof $\alpha_1 \to \cdots \to \alpha_n \to \beta$ is a tautology, and hence a logical axiom. Apply modus ponens n times. ■

Deduction Theorem If $\Gamma ; \gamma \vdash \varphi$, then $\Gamma \vdash (\gamma \to \varphi)$.

(The converse clearly holds also; in fact, the converse is essentially the rule modus ponens.)

First proof

$$\begin{aligned} \Gamma ; \gamma \vdash \varphi \quad & \text{iff } (\Gamma ; \gamma) \cup \Lambda \text{ tautologically implies } \varphi, \\ & \text{iff } \Gamma \cup \Lambda \text{ tautologically implies } (\gamma \to \varphi), \\ & \text{iff } \Gamma \vdash (\gamma \to \varphi). \quad ■ \end{aligned}$$

Second proof The second proof does not use the compactness theorem of sentential logic as does the first proof. It shows in a direct way how to transform a deduction of φ from $\Gamma ; \gamma$ to obtain a deduction of $(\gamma \to \varphi)$ from Γ. We show by induction that for every theorem φ of $\Gamma ; \gamma$ the formula $(\gamma \to \varphi)$ is a theorem of Γ.

Case 1: $\varphi = \gamma$. Then obviously $\vdash (\gamma \to \varphi)$.

Case 2: φ is a logical axiom or a member of Γ. Then $\Gamma \vdash \varphi$. And φ tautologically implies $(\gamma \to \varphi)$, whence by rule T we have $\Gamma \vdash (\gamma \to \varphi)$.

Case 3: φ is obtained by modus ponens from ψ and $\psi \to \varphi$. By the inductive hypothesis, $\Gamma \vdash (\gamma \to \psi)$ and $\Gamma \vdash (\gamma \to (\psi \to \varphi))$. And the set $\{\gamma \to \psi, \gamma \to (\psi \to \varphi)\}$ tautologically implies $\gamma \to \varphi$. Thus, by rule T, $\Gamma \vdash (\gamma \to \varphi)$.

So by induction the conclusion holds for any φ deducible from $\Gamma ; \gamma$. ■

Corollary 24D (contraposition) $\Gamma ; \varphi \vdash \neg\psi$ iff $\Gamma ; \psi \vdash \neg\varphi$.

Proof

$$\begin{aligned} \Gamma ; \varphi \vdash \neg\psi & \Rightarrow \Gamma \vdash \varphi \to \neg\psi && \text{by the deduction theorem,} \\ & \Rightarrow \Gamma \vdash \psi \to \neg\varphi && \text{by rule T,} \\ & \Rightarrow \Gamma ; \psi \vdash \neg\varphi && \text{by modus ponens.} \end{aligned}$$

(In the second step we use the fact that $\varphi \rightarrow \neg\psi$ tautologically implies $\psi \rightarrow \neg\varphi$.) By symmetry, the converse holds also. ■

Say that a set of formulas is *inconsistent* iff for some β, both β and $\neg\beta$ are theorems of the set. (In this event, any formula α is a theorem of the set, since $\beta \rightarrow \neg\beta \rightarrow \alpha$ is a tautology.)

Corollary 24E (reductio ad absurdum) If $\Gamma ; \varphi$ is inconsistent, then $\Gamma \vdash \neg\varphi$.

Proof From the deduction theorem we have $\Gamma \vdash (\varphi \rightarrow \beta)$ and $\Gamma \vdash (\varphi \rightarrow \neg\beta)$. And $\{\varphi \rightarrow \beta, \varphi \rightarrow \neg\beta\}$ tautologically implies $\neg\varphi$. ■

EXAMPLE $\vdash \exists x \forall y\, \varphi \rightarrow \forall y\, \exists x\, \varphi$.

There are strategic advantages to working backward.

It suffices to show that $\exists x \forall y\, \varphi \vdash \forall y\, \exists x\, \varphi$, by the deduction theorem.

It suffices to show that $\exists x \forall y\, \varphi \vdash \exists x\, \varphi$, by the generalization theorem.

It suffices to show that $\neg\forall x\, \neg\forall y\, \varphi \vdash \neg\forall x\, \neg\varphi$, as this is the same as the preceding.

It suffices to show that $\forall x\, \neg\varphi \vdash \forall x\, \neg\forall y\, \varphi$, by contraposition (and rule T).

It suffices to show that $\forall x\, \neg\varphi \vdash \neg\forall y\, \varphi$, by generalization.

It suffices to show that $\{\forall x\, \neg\varphi, \forall y\, \varphi\}$ is inconsistent, by reductio ad absurdum.

And this is easy:

1. $\forall x\, \neg\varphi \vdash \neg\varphi$ by axiom group 2 and modus ponens.
2. $\forall y\, \varphi \vdash \varphi$ for the same reason.

Lines 1 and 2 show that $\{\forall x\, \neg\varphi, \forall y\, \varphi\}$ is inconsistent.

Strategy

As the preceding example indicates, the generalization and deduction theorems (and to a smaller extent the corollaries) will be very useful in showing that certain formulas are deducible. But there is still the matter of strategy: For a given Γ and φ, where should one begin in order to show that $\Gamma \vdash \varphi$? One could, in principle, start enumerating all finite sequences of wffs until one encountered a deduction of φ from Γ. Although this would be an effective procedure (for reasonable languages) for locating a deduction if one exists, it is too inefficient to have more than theoretical interest.

One technique is to abandon formality and to give in English a proof that the truth of Γ implies the truth of φ. Then the proof in English can be formalized into a legal deduction. (In the coming pages we will see techniques for carrying out such a formalization in a reasonably natural way.)

There are also useful methods based solely on the syntactical form of φ. Assume then that φ is indeed deducible from Γ but that you are seeking a proof of this fact. There are several cases:

1. Suppose that φ is $(\psi \to \theta)$. Then it will suffice to show that $\Gamma ; \psi \vdash \theta$ (and this will always be possible).

2. Suppose that φ is $\forall x\, \psi$. If x does not occur free in Γ, then it will suffice to show that $\Gamma \vdash \psi$. (Even if x should occur free in Γ, the difficulty can be circumvented. There will always be a variable y such that $\Gamma \vdash \forall y\, \psi^x_y$ and $\forall y\, \psi^x_y \vdash \forall x\, \psi$. See the re-replacement lemma, Exercise 9.)

3. Finally, suppose that φ is the negation of another formula.

3a. If φ is $\neg(\psi \to \theta)$, then it will suffice to show that $\Gamma \vdash \psi$ and $\Gamma \vdash \neg\theta$ (by rule T). And this will always be possible.

3b. If φ is $\neg\neg\psi$, then of course it will suffice to show that $\Gamma \vdash \psi$.

3c. The remaining case is where φ is $\neg\forall x\, \psi$. It would suffice to show that $\Gamma \vdash \neg\psi^x_t$, where t is substitutable for x in ψ. (Why?) Unfortunately this is not always possible. There are cases in which

$$\Gamma \vdash \neg\forall x\, \psi,$$

and yet for every term t,

$$\Gamma \nvdash \neg\psi^x_t.$$

(One such example is $\Gamma = \varnothing$, $\psi = \neg(Px \to \forall y\, Py)$.) Contraposition is handy here;

$$\Gamma ; \alpha \vdash \neg\forall x\, \psi$$

iff

$$\Gamma ; \forall x\, \psi \vdash \neg\alpha.$$

(A variation on this is: $\Gamma ; \forall y\, \alpha \vdash \neg\forall x\, \psi$ if $\Gamma ; \forall x\, \psi \vdash \neg\alpha$.) If all else fails, one can try reductio ad absurdum.

EXAMPLE (Q2a) If x does not occur free in α, then

$$\vdash (\alpha \to \forall x\, \beta) \leftrightarrow \forall x(\alpha \to \beta).$$

To prove this, it suffices (by rule T) to show that

$$\vdash (\alpha \to \forall x\, \beta) \to \forall x(\alpha \to \beta)$$

and

$$\vdash \forall x(\alpha \to \beta) \to (\alpha \to \forall x\, \beta).$$

For the first of these, it suffices (by the deduction and generalization theorems) to show that

$$\{(\alpha \to \forall x\, \beta),\ \alpha\} \vdash \beta.$$

But this is easy; $\forall x\, \beta \to \beta$ is an axiom.

To obtain the converse,

$$\vdash \forall x(\alpha \to \beta) \to (\alpha \to \forall x\, \beta),$$

it suffices (by the deduction and generalization theorems) to show that

$$\{\forall x(\alpha \to \beta),\ \alpha\} \vdash \beta.$$

This again is easy.

In the above example we can replace α by $\neg\alpha$, β by $\neg\beta$, and use the contraposition tautology to obtain the corollary:

(Q3b) If x does not occur free in α, then

$$\vdash (\exists x\, \beta \to \alpha) \leftrightarrow \forall x(\beta \to \alpha).$$

The reader might want to convince himself that the above formula *is* valid.

Frequently an abbreviated style is useful in writing down a proof of deducibility, as in the following example.

EXAMPLE (Eq2) $\vdash \forall x\, \forall y(x \approx y \to y \approx x)$.

Proof

1. $\vdash x \approx y \to x \approx x \to y \approx x$. Ax 6.
2. $\vdash x \approx x$. Ax 5.
3. $\vdash x \approx y \to y \approx x$. 1, 2; T.
4. $\vdash \forall x\, \forall y(x \approx y \to y \approx x)$. 3; gen^2. ■

In line 1, "Ax 6" means that the formula belongs to axiom group 6. In line 3, "1, 2; T" means that this line is obtained from lines 1 and 2 by rule T. In line 4, "3; gen^2" means that the generalization theorem can be applied twice to line 3 to yield line 4. In the same spirit we write "MP," "ded," and "RAA" to refer to modus ponens, the deduction theorem, and reductio ad absurdum, respectively.

It must be emphasized that the four numbered lines above do *not* constitute a deduction of $\forall x\, \forall y(x \approx y \to y \approx x)$. Instead they form a proof (in the metalanguage we continue, without justification, to call English) that such a deduction exists. The shortest deduction of $\forall x\, \forall y(x \approx y \to y \approx x)$ known to the author is a sequence of seventeen formulas.

EXAMPLE $\vdash x \approx y \rightarrow \forall z\, Pxz \rightarrow \forall z\, Pyz.$

Proof

1. $\vdash x \approx y \rightarrow Pxz \rightarrow Pyz.$ Ax 6.
2. $\vdash \forall z\, Pxz \rightarrow Pxz.$ Ax 2.
3. $\vdash x \approx y \rightarrow \forall z\, Pxz \rightarrow Pyz.$ 1, 2; T.
4. $\{x \approx y, \forall z\, Pxz\} \vdash Pyz.$ 3; MP².
5. $\{x \approx y, \forall z\, Pxz\} \vdash \forall z\, Pyz.$ 4; gen.
6. $\vdash x \approx y \rightarrow \forall z\, Pxz \rightarrow \forall z\, Pyz.$ 5; ded². ■

EXAMPLE (Eq5) Let f be a two-place function symbol. Then

$$\vdash \forall x_1 \forall x_2 \forall y_1 \forall y_2 (x_1 \approx y_1 \rightarrow x_2 \approx y_2 \rightarrow fx_1x_2 \approx fy_1y_2).$$

Proof Two members of axiom group 6 are

$$x_1 \approx y_1 \rightarrow fx_1x_2 \approx fx_1x_2 \rightarrow fx_1x_2 \approx fy_1x_2,$$
$$x_2 \approx y_2 \rightarrow fx_1x_2 \approx fy_1x_2 \rightarrow fx_1x_2 \approx fy_1y_2.$$

From $\forall x\, x \approx x$ (in axiom group 5) we deduce

$$fx_1x_2 \approx fx_1x_2.$$

The three displayed formulas tautologically imply

$$x_1 \approx y_1 \rightarrow x_2 \approx y_2 \rightarrow fx_1x_2 \approx fy_1y_2. \quad ■$$

EXAMPLE (a) $\{\forall x(Px \rightarrow Qx), \forall z\, Pz\} \vdash Qc$. It is not hard to show that such a deduction exists. The deduction itself consists of seven formulas.

(b) $\{\forall x(Px \rightarrow Qx), \forall z\, Pz\} \vdash Qy$. This is just like (a). The point we are interested in here is that we can use the *same* seven-element deduction, with c replaced throughout by y.

(c) $\{\forall x(Px \rightarrow Qx), \forall z\, Pz\} \vdash \forall y\, Qy$. This follows from (b) by the generalization theorem.

(d) $\{\forall x(Px \rightarrow Qx), \forall z\, Pz\} \vdash \forall x\, Qx$. This follows from (c) by use of the fact that $\forall y\, Qy \vdash \forall x\, Qx$.

Parts (a) and (b) of the foregoing example illustrate a sort of interchangeability of constant symbols with free variables. This interchangeability is the basis for the following variation on the generalization theorem, for which part (c) is an example. Part (d) is covered by Corollary 24G. φ^c_y is, of course, the result of replacing c by y in φ.

Theorem 24F (generalization on constants) Assume that $\Gamma \vdash \varphi$ and that c is a constant symbol which does not occur in Γ. Then there is a variable y (which does not occur in φ) such that $\Gamma \vdash \forall y\, \varphi^c_y$. Furthermore, there is a deduction of $\forall y\, \varphi^c_y$ from Γ in which c does not occur.

Proof Let $\langle \alpha_0, \ldots, \alpha_n \rangle$ be a deduction of φ from Γ. (Thus $\alpha_n = \varphi$.) Let y be the first variable which does not occur in any of the α_i's. We claim that

$$(*) \qquad \langle (\alpha_0)^c_y, \ldots, (\alpha_n)^c_y \rangle$$

is a deduction from Γ of φ^c_y. So we must check that each $(\alpha_k)^c_y$ is in $\Gamma \cup \Lambda$ or is obtained from earlier formulas by modus ponens.

Case 1: $\alpha_k \in \Gamma$. Then c does not occur in α_k. So $(\alpha_k)^c_y = \alpha_k$, which is in Γ.

Case 2: α_k is a logical axiom. Then $(\alpha_k)^c_y$ is also a logical axiom. (Read the list of logical axioms and note that introducing a new variable will transform a logical axiom into another one.)

Case 3: α_k is obtained by modus ponens from α_i and α_j $(= (\alpha_i \rightarrow \alpha_k))$ for i, j less than k. Then $(\alpha_j)^c_y = ((\alpha_i)^c_y \rightarrow (\alpha_k)^c_y)$. So $(\alpha_k)^c_y$ is obtained by modus ponens from $(\alpha_i)^c_y$ and $(\alpha_j)^c_y$.

This completes the proof that $(*)$ above is a deduction of φ^c_y. Let Φ be the finite subset of Γ which is used in $(*)$. Thus $(*)$ is a deduction of φ^c_y from Φ, and y does not occur in Φ. So by the generalization theorem, $\Phi \vdash \forall y\, \varphi^c_y$. Furthermore, there is a deduction of $\forall y\, \varphi^c_y$ from Φ in which c does not appear. (For the proof to the generalization theorem did not add any new symbols to a deduction.) This is also a deduction from Γ of $\forall y\, \varphi^c_y$. ■

We will sometimes want to apply this theorem in circumstances in which not just any variable will do. In the following version, there is a variable x selected in advance.

Corollary 24G Assume that $\Gamma \vdash \varphi^x_c$, where the constant symbol c does not occur in Γ or in φ. Then $\Gamma \vdash \forall x\, \varphi$, and there is a deduction of $\forall x\, \varphi$ from Γ in which c does not occur.

Proof By the above theorem we have a deduction (without c) from Γ of $\forall y((\varphi^x_c)^c_y)$, where y does not occur in φ^x_c. But since c does not occur in φ,

$$(\varphi^x_c)^c_y = \varphi^x_y.$$

It remains to show that $\forall y\, \varphi^x_y \vdash \forall x\, \varphi$. We can easily do this if we know that

$$(\forall y\, \varphi^x_y) \to \varphi$$

is an axiom. That is, x must be substitutable for y in φ^x_y, and $(\varphi^x_y)^y_x$ must be φ. This is reasonably clear; for details see the re-replacement lemma (Exercise 9). ∎

Corollary 24H (rule EI) Assume that the constant symbol c does not occur in φ, ψ, or Γ, and that

$$\Gamma ; \varphi^x_c \vdash \psi.$$

Then

$$\Gamma ; \exists x\, \varphi \vdash \psi$$

and there is a deduction of ψ from $\Gamma ; \exists x\, \varphi$ in which c does not occur.

Proof By contraposition we have

$$\Gamma ; \neg\psi \vdash \neg\varphi^x_c.$$

So by the preceding corollary we obtain

$$\Gamma ; \neg\psi \vdash \forall x\, \neg\varphi.$$

Applying contraposition again, we have the desired result. ∎

"EI" stands for "existential instantiation," a bit of traditional terminology.

We will not have occasion to use rule EI in any of our proofs, but it may be handy in exercises. It is the formal counterpart to the reasoning: "We know there is an x such that __x__. So call it c. Now from __c__ we can prove ψ." But notice that rule EI does *not* claim that $\exists x\, \varphi \vdash \varphi^x_c$, which is in fact usually false.

EXAMPLE, revisited $\vdash \exists x\, \forall y\, \varphi \to \forall y\, \exists x\, \varphi$. By the deduction theorem, it suffices to show that

$$\exists x\, \forall y\, \varphi \vdash \forall y\, \exists x\, \varphi.$$

By rule EI it suffices to show that

$$\forall y\, \varphi^x_c \vdash \forall y\, \exists x\, \varphi,$$

where c is new to the language. By the generalization theorem it suffices to show that

$$\forall y\, \varphi^x_c \vdash \exists x\, \varphi.$$

Since $\forall y\, \varphi^x_c \vdash \varphi^x_c$, it suffices to show that

$$\varphi^x_c \vdash \exists x\, \varphi.$$

By contraposition this is equivalent to

$$\forall x\, \neg\varphi \vdash \neg\varphi^x_c,$$

which is trivial (by axiom group 2 and modus ponens).

We can now see roughly how our particular list of logical axioms was formed. The tautologies were included to handle the sentential connective symbols. (We could economize considerably at this point by using only some of the tautologies.) Axiom group 2 reflects the intended meaning of the quantifier symbol. Then in order to be able to prove the generalization theorem we added axiom groups 3 and 4 and arranged for generalizations of axioms to be axioms.

Axiom groups 5 and 6 will turn out to be just enough to prove the crucial properties of equality; see the subsection on equality.

As we will prove in Section 2.5, every logical axiom is a valid formula. It might seem simpler to use as logical axioms the set of *all* valid formulas. But there are two (related) objections to doing this. For one, the notion of validity was defined *semantically*. That is, the definition referred to possible meanings (i.e., structures) for the language and to the notion of truth in a structure. For our present purposes (e.g., proving that the validities are effectively enumerable) we need a class Λ with a finitary, *syntactical* definition. That is, the definition of Λ involves only matters concerning the arrangement of the symbols in the logical axioms; there is no reference to matters of truth in structures. A second objection to the inclusion of all valid formulas as axioms is that we prefer a decidable set Λ, and the set of validities fails to be decidable.

Alphabetic variants

Often when we are discussing a formula such as

$$\forall x(x \not\approx \mathbf{0} \to \exists y\, x \approx \mathbf{S}y)$$

we are not interested in the particular choice of the variables x and y. We want $\langle x, y\rangle$ to be a pair of distinct variables, but often it makes no difference whether the pair is $\langle \boldsymbol{v}_4, \boldsymbol{v}_9\rangle$ or $\langle \boldsymbol{v}_8, \boldsymbol{v}_1\rangle$.

But when it comes time to substitute a term t into a formula, then the choice of quantified variables can make the difference between the substitutability of t and its failure. In this subsection we will discuss what to do when substitutability fails. As will be seen, the difficulty can always be surmounted by suitably juggling the quantified variables.

For example, suppose we want to show that

$$\vdash \forall x\, \forall y\, Pxy \rightarrow \forall y\, Pyy.$$

There is the difficulty that y is not substitutable for x in $\forall y\, Pxy$, so the above sentence is not in axiom group 2. This is a nuisance resulting from an unfortunate choice of variables. For example, showing that

$$\vdash \forall x\, \forall z\, Pxz \rightarrow \forall y\, Pyy$$

involves no such difficulties. So we can solve our original problem if we know that

$$\vdash \forall x\, \forall y\, Pxy \rightarrow \forall x\, \forall z\, Pxz,$$

which, again, involves no difficulties.

This slightly circuitous strategy (of interpolating $\forall x\, \forall z\, Pxz$ between $\forall x\, \forall y\, Pxy$ and $\forall y\, Pyy$) is typical of a certain class of problems. Say that we desire to substitute a term t for x in a wff φ. If t is, in fact, not so substitutable, then we replace $\forall x\, \varphi$ by $\forall x\, \varphi'$, where t is substitutable for x in φ'. In the above example φ is $\forall y\, Pxy$ and φ' is $\forall z\, Pxz$. In general φ' will differ from φ only in the choice of quantified variables. But φ' must be formed in a reasonable way so as to be logically equivalent to φ. For example, it would be unreasonable to replace $\forall y\, Pxy$ by $\forall x\, Pxx$, or $\forall y\, \forall z\, Qxyz$ by $\forall z\, \forall z\, Qxzz$.

Theorem 24I (existence of alphabetic variants) Let φ be a formula, t a term, and x a variable. Then we can find a formula φ' (which differs from φ only in the choice of quantified variables) such that

(a) $\varphi \vdash \varphi'$ and $\varphi' \vdash \varphi$;

(b) t is substitutable for x in φ'.

Proof We consider fixed t and x, and construct φ' by recursion on φ. The first cases are simple: For atomic φ we take $\varphi' = \varphi$, and then $(\neg\varphi)' = (\neg\varphi')$, $(\varphi \rightarrow \psi)' = (\varphi' \rightarrow \psi')$. But now consider the choice of $(\forall y\, \varphi)'$.

If y does not occur in t, or if $y = x$, then we can just take $(\forall y\,\varphi)' = \forall y\,\varphi'$. But for the general case we must change the variable.

Choose a variable z which does not occur in φ' or t or x. Then define $(\forall y\,\varphi)' = \forall z(\varphi')^y_z$. To verify that (b) holds, we note that z does not occur in t and t is substitutable for x in φ' (by the inductive hypothesis). Hence (since $x \neq z$) t is also substitutable for x in $(\varphi')^y_z$. To verify that (a) holds, we calculate:

$\varphi \vdash \varphi'$	by the inductive hypothesis;
$\therefore \forall y\,\varphi \vdash \forall y\,\varphi'$.	
$\forall y\,\varphi' \vdash (\varphi')^y_z$	since z does not occur in φ';
$\therefore \forall y\,\varphi' \vdash \forall z(\varphi')^y_z$	by generalization;
$\therefore \forall y\,\varphi \vdash \forall z(\varphi')^y_z$.	

In the other direction,

$\forall z(\varphi')^y_z \vdash ((\varphi')^y_z)^z_y$,	which is φ' by Exercise 9;
$\varphi' \vdash \varphi$	by the inductive hypothesis;
$\therefore \forall z(\varphi')^y_z \vdash \varphi$;	
$\therefore \forall z(\varphi')^y_z \vdash \forall y\,\varphi$	by generalization.

The last step uses the fact that y does not occur free in $(\varphi')^y_z$ unless $y = z$, and so does not occur free in $\forall z(\varphi')^y_z$ in any case. ■

The formulas φ' constructed as in the proof of this theorem will be called *alphabetic variants* of φ. The moral of the theorem is: One should not be daunted by failure of substitutability; the right alphabetic variant will avoid the difficulty.

Equality

We list here (assuming that our language includes $\approx$) the facts about equality which will be needed in the next section. First, the relation defined by $\boldsymbol{v}_1 \approx \boldsymbol{v}_2$ is reflexive, symmetric, and transitive (i.e., is an equivalence relation):

Eq1: $\vdash \forall x\, x \approx x$.

Proof Axiom group 5. ■

Eq2: $\vdash \forall x\,\forall y(x \approx y \to y \approx x)$.

Proof Page 114. ■

Eq3: $\vdash \forall x \, \forall y \, \forall z(x \approx y \to y \approx z \to x \approx z)$.

Proof Exercise 11. ■

In addition, we will need to know that equality is compatible with the predicate and function symbols:

Eq4 (for a two-place predicate symbol P):

$$\vdash \forall x_1 \forall x_2 \forall y_1 \forall y_2 (x_1 \approx y_1 \to x_2 \approx y_2 \to Px_1x_2 \to Py_1y_2).$$

Similarly for n-place predicate symbols.

Proof It suffices to show that

$$\{x_1 \approx y_1,\ x_2 \approx y_2,\ Px_1x_2\} \vdash Py_1y_2.$$

This is obtained by application of modus ponens to the two members of axiom group 6:

$$x_1 \approx y_1 \to Px_1x_2 \to Py_1x_2,$$
$$x_2 \approx y_2 \to Py_1x_2 \to Py_1y_2. \quad ■$$

Eq5 (for a two-place function symbol f):

$$\vdash \forall x_1 \forall x_2 \forall y_1 \forall y_2 (x_1 \approx y_1 \to x_2 \approx y_2 \to fx_1x_2 \approx fy_1y_2).$$

Similarly for n-place function symbols.

Proof Page 115. ■

Final comments

A logic book in the bootstrap tradition might well begin with this section on a deductive calculus. Such a book would first state the logical axioms and the rules of inference and would explain that they are acceptable to reasonable people. Then it would proceed to show that many formulas were deducible (or deducible from certain nonlogical axioms such as axioms for set theory).

Our viewpoint is very different. We study, among other things, the facts about the procedure described in the preceding paragraph. And we employ in this any correct mathematical reasoning, whether or not such reasoning is known to have counterparts in the deductive calculus under study.

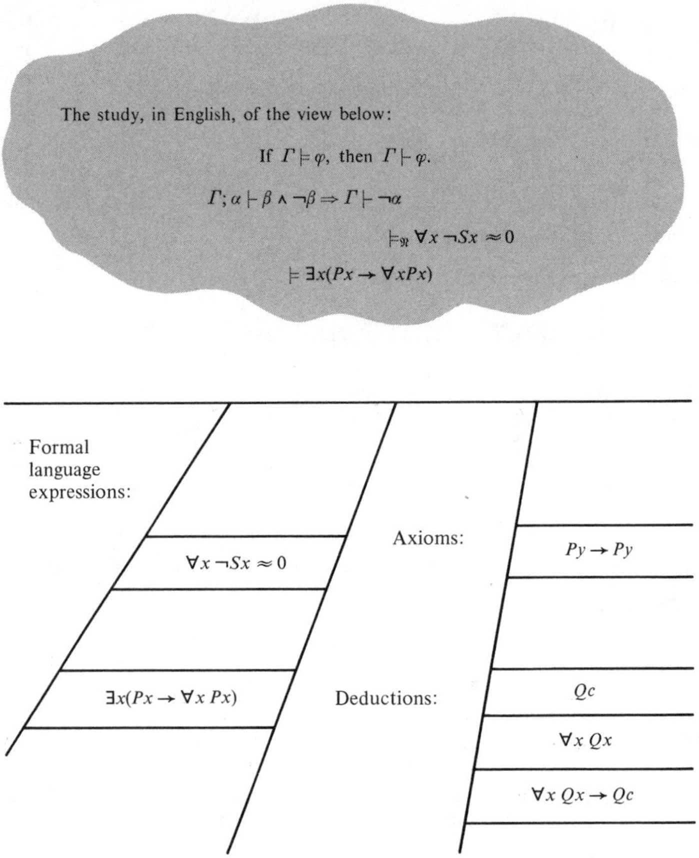

Figure 8. The metalanguage above, in which we study the object language below.

Figure 8 is intended to illustrate the separation between (a) the level at which we carry out our reasoning and prove our results, and (b) the level of the deductive calculus which we study.

EXERCISES

1. For a term u, let u_t^x be the expression obtained from u by replacing the variable x by the term t. Restate this definition without using any form of the word "replace" or its synonyms. *Suggestion*: Use recursion on u. (Observe that from the new definition it is obvious that u_t^x is itself a term.)

2. To which axiom groups, if any, do each of the following formulas belong?

(a) $[(\forall x\, Px \to \forall y\, Py) \to Pz] \to [\forall x\, Px \to (\forall y\, Py \to Pz)]$.
(b) $\forall y[\forall x(Px \to Px) \to (Pc \to Pc)]$.
(c) $\forall x\, \exists y\, Pxy \to \exists y\, Pyy$.

3. (a) Let $\mathfrak{A}$ be a structure and let $s : V \to |\mathfrak{A}|$. Define a truth assignment v on the set of prime formulas by

$$v(\alpha) = T \qquad \text{iff } \models_{\mathfrak{A}} \alpha\, [s].$$

Show that for any formula (prime or not),

$$\bar{v}(\alpha) = T \qquad \text{iff } \models_{\mathfrak{A}} \alpha\, [s].$$

(b) Conclude that if Γ tautologically implies φ, then Γ logically implies φ.

4. Give a deduction (from $\varnothing$) of $\forall x\, \varphi \to \exists x\, \varphi$. (Note that you should not merely prove that such a deduction exists. You are instead asked to write out the entire deduction.)

5. Find a function f such that if a formula φ has a deduction of length n from a set Γ, and if x does not occur free in Γ, then $\forall x\, \varphi$ has a deduction from Γ of length $f(n)$. The more slowly your function grows, the better.

6. (a) Show that if $\vdash \alpha \to \beta$, then $\vdash \forall x\, \alpha \to \forall x\, \beta$.
(b) Show that it is not in general true that $\alpha \to \beta \models \forall x\, \alpha \to \forall x\, \beta$.

7. (a) Show that $\vdash \exists x(Px \to \forall x\, Px)$.
(b) Show that $\{Qx, \forall y(Qy \to \forall z\, Pz)\} \vdash \forall x\, Px$.

8. (Q2b) Assume that x does not occur free in α. Show that

$$\vdash (\alpha \to \exists x\, \beta) \leftrightarrow \exists x(\alpha \to \beta).$$

As a corollary conclude that, under the same assumption, we have Q3a:

$$\vdash (\forall x\, \beta \to \alpha) \leftrightarrow \exists x(\beta \to \alpha).$$

9. (Re-replacement lemma) (a) Show by example that $(\varphi^x_y)^y_x$ is not in general equal to φ. And that it is possible both for x to occur in $(\varphi^x_y)^y_x$ at a place where it does not occur in φ, and for x to occur in φ at a place where it does not occur in $(\varphi^x_y)^y_x$.
(b) Show that if y does not occur at all in φ, then x is substitutable for y in φ^x_y and $(\varphi^x_y)^y_x = \varphi$. *Suggestion*: Use induction on φ.

10. Show that

$$\forall x\, \forall y\, Pxy \vdash \forall y\, \forall x\, Pyx.$$

11. (Eq3) Show that

$$\vdash \forall x\, \forall y\, \forall z(x \approx y \rightarrow y \approx z \rightarrow x \approx z).$$

12. Show that any consistent set Γ of formulas can be extended to a consistent set Δ having the property that for any formula α, either $\alpha \in \Delta$ or $(\neg\alpha) \in \Delta$. (Assume that the language is countable. Do not use the compactness theorem of sentential logic.)

13. Form a deductive calculus for sentential logic by taking as logical axioms the set of tautologies, and taking modus ponens as the rule of inference. Show that for a set Γ ; α of wffs of sentential logic, α is a theorem of Γ iff Γ tautologically implies α.

14. Show that $\vdash Py \leftrightarrow \forall x(x \approx y \rightarrow Px)$.

15. Show that deductions (from $\varnothing$) of the following formulas exist:

(a) $\exists x(\alpha \vee \beta) \leftrightarrow \exists x\, \alpha \vee \exists x\, \beta$.
(b) $\forall x\, \alpha \vee \forall x\, \beta \rightarrow \forall x(\alpha \vee \beta)$.
(c) $\exists x(\alpha \wedge \beta) \rightarrow \exists x\, \alpha \wedge \exists x\, \beta$.
(d) $\forall x(\alpha \rightarrow \beta) \rightarrow (\exists x\, \alpha \rightarrow \exists x\, \beta)$.
(e) $\exists x(Py \wedge Qx) \leftrightarrow Py \wedge \exists x\, Qx$.

§ 2.5 SOUNDNESS AND COMPLETENESS THEOREMS

In this section we establish two major theorems: the soundness of our deductive calculus ($\Gamma \vdash \varphi \Rightarrow \Gamma \models \varphi$) and its completeness ($\Gamma \models \varphi \Rightarrow \Gamma \vdash \varphi$). We will then be able to draw a number of interesting conclusions (including the compactness and enumeration theorems). Although our deductive calculus was chosen in a somewhat arbitrary way, the significant fact is that *some* such deductive calculus is sound and complete. This should be encouraging to the "working mathematician" concerned about the existence of proofs from axioms; see the Retrospectus subsection of Section 2.6.

Soundness Theorem If $\Gamma \vdash \varphi$, then $\Gamma \models \varphi$.

The idea of the proof is that the logical axioms are logically implied by anything, and that modus ponens preserves logical implications.

Lemma 25A Every logical axiom is valid.

Proof of the soundness theorem, assuming the lemma We show by induction that any formula φ deducible from Γ is logically implied by Γ.

Case 1: φ is a logical axiom. Then by the lemma $\models \varphi$, so *a fortiori* $\Gamma \models \varphi$.

Case 2: $\varphi \in \Gamma$. Then clearly $\Gamma \models \varphi$.

Case 3: φ is obtained by modus ponens from ψ and $\psi \to \varphi$, where (by the inductive hypothesis) $\Gamma \models \psi$ and $\Gamma \models (\psi \to \varphi)$. It then follows at once that $\Gamma \models \varphi$ (cf. Exercise 3 of the preceding section). ■

It remains, of course, to prove that lemma. We know from Exercise 6 of Section 2.2 that any generalization of a valid formula is valid. So it suffices to consider only logical axioms which are not themselves generalizations of other axioms. We will examine the various axiom groups in order of complexity.

Axiom group 3: See Exercise 3 of Section 2.2.

Axiom group 4: See Exercise 4 of Section 2.2.

Axiom group 5: Trivial. $\mathfrak{A}$ satisfies $x \approx x$ with s iff $s(x) = s(x)$, which is always true.

Axiom group 1: We know from Exercise 3 of the preceding section that if $\varnothing$ tautologically implies α, then $\varnothing \models \alpha$. And that is just what we need.

Axiom group 6 (for an example, see Exercise 5 of Section 2.2.): Assume that α is atomic and α' is obtained from α by replacing x at some places by y. It suffices to show that

$$\{x \approx y, \alpha\} \models \alpha'.$$

So take any $\mathfrak{A}$, s such that

$$\models_{\mathfrak{A}} x \approx y\ [s], \qquad \text{i.e., } s(x) = s(y).$$

Then any term t has the property that if t' is obtained from t by replacing x at some places by y, then $\bar{s}(t) = \bar{s}(t')$. This is obvious; a full proof would use induction on t.

If α is $t_1 \approx t_2$, then α' must be $t_1' \approx t_2'$, where t_i' is obtained from t_i as described:

$$\begin{aligned} \models_{\mathfrak{A}} \alpha\ [s] \quad & \text{iff } \bar{s}(t_1) = \bar{s}(t_2), \\ & \text{iff } \bar{s}(t_1') = \bar{s}(t_2'), \\ & \text{iff } \models_{\mathfrak{A}} \alpha'\ [s]. \end{aligned}$$

Similarly, if α is $Pt_1 \cdots t_n$, then α' is $Pt_1' \cdots t_n'$ and an analogous argument applies.

Finally, we come to axiom group 2. It will be helpful to consider first a simple example; we will show that $\forall x\, Px \to Pt$ is valid. Assume that

$$\models_{\mathfrak{A}} \forall x\, Px\ [s].$$

Then for any d in $|\mathfrak{A}|$,

$$\models_{\mathfrak{A}} Px\,[s(x|d)].$$

So in particular we may take $d = \bar{s}(t)$:

(a) $$\models_{\mathfrak{A}} Px\,[s(x|\bar{s}(t))].$$

This is equivalent (by the definition of satisfaction of atomic formulas) to

$$\bar{s}(t) \in P^{\mathfrak{A}},$$

which in turn is equivalent to

(b) $$\models_{\mathfrak{A}} Pt\,[s].$$

For this argument to be applicable to the nonatomic case, we need a way of passing from (a) to (b). This will be provided by the substitution lemma below, which states that

$$\models_{\mathfrak{A}} \varphi\,[s(x|\bar{s}(t))] \qquad \text{iff } \models_{\mathfrak{A}} \varphi_t^x\,[s]$$

whenever t is substitutable for x in φ.

Consider a fixed $\mathfrak{A}$ and s. For any term u, let u_t^x be the result of replacing the variable x in u by the term t.

Lemma 25B $\bar{s}(u_t^x) = \overline{s(x|\bar{s}(t))}(u)$.

This asserts that a substitution can be carried out either in the term u or in s, with equivalent results. The corresponding diagram is

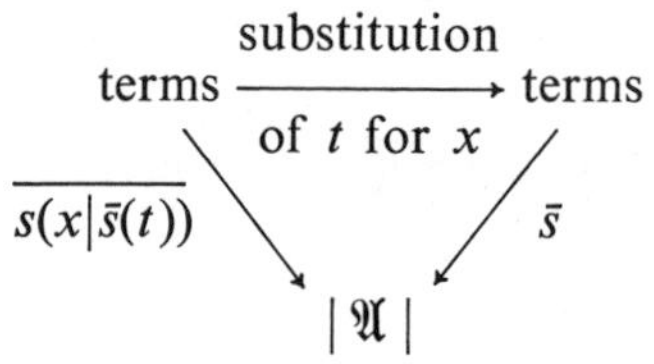

Proof By induction on the term u. If u is a constant symbol or a variable other than x, then $u_t^x = u$ and the desired equation reduces to $\bar{s}(u) = \bar{s}(u)$. If $u = x$, then the equation reduces to $\bar{s}(t) = \bar{s}(t)$. The inductive step, although cumbersome to write, is mathematically trivial. ∎

The substitution lemma is similar in spirit; it states that a substitution can be carried out either within φ or in s, with equivalent results. For an example see Exercise 10 of Section 2.2.

Substitution Lemma If the term t is substitutable for the variable x in the wff φ, then

$$\models_{\mathfrak{A}} \varphi^x_t\,[s] \qquad \text{iff } \models_{\mathfrak{A}} \varphi\,[s(x|\bar{s}(t))].$$

Proof We use induction on φ to show that the above holds for every s.

Case 1: φ is atomic. Then the conclusion follows from the preceding lemma. For example, if φ is Pu for some term u, then

$$\begin{aligned} \models_{\mathfrak{A}} Pu^x_t\,[s] \quad & \text{iff } \bar{s}(u^x_t) \in P^{\mathfrak{A}}, \\ & \text{iff } \overline{s(x|\bar{s}(t))}(u) \in P^{\mathfrak{A}} \qquad \text{by Lemma 25B}, \\ & \text{iff } \models_{\mathfrak{A}} Pu\,[s(x|\bar{s}(t))]. \end{aligned}$$

Case 2: φ is $\neg\psi$ or $\psi \rightarrow \theta$. Then the conclusion for φ follows at once from the inductive hypotheses for ψ and θ.

Case 3: φ is $\forall y\,\psi$, and x does not occur free in φ. Then s and $s(x|\bar{s}(t))$ agree on all variables which occur free in φ. And also φ^x_t is just φ. So the conclusion is immediate.

Case 4: φ is $\forall y\,\psi$, and x does occur free in φ. Because t is substitutable for x in φ, we know that y does not occur in t and t is substitutable for x in ψ (see the definition of "substitutable"). By the first of these,

$$(*) \qquad \bar{s}(t) = \overline{s(y|d)}(t)$$

for any d in $|\mathfrak{A}|$. Since $x \neq y$, $\varphi^x_t = \forall y\,\psi^x_t$.

$$\begin{aligned} \models_{\mathfrak{A}} \varphi^x_t\,[s] \quad & \text{iff for every } d, \models_{\mathfrak{A}} \psi^x_t\,[s(y|d)], \\ & \text{iff for every } d, \models_{\mathfrak{A}} \psi\,[s(y|d)(x|\bar{s}(t))] \text{ by the inductive hypothesis and } (*); \\ & \text{iff } \models_{\mathfrak{A}} \varphi\,[s(x|\bar{s}(t))]. \end{aligned}$$

So by induction the lemma holds for all φ. ■

Axiom group 2: Assume that t is substitutable for x in φ. Assume that $\mathfrak{A}$ satisfies $\forall x\,\varphi$ with s. We need to show that $\models_{\mathfrak{A}} \varphi^x_t\,[s]$. We know that for any d in $|\mathfrak{A}|$,

$$\models_{\mathfrak{A}} \varphi\,[s(x|d)],$$

In particular, let $d = \bar{s}(t)$:

$$\models_{\mathfrak{A}} \varphi\,[s(x|\bar{s}(t))].$$

So, by the substitution lemma,

$$\models_{\mathfrak{A}} \varphi_t^x\,[s].$$

Hence $\forall x\,\varphi \rightarrow \varphi_t^x$ is valid.

This completes the proof that all logical axioms are valid. And so the soundness theorem is proved.

Corollary 25C If $\vdash(\varphi \leftrightarrow \psi)$, then φ and ψ are logically equivalent.

Corollary 25D If φ' is an alphabetic variant of φ (see Theorem 24I), then φ and φ' are logically equivalent.

Recall that a set Γ is consistent iff there is no formula φ such that both $\Gamma \vdash \varphi$ and $\Gamma \vdash \neg\varphi$. Define Γ to be *satisfiable* iff there is some $\mathfrak{A}$ and s such that $\mathfrak{A}$ satisfies every member of Γ with s.

Corollary 25E If Γ is satisfiable, then Γ is consistent.

This corollary is actually equivalent to the soundness theorem, as the reader is invited to verify.

The completeness theorem is the converse to the soundness theorem and is a deeper result.

Completeness Theorem (Gödel, 1930)

(a) If $\Gamma \models \varphi$, then $\Gamma \vdash \varphi$.

(b) Any consistent set of formulas is satisfiable.

Actually parts (a) and (b) are equivalent; cf. Exercise 2. So it suffices to prove part (b). We will give a proof for a countable language; later we will indicate what alterations are needed for languages of larger cardinality. (A countable language is one with countably many symbols, or equivalently (by Theorem 0B) one with countably many wffs.)

The ideas of the proof are related to those in the proof of the compactness theorem for sentential logic. We begin with a consistent set Γ. In steps 1–3 we extend Γ to a set Δ of formulas for which

(i) $\Gamma \subseteq \Delta$.

(ii) Δ is consistent and is maximal in the sense that for any formula α, either $\alpha \in \Delta$ or $(\neg\alpha) \in \Delta$.

(iii) For any formula φ and variable x, there is a c such that

$$(\neg\forall x\,\varphi \rightarrow \neg\varphi_c^x) \in \Delta.$$

Then in step 4 we form a structure $\mathfrak{A}$ in which members of Γ not containing $\approx$ can be satisfied. $|\mathfrak{A}|$ is the set of terms, and for a predicate symbol P,

$$\langle t_1, \ldots, t_n \rangle \in P^{\mathfrak{A}} \qquad \text{iff } Pt_1 \cdots t_n \in \Delta.$$

Finally, in step 5 we change $\mathfrak{A}$ to accommodate formulas containing the equality symbol.

It is suggested that on a first reading the details which are provided for most of the steps be omitted. Once the outline is clearly in mind, the entire proof should be read. (The nondetails are marked with a stripe in the left margin.)

Proof Let Γ be a consistent set of wffs in a countable language.

Step 1: Expand the language by adding a countably infinite set of new constant symbols. Then Γ remains consistent as a set of wffs in the new language.

Details: If not then for some β, there is a deduction (in the expanded language) of $(\beta \wedge \neg\beta)$ from Γ. This deduction contains only finitely many of the new constant symbols. By the theorem for generalization on constants (Theorem 24F), each can be replaced by a variable. We then have a deduction (in the original language) of $(\beta' \wedge \neg\beta')$ from Γ. This contradicts our assumption that Γ was consistent.

Step 2: For each wff φ (in the new language) and each variable x, we want to add to Γ the wff

$$\neg\forall x\, \varphi \rightarrow \neg\varphi^x_c,$$

where c is one of the new constant symbols. (The idea is that c volunteers to provide a counterexample to φ, if there is any.) We can do this in such a way that Γ together with the set Θ of all the added wffs is still a consistent set.

Details: Adopt a fixed enumeration of the pairs $\langle \varphi, x \rangle$, where φ is a wff (of the expanded language) and x is a variable:

$$\langle \varphi_1, x_1 \rangle, \langle \varphi_2, x_2 \rangle, \langle \varphi_3, x_3 \rangle, \ldots.$$

This is possible since the language is countable. Let θ_1 be

$$\neg\forall x_1\, \varphi_1 \rightarrow \neg{\varphi_1}^{x_1}_{c_1},$$

where c_1 is the first of the new constant symbols not occurring in φ_1. Then go on to $\langle \varphi_2, x_2 \rangle$ and define θ_2. In general, θ_n is

$$\neg\forall x_n \varphi_n \rightarrow \neg\varphi_n{}^{x_n}_{c_n},$$

where c_n is the first of the new constant symbols not occurring in φ_n or in θ_k for any $k < n$.

Let Θ be the set $\{\theta_1, \theta_2, \ldots\}$. We claim that $\Gamma \cup \Theta$ is consistent. If not, then (because deductions are finite) for some $m \geq 0$,

$$\Gamma \cup \{\theta_1, \ldots, \theta_m, \theta_{m+1}\}$$

is inconsistent. Take the least such m. Then by RAA

$$\Gamma \cup \{\theta_1, \ldots, \theta_m\} \vdash \neg\theta_{m+1}.$$

Now θ_{m+1} is

$$\neg\forall x\, \varphi \rightarrow \neg\varphi^x_c$$

for some x, φ, and c. So by rule T, we obtain the two facts:

$$(*) \qquad \begin{aligned} &\Gamma \cup \{\theta_1, \ldots, \theta_m\} \vdash \neg\forall x\varphi, \\ &\Gamma \cup \{\theta_1, \ldots, \theta_m\} \vdash \varphi^x_c. \end{aligned}$$

Since c does not appear in any formula on the left side, we can apply the Corollary 24G to the second of these, obtaining

$$\Gamma \cup \{\theta_1, \ldots, \theta_m\} \vdash \forall x\, \varphi.$$

This and (*) contradict the leastness of m (or the consistency of Γ, if $m = 0$).

Step 3: We now extend the consistent set $\Gamma \cup \Theta$ to a consistent set Δ which is maximal in the sense that for any wff φ either $\varphi \in \Delta$ or $(\neg\varphi) \in \Delta$.

Details: We can imitate the proof used at the analogous place in the proof of sentential compactness in Section 1.7. Or we can argue as follows: Let Λ be the set of logical axioms for the expanded language. Since $\Gamma \cup \Theta$ is consistent, there is no formula β such that $\Gamma \cup \Theta \cup \Lambda$ tautologically implies both β and $\neg\beta$. (This is by Theorem 24B; the compactness theorem of sentential logic is here used.) Hence there is a truth assignment v for the set of all prime formulas which satisfies (every member of) $\Gamma \cup \Theta \cup \Lambda$. Let

$$\Delta = \{\varphi : \bar{v}(\varphi) = T\}.$$

Clearly for any φ either $\varphi \in \Delta$ or $(\neg\varphi) \in \Delta$ but not both. Also we have

$$\begin{aligned} \Delta \vdash \varphi &\Rightarrow \Delta \text{ tautologically implies } \varphi && (\text{since } \Lambda \subseteq \Delta),\\ &\Rightarrow v(\varphi) = T && \text{since } v \text{ satisfies } \Delta,\\ &\Rightarrow \varphi \in \Delta. \end{aligned}$$

Consequently, Δ is consistent, lest both φ and $(\neg\varphi)$ belong to Δ.

Actually, regardless of how Δ is constructed, it must be deductively closed. For

$$\begin{aligned} \Delta \vdash \varphi &\Rightarrow \Delta \nvdash \neg\varphi && \text{by consistency,}\\ &\Rightarrow (\neg\varphi) \notin \Delta, \\ &\Rightarrow \varphi \in \Delta && \text{by maximality.} \end{aligned}$$

Step 4: We now make from Δ a structure $\mathfrak{A}$ for the new language, but with the equality symbol (if any) replaced by a new two-place predicate symbol E. $\mathfrak{A}$ will not itself be the structure in which Γ will be satisfied but will be a preliminary structure.

(a) $|\mathfrak{A}|$ = the set of all terms of the new language.
(b) Define the binary relation $E^{\mathfrak{A}}$ by

$$\langle u, t\rangle \in E^{\mathfrak{A}} \qquad \text{iff the formula } u \approx t \text{ belongs to } \Delta.$$

(c) For each n-place predicate parameter P, define the n-ary relation $P^{\mathfrak{A}}$ by

$$\langle t_1, \ldots, t_n\rangle \in P^{\mathfrak{A}} \qquad \text{iff } Pt_1 \cdots t_n \in \Delta.$$

(d) For each n-place function symbol f, let $f^{\mathfrak{A}}$ be the function defined by

$$f^{\mathfrak{A}}(t_1, \ldots, t_n) = ft_1 \cdots t_n.$$

This includes the $n = 0$ case; for a constant symbol c we take $c^{\mathfrak{A}} = c$. Define also a function $s : V \to |\mathfrak{A}|$, namely the identity function on V.

It then follows that for any term t, $\bar{s}(t) = t$. For any wff φ, let φ^* be the result of replacing the equality symbol in φ by E. Then

$$\models_{\mathfrak{A}} \varphi^* [s] \qquad \text{iff } \varphi \in \Delta.$$

Details: That $\bar{s}(t) = t$ can be proved by induction on t, but the proof is entirely straightforward.

The other claim, that

$$\models_{\mathfrak{A}} \varphi^* [s] \qquad \text{iff } \varphi \in \Delta,$$

we prove by induction on the number of places at which connective or quantifier symbols appear in φ.

Case 1: Atomic formulas. We defined $\mathfrak{A}$ in such a way as to make this case immediate. For example, if φ is Pt, then

$$\begin{aligned} \models_{\mathfrak{A}} Pt\,[s] \qquad & \text{iff } \bar{s}(t) \in P^{\mathfrak{A}}, \\ & \text{iff } t \in P^{\mathfrak{A}}, \\ & \text{iff } Pt \in \Delta. \end{aligned}$$

Similarly,

$$\begin{aligned} \models_{\mathfrak{A}} uEt\,[s] \qquad & \text{iff } \langle \bar{s}(u), \bar{s}(t) \rangle \in E^{\mathfrak{A}}, \\ & \text{iff } \langle u, t \rangle \in E^{\mathfrak{A}}, \\ & \text{iff } u \approx t \in \Delta. \end{aligned}$$

Case 2: Negation.

$$\begin{aligned} \models_{\mathfrak{A}} (\neg\varphi)^* [s] \qquad & \text{iff } \not\models_{\mathfrak{A}} \varphi^* [s], \\ & \text{iff } \varphi \notin \Delta \qquad \text{by inductive hypothesis,} \\ & \text{iff } (\neg\varphi) \in \Delta \qquad \text{by properties of } \Delta. \end{aligned}$$

Case 3: Conditional.

$$\begin{aligned} \models_{\mathfrak{A}} (\varphi \to \psi)^* [s] \qquad & \text{iff } \not\models_{\mathfrak{A}} \varphi^* [s] \quad \text{or} \quad \models_{\mathfrak{A}} \psi^* [s], \\ & \text{iff } \varphi \notin \Delta \quad \text{or} \quad \psi \in \Delta \qquad \text{by inductive hypothesis,} \\ & \text{iff } (\neg\varphi) \in \Delta \quad \text{or} \quad \psi \in \Delta, \\ & \Rightarrow \Delta \vdash (\varphi \to \psi), \qquad \text{in fact tautologically,} \\ & \Rightarrow \varphi \notin \Delta \quad \text{or} \quad [\varphi \in \Delta \quad \text{and} \quad \Delta \vdash \psi], \\ & \Rightarrow (\neg\varphi) \in \Delta \quad \text{or} \quad \psi \in \Delta, \end{aligned}$$

which closes the loop. And

$$\Delta \vdash (\varphi \to \psi) \qquad \text{iff } (\varphi \to \psi) \in \Delta.$$

(This should be compared with Exercise 2 of Section 1.7.)

Case 4: Quantification. We want to show that

$$\models_{\mathfrak{A}} \forall x\, \varphi^* [s] \qquad \text{iff } \forall x\, \varphi \in \Delta.$$

(The notational ambiguity is harmless since $\forall x(\varphi^*)$ is the same as $(\forall x\,\varphi)^*$.) Δ includes the wff θ:

$$\neg\forall x\,\varphi \rightarrow \neg\varphi^x_c.$$

To show that

$$\models_{\mathfrak{A}} \forall x\,\varphi^*\,[s] \Rightarrow \forall x\,\varphi \in \Delta,$$

we can argue: If φ^* is true of everything, then it is true of c, whence by the inductive hypothesis $\varphi^x_c \in \Delta$. But then $\forall x\,\varphi \in \Delta$, because c was chosen to be a counterexample to φ if there was one. In more detail:

$$\begin{aligned}
\models_{\mathfrak{A}} \forall x\,\varphi^*\,[s] &\Rightarrow\ \models_{\mathfrak{A}} \varphi^*\,[s(x|c)] && \\
&\Rightarrow\ \models_{\mathfrak{A}} (\varphi^*)^x_c\,[s] && \text{by the substitution lemma} \\
&\Rightarrow\ \models_{\mathfrak{A}} (\varphi^x_c)^*\,[s], && \text{this being the same formula} \\
&\Rightarrow \varphi^x_c \in \Delta && \text{by the inductive hypothesis} \\
&\Rightarrow (\neg\varphi^x_c) \notin \Delta && \text{by consistency} \\
&\Rightarrow (\neg\forall x\,\varphi) \notin \Delta && \text{since } \theta \in \Delta \text{ and } \Delta \text{ is deductively closed} \\
&\Rightarrow \forall x\,\varphi \in \Delta. &&
\end{aligned}$$

(This is our only use of Θ. We needed to know that if $(\neg\forall x\,\varphi) \in \Delta$, then for a particular c we would have $(\neg\varphi^x_c) \in \Delta$.)

We turn now to the converse. We can *almost* argue as follows:

$$\begin{aligned}
\not\models_{\mathfrak{A}} \forall x\,\varphi^*\,[s] &\Rightarrow\ \not\models_{\mathfrak{A}} \varphi^*\,[s(x|t)] && \text{for some } t \\
&\rightsquigarrow\ \not\models_{\mathfrak{A}} (\varphi^x_t)^*\,[s] && \text{by the substitution lemma} \\
&\Rightarrow \varphi^x_t \notin \Delta && \text{by the inductive hypothesis} \\
&\rightsquigarrow \forall x\,\varphi \notin \Delta && \text{since } \Delta \text{ is deductively closed.}
\end{aligned}$$

The flaw here is that the two wavy implications require that t be substitutable for x in φ. This may not be the case, but we can use the usual repair: We change to an alphabetic variant ψ of φ in which t is substitutable for x. Then

$$\begin{aligned}
\not\models_{\mathfrak{A}} \forall x\,\varphi^*\,[s] &\Rightarrow\ \not\models_{\mathfrak{A}} \varphi^*\,[s(x|t)] && \text{for some } t\text{, henceforth fixed} \\
&\Rightarrow\ \not\models_{\mathfrak{A}} \psi^*\,[s(x|t)] && \text{by the semantical equivalence of alphabetic variants (Corollary 25D)} \\
&\Rightarrow\ \not\models_{\mathfrak{A}} (\psi^x_t)^*\,[s] && \text{by the substitution lemma} \\
&\Rightarrow \psi^x_t \notin \Delta && \text{by the inductive hypothesis} \\
&\Rightarrow \forall x\,\psi \notin \Delta && \text{since } \Delta \text{ is deductively closed} \\
&\Rightarrow \forall x\,\varphi \notin \Delta && \text{by the syntactical equivalence of alphabetic variants (Theorem 24I).}
\end{aligned}$$

This completes the list of possible cases; it now follows by induction that for any φ,

$$\models_{\mathfrak{A}} \varphi^* [s] \qquad \text{iff } \varphi \in \Delta.$$

If our original language did not include the equality symbol, then we are done. For we need only restrict $\mathfrak{A}$ to the original language to obtain a structure which satisfies every member of Γ with the identity function. But now assume that the equality symbol is in the language. Then $\mathfrak{A}$ will no longer serve. For example, if Γ contains the sentence $c \approx d$ (where c and d are distinct constant symbols), then we need a structure $\mathfrak{B}$ in which $c^{\mathfrak{B}} = d^{\mathfrak{B}}$. We obtain $\mathfrak{B}$ as the quotient structure $\mathfrak{A}/E$ of $\mathfrak{A}$ modulo $E^{\mathfrak{A}}$.

Step 5: $E^{\mathfrak{A}}$ is an equivalence relation on $|\mathfrak{A}|$. For each t in $|\mathfrak{A}|$ let $[t]$ be its equivalence class. $E^{\mathfrak{A}}$ is, in fact, a *congruence relation* for $\mathfrak{A}$. This means that the following conditions are met:

(i) $E^{\mathfrak{A}}$ is an equivalence relation on $|\mathfrak{A}|$.
(ii) $P^{\mathfrak{A}}$ is compatible with $E^{\mathfrak{A}}$ for each predicate symbol P:

$$\langle t_0, \ldots, t_n\rangle \in P^{\mathfrak{A}} \quad \text{and} \quad t_i E^{\mathfrak{A}} t_i' \qquad \text{for } i \leq n \Rightarrow \langle t_0', \ldots, t_n'\rangle \in P^{\mathfrak{A}}.$$

(iii) $f^{\mathfrak{A}}$ is compatible with $E^{\mathfrak{A}}$ for each function symbol f:

$$t_i E^{\mathfrak{A}} t_i' \qquad \text{for } i \leq n \Rightarrow f^{\mathfrak{A}}(t_0, \ldots, t_n)\, E^{\mathfrak{A}}\, f^{\mathfrak{A}}(t_0', \ldots, t_n').$$

Under these circumstances we can form the quotient structure $\mathfrak{A}/E$, defined as follows:

(a) $|\mathfrak{A}/E|$ is the set of all equivalence classes of members of $|\mathfrak{A}|$.
(b) For each n-place predicate symbol P,

$$\langle [t_1], \ldots, [t_n]\rangle \in P^{\mathfrak{A}/E} \qquad \text{iff } \langle t_1, \ldots, t_n\rangle \in P^{\mathfrak{A}}.$$

(c) For each n-place function symbol f,

$$f^{\mathfrak{A}/E}([t_1], \ldots, [t_n]) = [f^{\mathfrak{A}}(t_1, \ldots, t_n)].$$

This includes the $n = 0$ cases:

$$c^{\mathfrak{A}/E} = [c^{\mathfrak{A}}].$$

Let $h : |\mathfrak{A}| \to |\mathfrak{A}/E|$ be the natural map:

$$h(t) = [t].$$

Then h is a homomorphism of $\mathfrak{A}$ onto $\mathfrak{A}/E$. Furthermore, $E^{\mathfrak{A}/E}$ is the equality relation on $|\mathfrak{A}/E|$. Consequently, for any φ:

$$\begin{aligned}\varphi \in \Delta &\Leftrightarrow \models_{\mathfrak{A}} \varphi^* [s] \\ &\Leftrightarrow \models_{\mathfrak{A}/E} \varphi^* [h \circ s] \\ &\Leftrightarrow \models_{\mathfrak{A}/E} \varphi [h \circ s].\end{aligned}$$

So $\mathfrak{A}/E$ satisfies every member of Δ (and hence every member of Γ) with $h \circ s$.

Details: Recall that

$$\begin{aligned} tE^{\mathfrak{A}}t' \quad & \text{iff } (t \approx t') \in \Delta, \\ & \text{iff } \Delta \vdash t \approx t'. \end{aligned}$$

(i) $E^{\mathfrak{A}}$ is an equivalence relation on $\mathfrak{A}$ by properties Eq1, Eq2, and Eq3 of equality.

(ii) $P^{\mathfrak{A}}$ is compatible with $E^{\mathfrak{A}}$ by property Eq4 of equality.

(iii) $f^{\mathfrak{A}}$ is compatible with $E^{\mathfrak{A}}$ by property Eq5 of equality.

It then follows from the compatibility of $P^{\mathfrak{A}}$ with $E^{\mathfrak{A}}$ that $P^{\mathfrak{A}/E}$ is well defined. Similarly, $f^{\mathfrak{A}/E}$ is well defined because $f^{\mathfrak{A}}$ is compatible with $E^{\mathfrak{A}}$.

It is immediate from the construction that h is a homomorphism of $\mathfrak{A}$ onto $\mathfrak{A}/E$. And

$$\begin{aligned} [t]E^{\mathfrak{A}/E}[t'] \quad & \text{iff } tE^{\mathfrak{A}}t', \\ & \text{iff } [t] = [t']. \end{aligned}$$

Finally,

$$\begin{aligned}\varphi \in \Delta &\Leftrightarrow \models_{\mathfrak{A}} \varphi^* [s] && \text{by step 4} \\ &\Leftrightarrow \models_{\mathfrak{A}/E} \varphi^* [h \circ s] && \text{by the homomorphism theorem} \\ &\Leftrightarrow \models_{\mathfrak{A}/E} \varphi [h \circ s],\end{aligned}$$

the last step being justified by the fact that $E^{\mathfrak{A}/E}$ is the equality relation on $|\mathfrak{A}/E|$.

Step 6: Restrict the structure $\mathfrak{A}/E$ to the original language. This restriction of $\mathfrak{A}/E$ satisfies every member of Γ with $h \circ s$. ■

For an uncountable language, a few modifications to the foregoing proof of the completeness theorem are needed. Say that the language has cardinality $\varkappa$. (By this we mean that it has $\varkappa$ symbols or, equivalently, $\varkappa$ formulas.) We will describe the modifications needed, assuming the reader has a substantial knowledge of set theory. In step 1 we add $\varkappa$ new constant symbols; the details remain unchanged. In step 2, only the details change. The cardinal $\varkappa$ is an initial ordinal. (We have tacitly well-ordered the language here.) "Enumerate" the pairs

$$\langle \varphi_\alpha, x_\alpha \rangle_{\alpha < \varkappa}$$

indexed by ordinals less than $\varkappa$. For $\alpha < \varkappa$, θ_α is

$$\neg\forall x_\alpha \varphi_\alpha \rightarrow (\neg\varphi)^{x_\alpha}_{c_\alpha},$$

where c_α is the first of the new constant symbols not in φ_α or in θ_β for any $\beta < \alpha$. (This excludes at most $\aleph_0 \cdot \operatorname{card}(\alpha)$ constant symbols, so there are some left.) Finally, in step 3, we can obtain the maximal set Δ by use of Zorn's lemma. The rest of the proof remains unchanged.

Compactness Theorem (a) If $\Gamma \models \varphi$, then for some finite $\Gamma_0 \subseteq \Gamma$ we have $\Gamma_0 \models \varphi$.

(b) If every finite subset Γ_0 of Γ is satisfiable, then Γ is satisfiable.

In particular, a set Σ of sentences has a model iff every finite subset has a model.

To prove part (a) of the compactness theorem, we simply observe that

$$\begin{aligned} \Gamma \models \varphi &\Rightarrow \Gamma \vdash \varphi \\ &\Rightarrow \Gamma_0 \vdash \varphi \qquad \text{for some finite } \Gamma_0 \subseteq \Gamma, \text{ deductions being finite} \\ &\Rightarrow \Gamma_0 \models \varphi. \end{aligned}$$

Part (b) has a similar proof. If every finite subset of Γ is satisfiable, then by soundness every finite subset of Γ is consistent. Thus Γ is consistent, since deductions are finite. So by completeness, Γ is satisfiable. (Actually parts (a) and (b) are equivalent; cf. Exercise 3 of Section 1.7.) ■

When a person first hears of the compactness theorem, his natural inclination is to try to combine (by some algebraic or set-theoretic operation) the structures in which the various finite subsets are satisfied, in such a way as to obtain a structure in which the entire set is satisfied. In fact, such a proof is possible; the operation to use is the *ultraproduct* construction. But we will refrain from digressing further into this intriguing possibility.

Notice that the compactness theorem involves only semantical notions of Section 2.3; it does not involve deductions at all. And there are proofs that avoid deductions. The same remarks apply to the following theorem.

***Enumerability Theorem** For a reasonable language, the set of valid wffs can be effectively enumerated.

By a reasonable language we mean one whose set of parameters can be effectively enumerated and such that the two relations

$$\{(P, n) : P \text{ is an } n\text{-place predicate symbol}\}$$

and

$$\{(f, n) : f \text{ is an } n\text{-place function symbol}\}$$

are decidable. For example, any language with only finitely many parameters is certainly reasonable. On the other hand, a reasonable language must be countable, since we cannot effectively enumerate an uncountable set.

A precise version of this theorem will be given in Section 3.4. (See especially item 20 there.) The proofs of the two versions are in essence the same.

Proof The essential fact is that Λ, and hence the set of deductions, are decidable.

Suppose that we are given some expression ε. (The assumption of reasonableness enters already here. There are only countably many things eligible to be given by one person to another.) We want to decide whether or not ε is in Λ. First we check that ε has the syntactical form necessary to be a formula. (For sentential logic we gave detailed instructions for such a check; see Section 1.4. Similar instructions can be given for first-order languages, by using Section 2.3.) If ε passes that test, we then check (by constructing a truth table) to see if ε is a generalization of a tautology. If not, then we proceed to see if ε has the syntactical form necessary to be in axiom group 2. And so forth. If ε has not been accepted by the time we finish with axiom group 6, then ε is not in Λ.

(The above is intended to convince the reader that he really can tell members of Λ from nonmembers. The reader who remains dubious can look forward to the rerun in Section 3.4.)

Since Λ is decidable, the set of tautological consequences of Λ are effectively enumerable; see Theorem 17G. But

$$\begin{aligned} &\{\alpha : \alpha \text{ is a tautological consequence of } \Lambda\} \\ &\quad = \{\alpha : \vdash \alpha\} \text{ by Theorem 24B,} \\ &\quad = \{\alpha : \alpha \text{ is valid}\}. \qquad \blacksquare \end{aligned}$$

An alternative to the last paragraph of this proof is the following argument, which is possibly more illuminating: First we claim that the set of deductions (from $\varnothing$) is decidable. For given a finite sequence $\alpha_0, \ldots, \alpha_n$ we can examine each α_i in turn to see if it is in Λ or is obtainable by modus ponens from earlier members of the sequence. Then to enumerate the validities, we begin by enumerating all finite sequences of wffs. We look at each sequence as it is produced and decide whether or not it is a deduction. If not, we discard it. But if it is, then we put its last member on the list of validities. Continuing in this way, we generate a list on which any valid formula will eventually appear.

★**Corollary 25F** Let Γ be a decidable set of formulas in a reasonable language.

(a) The set of theorems of Γ is effectively enumerable.

(b) The set $\{\varphi : \Gamma \models \varphi\}$ of formulas logically implied by Γ is effectively enumerable.

(Of course parts (a) and (b) refer to the same set. This corollary includes the enumeration theorem itself, in which $\Gamma = \varnothing$.)

Proof 1 Enumerate the validities; whenever you find one of the form

$$\alpha_n \to \cdots \to \alpha_1 \to \alpha_0,$$

check to see if $\alpha_n, \ldots, \alpha_1$ are in Γ. If so, then put α_0 on the list of theorems of Γ. In this way, any theorem of Γ is eventually listed. ■

Proof 2 $\Gamma \cup \Lambda$ is decidable, so its set of tautological consequences is effectively enumerable. And that is just the set we want. ■

For example, let Γ be the (decidable) set of axioms for any of the usual systems of set theory. Then this corollary tells us that the set of theorems of set theory is effectively enumerable.

★**Corollary 25G** Assume that Γ is a decidable set of formulas in a reasonable language, and for any sentence σ either $\Gamma \models \sigma$ or $\Gamma \models \neg\sigma$. Then the set of sentences implied by Γ is decidable.

Proof If Γ is inconsistent, then we have just the (decidable) set of all sentences. So assume that Γ is consistent. Suppose that we are given a sentence σ and asked to decide whether or not $\Gamma \models \sigma$. We can enumerate the theorems of Γ and look for σ or $\neg\sigma$. Eventually one will appear, and then we know the answer. ■

(Observe that this proof actually describes two decision procedures. One is correct when Γ is inconsistent, the other is correct when Γ is consistent. So in either case there is a decision procedure. But we cannot necessarily determine effectively, given a finite description of Γ, which one is to be used.)

It should be remarked that our proofs of enumerability cannot, in general, be strengthened to proofs of decidability. For almost all languages the set of validities is *not* decidable. (See Church's theorem, Section 3.5.)

Historical notes

The completeness theorem (for countable languages) was contained in the 1930 doctoral dissertation of Kurt Gödel. (It is not to be confused with the "Gödel incompleteness theorem," published in 1931. We will consider this latter result in Chapter 3.) The compactness theorem (for countable languages) was given as a corollary.

The compactness theorem for uncountable languages was implicit in a 1936 paper by Anatolii Mal′cev. His proof used Skolem functions (cf. Section 4.2) and the compactness theorem of sentential logic. The first explicit statement of the compactness theorem for uncountable languages was in a 1941 paper by Mal′cev.

The enumerability theorem, as well as following from Gödel's 1930 work, was also implicit in results published in 1928 by Thoralf Skolem.

The proof we have given for the completeness theorem is patterned after one given by Leon Henkin in his dissertation, published in 1949. Unlike Gödel's original proof, Henkin's proof generalizes easily to languages of any cardinality.

EXERCISES

1. (Semantical rule EI) Assume that the constant symbol c does not occur in φ, ψ, or Γ, and that $\Gamma ; \varphi^x_c \models \psi$. Show (without using the soundness and completeness theorems) that $\Gamma ; \exists x \, \varphi \models \psi$.

2. Prove the equivalence of parts (a) and (b) of the completeness theorem.

3. Assume that $\Gamma \vdash \varphi$ and that P is a predicate symbol which occurs neither in Γ nor in φ. Is there a deduction of φ from Γ in which P nowhere occurs?

4. Assume that the language has only finitely many parameters.

(a) Show that we can effectively decide, given a finite structure $\mathfrak{A}$ and a sentence σ, whether or not $\models_{\mathfrak{A}} \sigma$.

(b) Show that the set of sentences having finite models is effectively enumerable.

5. Assume that the language has only finitely many parameters.

(a) Let Σ be a set of sentences such that for any $\sigma \in \Sigma$, if σ has a counterexample (i.e., an $\mathfrak{A}$ in which σ is false), then σ has such a counterexample with $|\mathfrak{A}|$ finite. Find an effective procedure which, given any $\sigma \in \Sigma$, will decide whether or not σ is valid.

(b) Show that the set of valid $\forall_2$ sentences without function symbols is decidable. (See Exercises 18 and 19 of Section 2.2 for terminology and background. An $\forall_2$ formula is one of the form $\forall x_1 \cdots \forall x_n \varphi$, where φ is existential.)

6. Let $\Gamma = \{\neg\forall \mathbf{v}_1 \, P\mathbf{v}_1, P\mathbf{v}_1, P\mathbf{v}_2, P\mathbf{v}_3, \ldots\}$. Is Γ consistent? Is Γ satisfiable?

§ 2.6 MODELS OF THEORIES

In this section we will leave behind deductions and logical axioms. Instead we return to topics discussed in Section 2.2. But now, in the presence of the theorems of the preceding section, we will be able to answer more questions than we could before.

Size of models

In the completeness theorem we started with a set Γ (in a language of cardinality[1] $\varkappa$) and formed a structure $\mathfrak{A}/E$ in which it was satisfied. $\mathfrak{A}/E$ was constructed from a preliminary structure $\mathfrak{A}$. The universe of $\mathfrak{A}$ was the set of all terms in the language obtained by adding $\varkappa$ new constant symbols. So clearly $|\mathfrak{A}|$ contained at least $\varkappa$ terms. On the other hand, there are only $\varkappa$ expressions in the augmented language (by Theorem 0D), so $|\mathfrak{A}|$ could have no more than $\varkappa$ terms. Thus the cardinality of $\mathfrak{A}$ (by which we mean the cardinality of $|\mathfrak{A}|$) was $\varkappa$.

The universe of $\mathfrak{A}/E$ consisted of equivalence classes of members of $\mathfrak{A}$, so $\text{card}\,|\mathfrak{A}/E| \leq \text{card}\,|\mathfrak{A}|$. (We can map $|\mathfrak{A}/E|$ one-to-one into $|\mathfrak{A}|$ by assigning to each equivalence class some chosen member.) Thus, when the smoke had cleared, Γ was satisfied in a structure $\mathfrak{A}/E$ of cardinality $\leq \varkappa$.

[1] The reader who wishes to avoid uncountable cardinals is advised to skip from part (a) of the Löwenheim–Skolem theorem to Theorem 26B.

Löwenheim–Skolem Theorem (1915) (a) Let Γ be a satisfiable set of formulas in a countable language. Then Γ is satisfiable in some countable structure.

(b) Let Γ be a satisfiable set of formulas in a language of cardinality κ. Then Γ is satisfiable in some structure of cardinality $\leq \kappa$.

Proof (a) is a special case of (b), wherein $\kappa = \aleph_0$. To prove (b), first observe that Γ is consistent, by the soundness theorem. Then by the completeness theorem (plus the foregoing remarks) it can be satisfied in a structure of cardinality $\leq \kappa$. ■

(There is another, more direct proof of this theorem that will be indicated in Section 4.2; see especially Exercise 1 there. That proof, which does not use a deductive calculus, begins with an arbitrary structure $\mathfrak{A}$ in which Γ can be satisfied, and by various manipulations extracts from it a suitable substructure of cardinality κ or less.)

The Löwenheim–Skolem theorem was published by Leopold Löwenheim in 1915 for the case where Γ is a singleton; Thoralf Skolem in 1920 extended this to a possibly infinite Γ. The theorem marked a new phase in mathematical logic. Earlier work had been done in the direction of *formalizing* mathematics by means of formal languages and deductive calculi; this work was initiated largely by Gottlob Frege in 1879. For example, the *Principia Mathematica* (1910–1913) of Whitehead and Russell carried out such a formalization in great detail. But the modern phase began when logicians stepped back and began to prove results *about* the formal systems they had been constructing. Other early work in this trend was done by Kurt Gödel (as mentioned before), Alfred Tarski, and others.

For a sample application of the Löwenheim–Skolem theorem, let A_{ST} be your favorite set of axioms for set theory. Hopefully these axioms are consistent. And so they have some model. By the Löwenheim–Skolem theorem, the axioms have a countable model $\mathfrak{S}$. Of course, $\mathfrak{S}$ is also a model of all the sentences logically implied by A_{ST}. One of these sentences asserts (when translated back into English according to the intended translation) that there are uncountably many sets. There is no contradiction here, but the situation is sufficiently puzzling to be called "Skolem's paradox." It is true that *in the structure* $\mathfrak{S}$ there is no point which satisfies the formal definition of being a one-to-one map of the natural numbers onto the universe. But this in no way excludes the possibility of there being (outside $\mathfrak{S}$) some real function providing such a one-to-one correspondence.

For a structure $\mathfrak{A}$, define the *theory* of $\mathfrak{A}$, $\text{Th}\,\mathfrak{A}$, to be the set of all sentences true in $\mathfrak{A}$. Suppose that we have an uncountable structure $\mathfrak{A}$ for a countable language. By the Löwenheim–Skolem theorem (applied to Th $\mathfrak{A}$) there is a countable $\mathfrak{B}$ which is a model of Th $\mathfrak{A}$. It follows that $\mathfrak{A} \equiv \mathfrak{B}$, for

$$\models_{\mathfrak{A}} \sigma \Rightarrow \sigma \in \text{Th}\,\mathfrak{A} \Rightarrow \models_{\mathfrak{B}} \sigma$$

and

$$\not\models_{\mathfrak{A}} \sigma \Rightarrow \models_{\mathfrak{A}} \neg\sigma \Rightarrow (\neg\sigma) \in \text{Th}\,\mathfrak{A} \Rightarrow \models_{\mathfrak{B}} \neg\sigma.$$

Conversely, suppose that we start with a countable structure $\mathfrak{B}$. Is there an uncountable $\mathfrak{A}$ such that $\mathfrak{A} \equiv \mathfrak{B}$? If $\mathfrak{B}$ is finite (and the language includes equality), then this is impossible. But if $\mathfrak{B}$ is infinite, then there will be such an $\mathfrak{A}$, by the following "upward and downward Löwenheim–Skolem theorem." The upward part is due to Tarski, whence the "T" of "LST."

LST Theorem Let Γ be a set of formulas in a language of cardinality $\varkappa$, and assume that Γ is satisfiable in some infinite structure. Then for every cardinal $\lambda \geq \varkappa$, there is a structure of cardinality λ in which Γ is satisfiable.

Proof Let $\mathfrak{A}$ be the infinite structure in which Γ is satisfiable. Expand the language by adding a set C of λ new constant symbols. Let

$$\Sigma = \{c_1 \not\approx c_2 : c_1, c_2 \text{ distinct members of } C\}.$$

Then every finite subset of $\Sigma \cup \Gamma$ is satisfiable in the structure $\mathfrak{A}$, expanded to assign distinct objects to the finitely many new constant symbols in the subset. (Since $\mathfrak{A}$ is infinite, there is room to accommodate any finite number of these.) So by compactness $\Sigma \cup \Gamma$ is satisfiable, and by the Löwenheim–Skolem theorem it is satisfiable in a structure $\mathfrak{B}$ of cardinality $\leq \lambda$. (The expanded language has cardinality $\varkappa + \lambda = \lambda$.) But any model of Σ clearly has cardinality $\geq \lambda$. So $\mathfrak{B}$ has cardinality λ; restrict $\mathfrak{B}$ to the original language. ■

Corollary 26A (a) Let Σ be a set of sentences in a countable language. If Σ has some infinite model, then Σ has models of every infinite cardinality.

(b) Let $\mathfrak{A}$ be an infinite structure for a countable language. Then for any infinite cardinal λ, there is a structure $\mathfrak{B}$ of cardinality λ such that $\mathfrak{B} \equiv \mathfrak{A}$.

Proof (a) Take $\Gamma = \Sigma$, $\varkappa = \aleph_0$ in the theorem. (b) Take $\Sigma = \text{Th}\,\mathfrak{A}$ in part (a). ■

Consider a set Σ of sentences, to be thought of as nonlogical axioms. (For example, Σ might be a set of axioms for set theory or a set of axioms for number theory.) Call Σ *categorical* iff any two models of Σ are isomorphic. The above corollary implies that if Σ has any infinite models, then Σ is not categorical. There is, for example, no set of sentences whose models are exactly the structures isomorphic to $(N, 0, S, +, \cdot)$. This is indicative of a limitation in the expressiveness of first-order languages. (As will be seen in Section 4.1, there are categorical second-order sentences. But second-order sentences are peculiar objects, obtained at the cost of holding the notion of *subset* fixed, immune from interpretation by structures.)

It is possible to have sentences having only finite models. For example, any model of $\forall x \, \forall y \, x \approx y$ has cardinality 1. But if all models of Σ are finite, then there is a finite bound on the size of the models, by the following theorem.

Theorem 26B If a set Σ of sentences has arbitrarily large finite models, then it has an infinite model.

Proof For each integer $k \geq 2$, we can find a sentence λ_k which translates, "There are at least k things." For example,

$$\lambda_2 = \exists v_1 \, \exists v_2 \, v_1 \not\approx v_2,$$

$$\lambda_3 = \exists v_1 \, \exists v_2 \, \exists v_3 (v_1 \not\approx v_2 \wedge v_2 \not\approx v_3 \wedge v_1 \not\approx v_3).$$

Consider the set

$$\Sigma \cup \{\lambda_2, \lambda_3, \ldots\}.$$

By hypothesis any finite subset has a model. So by compactness the entire set has a model, which clearly must be infinite. ■

Both the proof to this theorem and the proof of the preceding theorem illustrate a useful method for obtaining a structure with given properties. One writes down sentences (possibly in an expanded language) stating the properties one wants. One then argues that any finite subset of the sentences has a model. The compactness theorem does the rest. We will see more examples of this method in the coming pages.

Corollary 26C The class of all finite structures (for a fixed language) is not EC_Δ. The class of all infinite structures is not EC.

Proof The first sentence follows immediately from the theorem. If the class of all infinite structures is Mod τ, then the class of all finite structures is Mod $\neg\tau$. But this class isn't even EC_Δ, much less EC. ■

The class of infinite structures *is* EC_Δ, being $\mathrm{Mod}\{\lambda_2, \lambda_3, \ldots\}$.

For example, it is a priori conceivable that there might be some very subtle equation of group theory that was true in every finite group but false in every infinite group. But by the above theorem, no such equation exists.

EXAMPLE Consider the structure

$$\mathfrak{N} = (N, 0, S, <, +, \cdot).$$

By the LST theorem there are structures elementarily equivalent to $\mathfrak{N}$ which are uncountable (and hence not isomorphic to $\mathfrak{N}$). We claim now that there is also a countable structure $\mathfrak{M}_0$ elementarily equivalent to, but not isomorphic to, the structure $\mathfrak{N}$.

Proof Expand the language by adding a new constant symbol c. Let

$$\Sigma = \{\mathbf{0} < c,\ \mathbf{S0} < c,\ \mathbf{SS0} < c, \ldots\}.$$

We claim that $\Sigma \cup \mathrm{Th}\,\mathfrak{N}$ has a model. For consider a finite subset. That finite subset is true in

$$\mathfrak{N}_k = (N, 0, S, <, +, \cdot, k)$$

(where $k = c^{\mathfrak{N}_k}$) for some large k. So by the compactness theorem $\Sigma \cup \mathrm{Th}\,\mathfrak{N}$ has a model.

By the Löwenheim–Skolem theorem, $\Sigma \cup \mathrm{Th}\,\mathfrak{N}$ has a countable model

$$\mathfrak{M} = (|\,\mathfrak{M}\,|, \mathbf{0}^{\mathfrak{M}}, \mathbf{S}^{\mathfrak{M}}, <^{\mathfrak{M}}, +^{\mathfrak{M}}, \cdot^{\mathfrak{M}}, c^{\mathfrak{M}}).$$

Let $\mathfrak{M}_0$ be the restriction of $\mathfrak{M}$ to the original language:

$$\mathfrak{M}_0 = (|\,\mathfrak{M}\,|, \mathbf{0}^{\mathfrak{M}}, \mathbf{S}^{\mathfrak{M}}, <^{\mathfrak{M}}, +^{\mathfrak{M}}, \cdot^{\mathfrak{M}}).$$

Since $\mathfrak{M}_0$ is a model of $\mathrm{Th}\,\mathfrak{N}$, we have $\mathfrak{M}_0 \equiv \mathfrak{N}$. We leave it to the reader to show that $\mathfrak{M}_0$ is not isomorphic to $\mathfrak{N}$. ■

Theories

We define a *theory* to be a set of sentences closed under logical implication. That is, T is a theory iff T is a set of sentences such that for any sentence σ of the language,

$$T \models \sigma \Rightarrow \sigma \in T.$$

(Note that we admit only sentences, not formulas with free variables.)

For example, there is always a smallest theory, consisting of the valid sentences of the language. At the other extreme there is the theory consisting of all the sentences of the language; it is the only unsatisfiable theory.

For a class $\mathscr{K}$ of structures (for the language), define the *theory* of $\mathscr{K}$ ($\mathrm{Th}\,\mathscr{K}$) by the equation

$$\mathrm{Th}\,\mathscr{K} = \{\sigma : \sigma \text{ is true in every member of } \mathscr{K}\}.$$

(This concept arose previously in the special case $\mathscr{K} = \{\mathfrak{A}\}$.)

Theorem 26D $\mathrm{Th}\,\mathscr{K}$ is indeed a theory.

Proof Any member of $\mathscr{K}$ is a model of $\mathrm{Th}\,\mathscr{K}$. Thus if σ is true in every model of $\mathrm{Th}\,\mathscr{K}$, then it is true in every member of $\mathscr{K}$. Whence it belongs to $\mathrm{Th}\,\mathscr{K}$. ∎

For example, if the parameters of the language are $\forall$, **0**, **1**, $+$, and $\cdot$, and $\mathscr{F}$ is the class of all fields, then $\mathrm{Th}\,\mathscr{F}$, the theory of fields, is simply the set of all sentences of the language which are true in all fields. If $\mathscr{F}_0$ is the class of fields of characteristic 0, then $\mathrm{Th}\,\mathscr{F}_0$ is the theory of fields of characteristic 0.

Recall that for a set Σ of sentences, we defined $\mathrm{Mod}\,\Sigma$ to be the class of all models of Σ. $\mathrm{Th}\,\mathrm{Mod}\,\Sigma$ is then the set of all sentences which are true in all models of Σ. But this is just the set of all sentences logically implied by Σ. Call this set the set of *consequences* of Σ, $\mathrm{Cn}\,\Sigma$. Thus

$$\begin{aligned}\mathrm{Cn}\,\Sigma &= \{\sigma : \Sigma \models \sigma\}\\ &= \mathrm{Th}\,\mathrm{Mod}\,\Sigma.\end{aligned}$$

For example, set theory is the set of consequences of a certain set of sentences, known, unsurprisingly, as axioms for set theory. A set T of sentences is a theory iff $T = \mathrm{Cn}\,T$.

A theory T is *complete* iff for every sentence σ, either $\sigma \in T$ or $(\neg\sigma) \in T$. For example, for any one structure $\mathfrak{A}$, $\mathrm{Th}\,\{\mathfrak{A}\}$ (written, as before, "$\mathrm{Th}\,\mathfrak{A}$") is always a complete theory. In fact, it is clear upon reflection that $\mathrm{Th}\,\mathscr{K}$ is a complete theory iff any two members of $\mathscr{K}$ are elementarily equivalent. And a theory T is complete iff any two models of T are elementarily equivalent.

For example, the theory of fields is not complete, since the sentences

$$\mathbf{1} + \mathbf{1} \approx \mathbf{0},$$
$$\exists x\, x \cdot x \approx \mathbf{1} + \mathbf{1}$$

are true in some fields but false in others. The theory of algebraically closed fields of characteristic 0 is complete, but this is by no means obvious. (See Theorem 26G.)

★Definition A theory T is *axiomatizable* iff there is a decidable set Σ of sentences such that $T = \text{Cn}\,\Sigma$.

Definition A theory T is *finitely axiomatizable* iff $T = \text{Cn}\,\Sigma$ for some finite set Σ of sentences.

In the latter case we have $T = \text{Cn}\,\{\sigma\}$ (written "$T = \text{Cn}\,\sigma$"), where σ is the conjunction of the finitely many members of Σ. For example, the theory of fields is finitely axiomatizable. For the class $\mathscr{F}$ of fields is Mod Φ, where Φ is the finite set of field axioms. And the theory of fields is Th Mod $\Phi = \text{Cn}\,\Phi$.

The theory of fields of characteristic 0 is axiomatizable, being Cn Φ_0, where Φ_0 consists of the (finitely many) field axioms together with the infinitely many sentences:

$$\mathbf{1} + \mathbf{1} \not\approx \mathbf{0},$$
$$\mathbf{1} + \mathbf{1} + \mathbf{1} \not\approx \mathbf{0},$$
$$\ldots .$$

This theory is not finitely axiomatizable. To prove this, first note that no finite subset of Φ_0 has the entire theory as its set of consequences. (For that finite subset would be true in some field of very large characteristic.) Then apply the following:

Theorem 26E If Cn Σ is finitely axiomatizable, then there is a finite $\Sigma_0 \subseteq \Sigma$ such that $\text{Cn}\,\Sigma_0 = \text{Cn}\,\Sigma$.

Proof Say that Cn Σ is finitely axiomatizable; then $\text{Cn}\,\Sigma = \text{Cn}\,\tau$ for some one sentence τ. In general $\tau \notin \Sigma$, but at least $\Sigma \models \tau$. ($\tau \in \text{Cn}\,\tau = \text{Cn}\,\Sigma$.) By the compactness theorem there is a finite $\Sigma_0 \subseteq \Sigma$ such that $\Sigma_0 \models \tau$. Then

$$\text{Cn}\,\tau \subseteq \text{Cn}\,\Sigma_0 \subseteq \text{Cn}\,\Sigma,$$

whence equality holds. ■

We can now restate Corollaries 25F and 25G in the present terminology:

★Corollary 26F (a) An axiomatizable theory (in a reasonable language) is effectively enumerable.

(b) A complete axiomatizable theory (in a reasonable language) is decidable.

We can represent the relationships among these concepts by means of a diagram (in which we have included the results of Exercise 5):

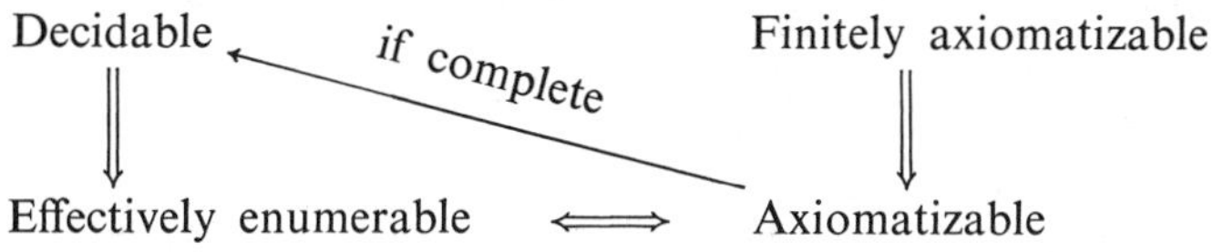

For example, a theory which is given in axiomatic form (such as Zermelo–Fraenkel set theory, which is Cn A_{ZF} for a certain set A_{ZF}) is effectively enumerable. We will argue in Section 3.6 that set theory is not decidable and not complete. Number theory, the theory of the structure $(\mathbb{N}, 0, S, <, +, \cdot, E)$, is complete but is not effectively enumerable and hence not axiomatizable (Section 3.5).

We can use part (b) of the preceding corollary to establish the decidability of an axiomatizable theory, provided we can show that the theory in question is complete. This can sometimes be done by means of the Łoś–Vaught test for completeness.

For a theory T and a cardinal λ, say that T is *λ-categorical* iff all models of T having cardinality λ are isomorphic.

Łoś–Vaught Test (1954) Let T be a theory in a countable language such that

1. T is λ-categorical for some infinite cardinal λ.
2. All models of T are infinite.

Then T is complete.

Proof It suffices to show for any two models $\mathfrak{A}$ and $\mathfrak{B}$ of T that $\mathfrak{A} \equiv \mathfrak{B}$. Since $\mathfrak{A}$ and $\mathfrak{B}$ are infinite, there exist (by the LST theorem) structures $\mathfrak{A}' \equiv \mathfrak{A}$ and $\mathfrak{B}' \equiv \mathfrak{B}$ having cardinality λ. $\mathfrak{A}'$ is isomorphic to $\mathfrak{B}'$, so we have

$$\mathfrak{A} \equiv \mathfrak{A}' \cong \mathfrak{B}' \equiv \mathfrak{B}.$$

Thus $\mathfrak{A} \equiv \mathfrak{B}$. ■

(If T is a theory in a language of cardinality $\varkappa$, then we must demand that $\varkappa \leq \lambda$.)

The converse to the Łoś–Vaught test is false. That is, there are complete theories which are not λ-categorical for any λ.

In Section 3.1 we will apply the Łoś–Vaught test to prove the decidability of the theory of the natural numbers with zero and successor. It can also be used to prove the decidability of the theory of the complex field.

Theorem 26G (a) The theory of algebraically closed fields of characteristic 0 is complete.

★(b) The theory of the complex field

$$\mathfrak{C} = (\mathbb{C}, 0, 1, +, \cdot)$$

is decidable.

Proof Let $\mathscr{A}$ be the class of algebraically closed fields of characteristic 0. Then $\mathscr{A} = \text{Mod}(\Phi_0 \cup \Gamma)$, where Φ_0 consists as before of the axioms for fields of characterisic 0, and Γ consists of the sentences

$$\forall a\, \forall b\, \forall c(a \not\approx 0 \rightarrow \exists x\; a \cdot x \cdot x + b \cdot x + c \approx 0),$$
$$\forall a\, \forall b\, \forall c\, \forall d(a \not\approx 0 \rightarrow \exists x\; a \cdot x \cdot x \cdot x + b \cdot x \cdot x + c \cdot x + d \approx 0),$$
$$\ldots .$$

The set $\Phi_0 \cup \Gamma$ is decidable and $\text{Th}\,\mathscr{A} = \text{Cn}(\Phi_0 \cup \Gamma)$, so this theory is axiomatizable. Part (a) of the theorem asserts that the theory is also complete, whence it is decidable.

Part (b) follows from part (a). For we have $\mathfrak{C} \in \mathscr{A}$, whence $\text{Th}\,\mathscr{A} \subseteq \text{Th}\,\mathfrak{C}$. The completeness of $\text{Th}\,\mathscr{A}$ implies that equality holds; see Exercise 1.

To prove part (a), we apply the Łoś–Vaught test. The models of $\text{Th}\,\mathscr{A}$ are exactly the members of $\mathscr{A}$. These are all infinite. We further claim that $\text{Th}\,\mathscr{A}$ is categorical in any uncountable cardinality. This is equivalent to saying that any two algebrically closed fields of characteristic 0 having the same uncountable cardinality are isomorphic.

This last assertion is a known result of algebra. We will sketch the proof, for the interest of those readers familiar with this topic. Any field $\mathfrak{F}$ is obtainable in the following way: (1) One begins with the prime subfield, which is determined within isomorphism by the characteristic of $\mathfrak{F}$. (2) One takes a transcendental extension, determined within isomorphism by the cardinality of the transcendence basis, i.e., by the transcendence degree of $\mathfrak{F}$ (over its prime subfield). (3) One finally takes some algebraic extension. We thus have a theorem of Steinitz: Two algebraically closed fields are isomorphic iff they have the same characteristic and the same transcendence degree.

If the transcendence degree of an infinite field $\mathfrak{F}$ is λ, then the cardinality of $\mathfrak{F}$ is the larger of λ and $\aleph_0$. Hence for an uncountable field, the cardinality equals the transcendence degree. So we conclude from Steinitz's theorem that two algebraically closed fields having same characteristic and the same uncountable cardinality are isomorphic. ■

The theory of the real field

$$(\mathbb{R}, 0, 1, +, \cdot)$$

is also decidable. But this result (which is due to Tarski) is much deeper than the above theorem. The theory of the real field is not categorical in any infinite cardinality, so the Łoś-Vaught test cannot be applied.

As a final application, we can show that the ordering of the rationals is elementarily equivalent to the ordering of the reals:

$$(\mathbb{Q}, <_Q) \equiv (\mathbb{R}, <_R),$$

where $\mathbb{Q}$ and $\mathbb{R}$ are the rationals and reals, respectively, and $<_Q$ and $<_R$ are the corresponding orderings. To show elementary equivalence, we show that both are models of some complete theory (which then must coincide with the theory of each structure). The key fact is provided by a theorem of Cantor: Any two countable dense linear orderings are isomorphic.

To give the details, we must back up a little. The language here has equality and the parameters $\forall$ and $<$. Let δ be the conjunction of the following sentences:

1. Ordering axioms (trichotomy and transitivity):

$$\forall x\, \forall y\, \forall z(x < y \vee x \approx y \vee y < x),$$
$$\forall x\, \forall y(x < y \rightarrow y \not< x),$$
$$\forall x\, \forall y\, \forall z(x < y \rightarrow y < z \rightarrow x < z).$$

2. Density:

$$\forall x\, \forall y(x < y \rightarrow \exists z(x < z < y)).$$

3. No endpoints:

$$\forall x\, \exists y\, \exists z(y < x < z).$$

The dense linear orderings without endpoints are, by definition, the structures for this language which are models of δ. It is clear that they are all infinite. Furthermore, we claim that the theory of these orderings, Cn δ, is categorical in cardinality $\aleph_0$. This is provided by the following fact.

Theorem 26H (Cantor) Any countable model of δ is isomorphic to $(\mathbb{Q}, <_Q)$.

We leave the proof to Exercise 3.

We can now apply the Łoś–Vaught test to conclude that Cn δ is complete. Hence any two models of δ are elementarily equivalent; in particular,

$$(\mathbb{Q}, <_Q) \equiv (\mathbb{R}, <_R).$$

We can also conclude that these structures have decidable theories.

Prenex normal form

It will at times be convenient to move all the quantifier symbols to the left of other symbols. For example,

$$\forall x(Ax \rightarrow \forall y\, Bxy)$$

is equivalent to

$$\forall x\, \forall y(Ax \rightarrow Bxy).$$

And

$$\forall x(Ax \rightarrow \exists y\, Bxy)$$

is equivalent to

$$\forall x\, \exists y(Ax \rightarrow Bxy).$$

Define a *prenex* formula to be one of the form (for some $n \geq 0$)

$$Q_1 x_1 \cdots Q_n x_n\, \alpha,$$

where Q_i is $\forall$ or $\exists$ and α is quantifier-free.

Prenex Normal Form Theorem For any formula we can find a logically equivalent prenex formula.

Proof We will make use of the following quantifier manipulation rules.

Q1a. $\neg \forall x\, \alpha \models\!\!\!\dashv \exists x\, \neg\alpha$.
Q1b. $\neg \exists x\, \alpha \models\!\!\!\dashv \forall x\, \neg\alpha$.
Q2a. $(\alpha \rightarrow \forall x\, \beta) \models\!\!\!\dashv \forall x(\alpha \rightarrow \beta)$ for x not free in α.
Q2b. $(\alpha \rightarrow \exists x\, \beta) \models\!\!\!\dashv \exists x(\alpha \rightarrow \beta)$ for x not free in α.
Q3a. $(\forall x\, \alpha \rightarrow \beta) \models\!\!\!\dashv \exists x(\alpha \rightarrow \beta)$ for x not free in β.
Q3b. $(\exists x\, \alpha \rightarrow \beta) \models\!\!\!\dashv \forall x(\alpha \rightarrow \beta)$ for x not free in β.

Q1 is clear; for the others see pages 113f. and Exercise 8 of Section 2.4.

We now show by induction that every formula has an equivalent prenex formula.

1. For atomic formulas this is vacuous, as any quantifier-free formula is trivially a prenex formula.

2. If α is equivalent to the prenex α', then $\forall x\, \alpha$ is equivalent to the prenex $\forall x\, \alpha'$.

3. If α is equivalent to the prenex α', then $\neg\alpha$ is equivalent to $\neg\alpha'$. Apply Q1 to $\neg\alpha'$ to obtain a prenex formula; for example,

$$\neg\forall x\, \exists y\, \exists z\, \beta \models\!\!\!\dashv \exists x\, \forall y\, \forall z\, \neg\beta.$$

4. Finally we come to the case of $\alpha \rightarrow \beta$. By inductive hypothesis we have prenex formulas α' and β' equivalent to α and β, respectively. By our theorems on alphabetic variants, we may further assume that any variable which occurs quantified in one of the formulas α' and β' does not occur at all in the other. We then use Q2 and Q3 to obtain a prenex formula equivalent to $\alpha' \rightarrow \beta'$ (and hence to $\alpha \rightarrow \beta$). Observe that there is some latitude in the order in which the rules Q2 and Q3 are applied. For example,

$$\forall x\, \exists y\, \varphi \rightarrow \exists u\, \psi$$

(where x and y do not occur in ψ, u does not occur in φ) is equivalent to any of the following:

$$\exists x\, \forall y\, \exists u(\varphi \rightarrow \psi),$$
$$\exists x\, \exists u\, \forall y(\varphi \rightarrow \psi),$$
$$\exists u\, \exists x\, \forall y(\varphi \rightarrow \psi). \qquad \blacksquare$$

Retrospectus

At the beginning of this book it was stated that symbolic logic is a mathematical model of deductive thought. This is as good a time as any to reflect on that statement, in the light of the material treated thus far.

As a first example, consider a mathematician working in set theory. He uses a language with an equality symbol, a symbol $\in$ for membership, and numerous defined symbols ($\varnothing$, $\cup$, etc.). In principle the defined symbols could be eliminated and any sentence replaced by an equivalent sentence in which the defined symbols did not appear. (In this connection see Section 2.7, where this topic is treated systematically.) He takes as primitive (or undefined) notions the notions of set and membership. He adopts some set A_{ST} of axioms involving these notions. He asserts that for certain sentences (his theorems), these sentences are true provided the axioms are true, regardless of what the undefined notions of set and membership actually mean. In support of these assertions he offers proofs, which are finitely long arguments intended to convince his colleagues of the correctness of the assertions.

In terms of first-order logic we can describe all this as follows: The language here is a first-order language with equality and a two-place predicate symbol $\in$. Thus $\forall$ and $\in$ are the only parameters open to interpretation. There is a certain set A_{ST} of sentences in this language singled out as being the set of (nonlogical) axioms. Then certain other sentences are logical consequences of A_{ST}, i.e., are true in any model of A_{ST}. If τ is a consequence of A_{ST} (and only then), there is a deduction of τ from A_{ST}.

Next consider a more typical case of the hypothetical working mathematician, that of the algebraist or analyst. The algebraist uses axioms for (say) group theory, but he also employs some amount of set theory. Similarly, the analyst deals with sentences that involve both numbers and sets of numbers. In both cases it is generally recognized that one could, in principle, convert the assertions of algebra and analysis to assertions of set theory. And then the remarks of the preceding paragraph again apply.

The interest that symbolic logic holds for the mathematician is largely due to the accuracy with which it mirrors mathematical deductions. In the long run, it will surely be useful to understand the fundamental processes of doing mathematics.

There remains the question of the accuracy with which first-order logic mirrors nonmathematical deductive thought. Logic, symbolic and nonsymbolic, has always formed a traditional part of the philosophical study of the process by which men come to hold certain ideas. Nonmathematical examples to which first-order logic applies are provided by a vast array of frivolous situations. Lewis Carroll gave such examples, one of which inferred that babies cannot manage crocodiles from the three hypotheses: (1) Babies are illogical. (2) Nobody is despised who can manage a crocodile. (3) Illogical persons are despised.

But what of nonfrivolous situations? Here the applicability is obscured by the fact that we usually do not make explicit the assumptions we use in drawing conclusions. There are specific areas (in diverse fields such as physics, medicine, and law) where assumptions not only can be made explicit but are being made explicit. In some cases it appears that less than the full versatility of first-order logic is required to formalize the real-life deductions. In the case of quantum mechanics, more features may be necessary.

EXERCISES

1. Let T_1 and T_2 be theories such that (i) $T_1 \subseteq T_2$, (ii) T_1 is complete, and (iii) T_2 is satisfiable. Show that $T_1 = T_2$.

2. Establish the following facts:

(a) $\Sigma_1 \subseteq \Sigma_2 \Rightarrow \text{Mod}\,\Sigma_2 \subseteq \text{Mod}\,\Sigma_1$.
$\mathscr{K}_1 \subseteq \mathscr{K}_2 \Rightarrow \text{Th}\,\mathscr{K}_2 \subseteq \text{Th}\,\mathscr{K}_1$.

(b) $\Sigma \subseteq \text{Th Mod}\,\Sigma$ and $\mathscr{K} \subseteq \text{Mod Th}\,\mathscr{K}$.

(c) $\text{Mod}\,\Sigma = \text{Mod Th Mod}\,\Sigma$ and $\text{Th}\,\mathscr{K} = \text{Th Mod Th}\,\mathscr{K}$. (Part (c) follows from (a) and (b).)

3. Prove that any two countable dense linear orderings without endpoints are isomorphic (Theorem 26H). *Suggestions*: Let $\mathfrak{A}$ and $\mathfrak{B}$ be such structures with $|\mathfrak{A}| = \{a_0, a_1, \ldots\}$ and $|\mathfrak{B}| = \{b_0, b_1, \ldots\}$. Construct an isomorphism in stages; at stage $2n$ be sure a_n is paired with some suitable b_j, and at stage $2n + 1$ be sure b_n is paired with some suitable a_i.

4. Find prenex formulas equivalent to

(a) $(\exists x\, Ax \wedge \exists x\, Bx) \rightarrow Cx$.

(b) $\forall x\, Ax \leftrightarrow \exists x\, Bx$.

*5. Prove the converse to part (a) of Corollary 26F: An effectively enumerable theory (in a reasonable language) is axiomatizable.

6. Consider a language with a two-place predicate symbol $<$, and let $\mathfrak{N} = (N, <)$ be the structure consisting of the natural numbers with their usual ordering. Show that there is some $\mathfrak{A}$ elementarily equivalent to $\mathfrak{N}$ such that $<^{\mathfrak{A}}$ has a descending chain. (That is, there must be $a_0, a_1, \ldots$ in $|\mathfrak{A}|$ such that $\langle a_{i+1}, a_i\rangle \in <^{\mathfrak{A}}$ for all i.)

7. Show that an infinite map can be colored with four colors iff every finite submap of it can be. *Suggestion*: Take a language having a constant symbol for each country and having four one-place predicate symbols for the colors. Use the compactness theorem.

8. Let Σ_1 and Σ_2 be sets of sentences such that nothing is a model of both Σ_1 and Σ_2. Show that there is a sentence τ such that $\text{Mod}\,\Sigma_1 \subseteq \text{Mod}\,\tau$ and $\text{Mod}\,\Sigma_2 \subseteq \text{Mod}\,\neg\tau$. (This can be stated: Disjoint EC_Δ classes can be separated by an EC class.)

9. Assume that σ is true in all infinite models of a theory T. Show that there is a finite number k such that σ is true in all models $\mathfrak{A}$ of T for which $|\mathfrak{A}|$ has k or more elements.

§ 2.7 INTERPRETATIONS BETWEEN THEORIES[1]

In some cases a theory T_1 can be shown to be every bit as powerful as another theory T_0. This is certainly the case if the theories are in the same language and $T_0 \subseteq T_1$. But even if the theories are in different languages, there may exist a way of translating from one language to the other in such a way that members of T_0 are translated as members of T_1. This sort of situation will be examined in this section.

We will begin by discussing the topic of defined symbols. This topic, as well as having significant interest of its own, will serve as an example for the situation of the preceding paragraph, wherein T_0 is constructed from T_1 by adding a new defined symbol. If the definition is done properly, the original theory T_1 should in principle be just as strong as the new T_0. We will consider only the case of defined function symbols, since the case of defined predicate symbols presents, in comparison, no real difficulties.

Defining functions

Frequently in mathematics it is useful to introduce definitions of new functions. For example, in set theory one defines the power-set operation $\mathscr{P}$ by a sentence like, "Let $\mathscr{P}x$ be the set whose members are the subsets of x." Or by a sentence in a formal language (here containing $\in$, $\subseteq$, and $\mathscr{P}$),

$$\forall \mathbf{v}_1 \forall \mathbf{v}_2[\mathscr{P}\mathbf{v}_1 \approx \mathbf{v}_2 \leftrightarrow \forall u(u \in \mathbf{v}_2 \leftrightarrow u \subseteq \mathbf{v}_1)].$$

Now definitions are unlike theorems and unlike axioms. Unlike theorems, definitions are not things we prove. We just declare them by fiat. But unlike axioms, we do not expect definitions to add substantive information. A definition is expected to add to our convenience, not to our knowledge.

If this expectation is to be realized, the definition must be made in a reasonable way. As an example of a most unreasonable definition in number theory, suppose that we introduce a new function symbol f by the "definition"

$$f(x) = y \qquad \text{iff } x < y.$$

(Or by the sentence in a formal language: $\forall \mathbf{v}_1 \forall \mathbf{v}_2(f\mathbf{v}_1 \approx \mathbf{v}_2 \leftrightarrow \mathbf{v}_1 < \mathbf{v}_2)$.) Since we know that $1 < 2$, we see that $f(1) = 2$. But also $1 < 3$, so we obtain $f(1) = 3$. And so we come to the conclusion (which does not itself involve f) that $2 = 3$.

[1] The results of this section will be used only in Section 3.6.

Obviously this definition of f was in some way very bad. It did not just make matters convenient for us; it enabled us to conclude that $2 = 3$, which we could not have done without the definition. The trouble was that the definition bestowed the name "$f(1)$" ambiguously upon many things (2 and 3 among them). Thus $f(1)$ was not "well defined." Names ought to designate unique objects.

In this subsection we want to consider conditions under which we can be assured that a definition will be satisfactory. To simplify the notation, we will consider only the definition of a one-place function symbol f, but the remarks will apply to n-place function symbols as well.

Consider then a theory T in a language not yet containing the one-place function symbol f. (For example, T might be the set of consequences of your favorite axioms for set theory.) We want to add f to the language, introducing it by the *definition*

$$(\delta) \qquad \forall \boldsymbol{v}_1 \forall \boldsymbol{v}_2 [f\boldsymbol{v}_1 \approx \boldsymbol{v}_2 \leftrightarrow \varphi],$$

where φ is a formula in the original language (i.e., a formula not containing f) in which only $\boldsymbol{v}_1$ and $\boldsymbol{v}_2$ may occur free.

Theorem 27A In the above situation, the following are equivalent:

(a) (The definition is noncreative.) For any sentence σ in the smaller language, if

$$T ; \delta \models \sigma$$

(in the augmented language), then already $T \models \sigma$.

(b) (f is well defined.) The sentence

$$(\varepsilon) \qquad \forall \boldsymbol{v}_1 \; \exists ! \boldsymbol{v}_2 \, \varphi$$

is in the theory T. (Here "$\exists ! \boldsymbol{v}_2 \, \varphi$" is an abbreviation for a longer formula; see Exercise 21 of Section 2.2.)

Proof To obtain (a) $\Rightarrow$ (b), simply note that $\delta \models \varepsilon$. So by taking $\sigma = \varepsilon$ in part (a), we obtain $T \models \varepsilon$.

Conversely, assume that $T \models \varepsilon$. Let $\mathfrak{A}$ be a model of T. ($\mathfrak{A}$ is a structure for the original language.) For $d \in |\mathfrak{A}|$, let $F(d)$ be the unique $e \in |\mathfrak{A}|$ such that $\models_{\mathfrak{A}} \varphi [\![d, e]\!]$. (There is a unique such e because $\models_{\mathfrak{A}} \varepsilon$.) Let $(\mathfrak{A}, F)$ be the structure for the augmented language which agrees with $\mathfrak{A}$ on the original parameters and which assigns F to the symbol f. Then it is easy to see that $(\mathfrak{A}, F)$ is a model of δ. Furthermore, $\mathfrak{A}$ and $(\mathfrak{A}, F)$ satisfy the same sentences of the original language. In particular $(\mathfrak{A}, F)$ is a model of T. Hence

$$T\,;\delta \models \sigma \Rightarrow \models_{(\mathfrak{A},\,F)}\sigma$$
$$\Rightarrow \models_{\mathfrak{A}}\sigma. \quad \blacksquare$$

(This argument can be stated more briefly by using second-order logic. ε is logically equivalent to the sentence $\exists f\,\delta$.)

Interpretations

The basic idea is that it is possible for one theory to be just as strong (in a sense to be made precise) as another theory in another language. In considering two languages simultaneously, there is no point in allowing them to conflict; e.g., the negation symbol of one language should not be a predicate symbol in the other. We can eliminate such conflicts by assuming that each of the languages is obtained from a third parent language by deleting some parameters (and perhaps equality).

For example, axiomatic set theory is at least as strong as the theory of the natural numbers with zero and successor, i.e., the theory of $(\mathbb{N}, 0, S)$. Any sentence in the language of $(\mathbb{N}, 0, S)$ can be translated in a natural way into a sentence of set theory. (This translation is sketched in Section 3.6.) If the original sentence was true in $(\mathbb{N}, 0, S)$, then the translation will be a consequence of the axioms of set theory. (This is not obvious; the proof uses facts to be developed in Section 3.1.)

Let us look more carefully at a second example. Consider on the one hand the theory of

$$(\mathbb{N}, 0, S)$$

in its language, and on the other hand the theory of

$$(\mathbb{Z}, +, \cdot)$$

in its language. (Here $\mathbb{Z}$ is the set of all integers, positive, negative, and zero.) We will shortly be in a position to claim that the second theory is as strong as the first. How might a sentence about the natural numbers $\mathbb{N}$ with 0 and S be translated into a sentence about the integers $\mathbb{Z}$ with addition and multiplication?

The first clue is that an integer is nonnegative iff it is the sum of four squares, so a quantifier $\forall x$ in the first language (where x is intended to range over $\mathbb{N}$) can be replaced by

$$\forall x(\exists y_1 \, \exists y_2 \, \exists y_3 \, \exists y_4 \, x \approx y_1 \cdot y_1 + y_2 \cdot y_2 + y_3 \cdot y_3 + y_4 \cdot y_4 \rightarrow$$

in the second language.

The second clue is that $\{0\}$ and the successor function (viewed as a relation) are definable in $(\mathbb{Z}, +, \cdot)$. The set $\{0\}$ is defined by

$$\mathbf{v}_1 + \mathbf{v}_1 \approx \mathbf{v}_1 .$$

The successor relation (extended to $\mathbb{Z}$) is defined by

$$\forall z(z \cdot z \approx z \wedge z + z \not\approx z \rightarrow \mathbf{v}_1 + z \approx \mathbf{v}_2).$$

Thus the sentence about $(\mathbb{N}, 0, S)$

$$\forall x \, \mathbf{S}x \not\approx \mathbf{0}$$

can be translated into

$$\begin{aligned}&\forall x[\exists y_1 \, \exists y_2 \, \exists y_3 \, \exists y_4 \, x \approx y_1 \cdot y_1 + y_2 \cdot y_2 + y_3 \cdot y_3 + y_4 \cdot y_4 \rightarrow \\ &\quad \neg \forall u(u + u \approx u \rightarrow \forall v(\forall z(z \cdot z \approx z \wedge z + z \not\approx z \rightarrow x + z \approx v) \rightarrow \\ &\quad v \approx u))].\end{aligned}$$

So much for examples. For our general discussion it will be helpful to introduce the notation

$$\varphi(t) = \varphi_t^{\mathbf{v}_1},$$
$$\varphi(t_1, t_2) = (\varphi_{t_1}^{\mathbf{v}_1})_{t_2}^{\mathbf{v}_2},$$

and so forth. Thus $\varphi = \varphi(\mathbf{v}_1) = \varphi(\mathbf{v}_1, \mathbf{v}_2)$. If we use "$\varphi(x)$" we will not worry too much about whether or not x is substitutable for $\mathbf{v}_1$ in φ. If it is not, then we actually want $\varphi(x)$ to be $\psi_x^{\mathbf{v}_1}$, where ψ is a suitable alphabetic variant of φ.

Assume now that we have the following general situation:

L_0 is a language. (A *language* can for all practical purposes be a set of parameters, possibly augmented by the equality symbol.)

T_1 is a theory in a (possibly different) language L_1, which includes equality.

Definition An *interpretation* π of L_0 into T_1 is a function on the set of parameters of L_0 such that

1. π assigns to $\forall$ a formula $\pi_\forall$ of L_1 in which at most $\mathbf{v}_1$ occurs free, such that

(i) $$T_1 \models \exists \mathbf{v}_1\, \pi_\forall .$$

(The idea is that in any model of T_1, the formula $\pi_\forall$ should define a nonempty set to be used as the universe of an L_0-structure.)

2. π assigns to each n-place predicate symbol P a formula π_P of L_1 in which at most the variables $\mathbf{v}_1, \ldots, \mathbf{v}_n$ occur free.

3. π assigns to each n-place function symbol f a formula π_f of L_1 in which at most $\mathbf{v}_1, \ldots, \mathbf{v}_n, \mathbf{v}_{n+1}$ occur free, such that

(ii) $$\begin{aligned} T_1 \models \forall \mathbf{v}_1 \cdots \forall \mathbf{v}_n(\pi_\forall(\mathbf{v}_1) \to \cdots \to \pi_\forall(\mathbf{v}_n) \\ \to \exists x(\pi_\forall(x) \wedge \forall \mathbf{v}_{n+1}(\pi_f(\mathbf{v}_1, \ldots, \mathbf{v}_{n+1}) \leftrightarrow \mathbf{v}_{n+1} \approx x))). \end{aligned}$$

(In English, this formula becomes, "For all $\vec{v}$ in the set defined by $\pi_\forall$, there is a unique x such that $\pi_f(\vec{v}, x)$, and furthermore x is in the set defined by $\pi_\forall$." The idea is to ensure that in any model of T_1, π_f defines a function on the universe defined by $\pi_\forall$. In the case of a constant symbol c, we have $n = 0$ and (ii) becomes

$$T_1 \models \exists x(\pi_\forall(x) \wedge \forall \mathbf{v}_1(\pi_c(\mathbf{v}_1) \leftrightarrow \mathbf{v}_1 \approx x)).$$

In other words, π_c defines a singleton whose one member is also in the set defined by $\pi_\forall$.)

For example, if L_0 is the language of $(\mathbb{N}, 0, S)$ and T_1 is the theory of $(\mathbb{Z}, +, \cdot)$, then we have

$$\begin{aligned} \pi_\forall(x) &= \exists y_1\, \exists y_2\, \exists y_3\, \exists y_4\, x \approx y_1 \cdot y_1 + y_2 \cdot y_2 + y_3 \cdot y_3 + y_4 \cdot y_4, \\ \pi_0(x) &= x + x \approx x, \\ \pi_S(x, y) &= \forall y(z \cdot z \approx z \wedge z + z \not\approx z \to x + z \approx y). \end{aligned}$$

(We are here exploiting the fact that in $(\mathbb{Z}, +, \cdot)$ we can, in effect, define the structure $(\mathbb{N}, 0, S)$.)

If L_0 coincides with L_1, there is trivially the identity interpretation π, for which

$$\begin{aligned} \pi_\forall &= \mathbf{v}_1 \approx \mathbf{v}_1, \\ \pi_P &= P\mathbf{v}_1 \cdots \mathbf{v}_n, \\ \pi_f &= f\mathbf{v}_1 \cdots \mathbf{v}_n \approx \mathbf{v}_{n+1}. \end{aligned}$$

The conditions (i) and (ii) are then met no matter what T_1 is.

Now assume that π is an interpretation and let $\mathfrak{B}$ be a model of T_1. There is a natural way to extract from $\mathfrak{B}$ a structure ${}^{\pi}\mathfrak{B}$ for L_0. Namely, let

$$\begin{aligned} |{}^{\pi}\mathfrak{B}| &= \text{the set defined in } \mathfrak{B} \text{ by } \pi_{\forall}, \\ P^{{}^{\pi}\mathfrak{B}} &= \text{the relation defined in } \mathfrak{B} \text{ by } \pi_P, \text{ restricted to } |{}^{\pi}\mathfrak{B}|, \\ f^{{}^{\pi}\mathfrak{B}}(a_1, \ldots, a_n) &= \text{the unique } b \text{ such that } \models_{\mathfrak{B}} \pi_f [\![a_1, \ldots, a_n, b]\!], \\ &\quad\ \text{where } a_1, \ldots, a_n \text{ are in } |{}^{\pi}\mathfrak{B}|. \end{aligned}$$

By condition (i) in the definition of interpretations, $|{}^{\pi}\mathfrak{B}| \neq \varnothing$. And, by condition (ii), the definition of $f^{{}^{\pi}\mathfrak{B}}$ makes sense; i.e., there is a unique b meeting the above condition. Hence ${}^{\pi}\mathfrak{B}$ is indeed a structure for the language L_0.

Define the set $\pi^{-1}[T_1]$ of L_0-sentences by the equation

$$\begin{aligned} \pi^{-1}[T_1] &= \text{Th}\{{}^{\pi}\mathfrak{B} : \mathfrak{B} \in \text{Mod } T_1\} \\ &= \{\sigma : \sigma \text{ is an } L_0\text{-sentence true in every structure } {}^{\pi}\mathfrak{B} \text{ obtainable} \\ &\qquad \text{from a model } \mathfrak{B} \text{ of } T_1\}. \end{aligned}$$

This *is* a theory, as is Th $\mathscr{K}$ for any class $\mathscr{K}$. It is a satisfiable theory iff T_1 is satisfiable.

EXAMPLE Earlier in this section we had a theory T containing the sentence

$$(\varepsilon) \qquad \forall v_1 \, \exists! v_2 \, \varphi.$$

We augmented the language to a larger language L^+ which contained a function symbol f. The "definition" of f was provided by the L^+-sentence

$$(\delta) \qquad \forall v_1 \forall v_2 (f v_1 \approx v_2 \leftrightarrow \varphi).$$

We showed that for a sentence σ in the original language of T, if $T ; \delta \models \sigma$, then $T \models \sigma$.

We have an interpretation π from L^+ into T. π is the identity interpretation on all parameters except f. The formula π_f is φ. The fact that $T \models \varepsilon$ is just what we need to verify that π is indeed an interpretation. For any model $\mathfrak{A}$ of T, ${}^{\pi}\mathfrak{A}$ is a structure previously called $(\mathfrak{A}, F)$; it is a model of $T ; \delta$.

We claim that

$$\pi^{-1}[T] = \text{Cn}(T ; \delta).$$

First observe that any model $\mathfrak{B}$ of $T\,;\,\delta$ equals ${}^{\pi}\mathfrak{A}$, where $\mathfrak{A}$ is the restriction of $\mathfrak{B}$ to the language of T. Hence for an L^{+}-sentence σ,

$$\begin{aligned}\sigma \in \pi^{-1}[T] &\Leftrightarrow \models_{\pi_{\mathfrak{A}}} \sigma \quad \text{for every model } \mathfrak{A} \text{ of } T\\ &\Leftrightarrow \models_{\mathfrak{B}} \sigma \quad \text{for every model } \mathfrak{B} \text{ of } T\,;\,\delta\\ &\Leftrightarrow T\,;\,\delta \models \sigma.\end{aligned}$$

Syntactical translation

In the preceding subsection on interpretations we talked about arbitrary models and such. But the reader may already have noticed that there is a much more down-to-earth thing to be said about an interpretation π of L_0 into T_1. Briefly: We can, given a formula φ of L_0, find a formula φ^{π} in L_1 which in some sense corresponds exactly to φ. We define φ^{π} by recursion on φ.

First, consider an atomic formula α of L_0. For example, if α is

$$Pfgx,$$

then α is logically equivalent to

$$\forall y(gx \approx y \rightarrow \forall z(fy \approx z \rightarrow Pz)).$$

And we can take for α^{π} the L_1-formula

$$\forall y(\pi_g(x, y) \rightarrow \forall z(\pi_f(y, z) \rightarrow \pi_P(z))).$$

In general, scan an atomic formula α from right to left. The rightmost place at which a function symbol occurs will initiate a segment of the form $gx_1 \cdots x_n$ for an n-place g. (In the example $n = 1$.) Replace this by some new variable y, and prefix $\forall y(\pi_g(x_1, \ldots, x_n, y) \rightarrow$. Continue to the next place at which a function symbol occurs. Finally, replace the predicate symbol P (if a parameter) by π_P (with the correct variables).

The definition of α^{π} can be stated more carefully by using recursion on the number of places at which function symbols occur in α. If that number is zero, then α is $Px_1 \cdots x_n$ and α^{π} is $\pi_P(x_1, \ldots, x_n)$. Otherwise, take the rightmost place at which a function symbol g occurs. If g is an n-place symbol, then that place initiates a segment $gx_1 \cdots x_n$. Replace this segment by some new variable y, obtaining a formula we can call $\alpha_y^{gx_1 \ldots x_n}$. Then α^{π} is

$$\forall y(\pi_g(x_1, \ldots, x_n, y) \rightarrow (\alpha_y^{gx_1 \ldots x_n})^{\pi}).$$

For example,

$$\begin{aligned}(Pfgx)^\pi &= \forall y(\pi_g(x, y) \to (Pfy)^\pi)\\ &= \forall y(\pi_g(x, y) \to \forall z(\pi_f(y, z) \to (Pz)^\pi))\\ &= \forall y(\pi_g(x, y) \to \forall z(\pi_f(y, z) \to \pi_P(z))).\end{aligned}$$

The interpretation of a nonatomic formula is defined in the obvious way. $(\neg\varphi)^\pi$ is $(\neg\varphi^\pi)$, $(\varphi \to \psi)^\pi$ is $(\varphi^\pi \to \psi^\pi)$, and $(\forall x\varphi)^\pi$ is $\forall x(\pi_\forall(x) \to \varphi^\pi)$. (Thus the quantifiers are "relativized" to $\pi_\forall$.)

The sense in which φ^π "says the same thing" as φ is made precise in the following basic lemma.

Lemma 27B Let π be an interpretation of L_0 into T_1, let $\mathfrak{B}$ be a model of T_1. For any formula φ of L_0 and any map s of the variables into $|\,^\pi\mathfrak{B}\,|$,

$$\models_{\pi_\mathfrak{B}} \varphi\,[s] \qquad \text{iff } \models_\mathfrak{B} \varphi^\pi\,[s].$$

This is not a deep fact. It just says that φ^π was defined correctly.

Proof We use induction on φ, but only the case of an atomic formula α is nontrivial. For α, we use induction on the number of places at which function symbols occur. It is easy if that number is zero. Otherwise,

$$\alpha^\pi = \forall y(\pi_g(x, y) \to \beta^\pi),$$

where $\beta^y_{gx} = \alpha$. (We have quietly assumed that g is a one-place symbol; the notation is bad enough already.) Let

$$\begin{aligned}b &= \text{the unique } b \text{ such that } \models_\mathfrak{B} \pi_g\,[\![s(x), b]\!]\\ &= g^{^\pi\mathfrak{B}}(s(x)).\end{aligned}$$

Then

$$\begin{aligned}\models_\mathfrak{B} \alpha^\pi\,[s] &\Leftrightarrow \models_\mathfrak{B} \beta^\pi\,[s(y|b)]\\ &\Leftrightarrow \models_{\pi_\mathfrak{B}} \beta\,[s(y|b)] && \text{by the inductive hypothesis}\\ &\Leftrightarrow \models_{\pi_\mathfrak{B}} \beta^y_{gx}\,[s] && \text{by the substitution lemma}\\ &\Leftrightarrow \models_{\pi_\mathfrak{B}} \alpha\,[s]. \qquad \blacksquare\end{aligned}$$

The following corollary justifies our choice of notation for $\pi^{-1}[T_1]$.

Corollary 27C For a sentence σ of L_0,

$$\sigma \in \pi^{-1}[T_1] \qquad \text{iff } \sigma^\pi \in T_1.$$

Proof Recall that by definition

$$\begin{aligned} \sigma \in \pi^{-1}[T_1] &\Leftrightarrow \text{for every model } \mathfrak{B} \text{ of } T_1, \models_{\pi_{\mathfrak{B}}} \sigma \\ &\Leftrightarrow \text{for every model } \mathfrak{B} \text{ of } T_1, \models_{\mathfrak{B}} \sigma^{\pi} \qquad \text{by Lemma 27B} \\ &\Leftrightarrow T_1 \models \sigma^{\pi}. \qquad \blacksquare \end{aligned}$$

Definition An *interpretation* π of a theory T_0 into a theory T_1 is an interpretation π of the language of T_0 into T_1 such that

$$T_0 \subseteq \pi^{-1}[T_1].$$

In other words, it is necessary that for an L_0-sentence σ,

$$\sigma \in T_0 \Rightarrow \sigma^{\pi} \in T_1.$$

$\pi^{-1}[T_1]$ is the largest theory which π interprets into T_1. If $T_0 = \pi^{-1}[T_1]$, then we have

$$\sigma \in T_0 \Leftrightarrow \sigma^{\pi} \in T_1.$$

In this case π is said to be a *faithful* interpretation of T_0 into T_1.

To return to an earlier example, consider the structures $(\mathbb{N}, 0, S)$ and $(\mathbb{Z}, +, \cdot)$. We had an interpretation π into Th $(\mathbb{Z}, +, \cdot)$, where

$$\begin{aligned} \pi_{\forall}(x) &= \exists y_1\, \exists y_2\, \exists y_3\, \exists y_4\, x \approx y_1 \cdot y_1 + y_2 \cdot y_2 + y_3 \cdot y_3 + y_4 \cdot y_4, \\ \pi_0(x) &= x + x \approx x, \\ \pi_S(x, y) &= \forall z(z \cdot z \approx z \wedge z + z \not\approx z \rightarrow x + z \approx y). \end{aligned}$$

We now claim that π is a faithful interpretation of Th $(\mathbb{N}, 0, S)$ into Th $(\mathbb{Z}, +, \cdot)$. For in this case, ${}^{\pi}(\mathbb{Z}, +, \cdot)$ *is* the structure $(\mathbb{N}, 0, S)$. Hence

$$\models_{(\mathbb{N},0,S)} \sigma \Leftrightarrow \models_{\pi_{(\mathbb{Z},+,\cdot)}} \sigma \Leftrightarrow \models_{(\mathbb{Z},+,\cdot)} \sigma^{\pi}.$$

In Chapter 3 we will be able to show that there is no interpretation of Th$(\mathbb{Z}, +, \cdot)$ into Th$(\mathbb{N}, 0, S)$. Thus the former theory is strictly stronger than the latter.

Finally, let us return to the situation with which we started this section. Assume that T is a theory containing the sentence ε, where

$$\begin{aligned} \varepsilon &= \forall \mathbf{v}_1\, \exists! \mathbf{v}_2\, \varphi; \\ \delta &= \forall \mathbf{v}_1 \forall \mathbf{v}_2 (f\mathbf{v}_1 \approx \mathbf{v}_2 \leftrightarrow \varphi); \end{aligned}$$

L^+ = the language obtained by adding the new function symbol f to the language of T;

π = the interpretation of L^+ into T which is the identity interpretation on all parameters except f, and $\pi_f = \varphi$.

In fact, π is a faithful interpretation of $\mathrm{Cn}(T\,;\,\delta)$ into T, since, as noted previously,

$$\pi^{-1}[T] = \mathrm{Cn}(T\,;\,\delta).$$

We can now draw an additional conclusion; the definition is *eliminable.*

Theorem 27D Assume that we have the situation described above. Then for any L^+-sentence σ we can find the sentence σ^π in the original language such that

(a) $T\,;\,\delta \models (\sigma \leftrightarrow \sigma^\pi)$.
(b) $T\,;\,\delta \models \sigma \Leftrightarrow T \models \sigma^\pi$.
(c) If f does not occur in σ, then $\models (\sigma \leftrightarrow \sigma^\pi)$.

Proof Part (c) follows from the fact that π is the identity interpretation on all parameters except f. Part (b) restates that π is a faithful interpretation of $\mathrm{Cn}(T\,;\,\delta)$ into T. Since π is faithful, for (a) it suffices to show that

$$T \models (\sigma \leftrightarrow \sigma^\pi)^\pi.$$

This follows from (c), since $(\sigma \leftrightarrow \sigma^\pi)^\pi$ is $(\sigma^\pi \leftrightarrow \sigma^{\pi\pi})$, which is valid. ■

EXERCISES

1. Assume that L_0 and L_1 are languages with the same parameters except that L_0 has an n-place function symbol f not in L_1 and L_1 has an $(n+1)$-place predicate symbol P not in L_0. Show that for any L_0-theory T there is a faithful interpretation of T into some L_1-theory.

2. Let L_0 be the language with equality and the two-place function symbols $+$ and $\cdot$. Let L_1 be the same, but with three-place *predicate* symbols for addition and multiplication. Let $\mathfrak{N}_i = (N\,+, \cdot)$ be the structure for L_i consisting of the natural numbers with addition and multiplication $(i = 0, 1)$. Show that any relation definable by an L_0-formula in $\mathfrak{N}_0$ is also definable by an L_1-formula in $\mathfrak{N}_1$.

3. Show that an interpretation of a complete theory into a satisfiable theory is faithful.

§ 2.8 NONSTANDARD ANALYSIS[1]

The differential and integral calculus was originally described by Leibniz and Newton in the seventeenth century in terms of quantities which were infinitely small yet nonzero. Newton used in his calculations a number *o* which, being infinitely small, could be multiplied by any finite number and still be negligible. But it was necessary to divide by *o*, so it had to be nonzero. Leibniz's *dx* was less than any assignable quantity, yet was nonzero.

These ideas were not easy to comprehend or to accept. Throughout the eighteenth century this business of working with infinitesimals was attacked (e.g., by Bishop Berkeley), distrusted (e.g., by D'Alembert), and used in enthusiastic experimentation (e.g., by Euler). While Euler was creating the mathematics which students now study in advanced calculus, he used infinitesimals in a loose, free-swinging manner that would not be tolerated in today's freshmen. Only in the nineteenth century were the foundations of calculus presented in the form now found in textbooks. The treatment of limits was then rigorous, and debate subsided.

In 1961 Abraham Robinson introduced a new method for treating limits, rescuing infinitesimals from their intellectual disrepute. This method combines the intuitive advantages of working with infinitely small quantities with modern standards of rigor. The basic idea is to utilize a nonstandard model of the theory of the real numbers.

Construction of $*\mathfrak{R}$

We will use a very large first-order language. In addition to symbols for $+$, $\cdot$, and $<$ we might as well add symbols for the exponentiation and absolute-value functions. And since there is no good reason to stop there, we go all the way and include a symbol for *every* operation on the set $\mathbb{R}$ of reals. We do the same for every relation on $\mathbb{R}$. Thus we have the language with equality and the following parameters:

0. $\forall$, intended to mean "for all real numbers."
1. An n-place predicate symbol $\boldsymbol{P}_R$ for each n-ary relation R on $\mathbb{R}$.
2. A constant symbol $\boldsymbol{c}_r$ for each $r \in \mathbb{R}$.
3. An n-place function symbol $\boldsymbol{f}_F$ for each n-ary operation F on $\mathbb{R}$.

For this language there is the standard structure $\mathfrak{R}$, with $|\mathfrak{R}| = \mathbb{R}$, $\boldsymbol{P}_R^{\mathfrak{R}} = R$, $\boldsymbol{c}_r^{\mathfrak{R}} = r$, and $\boldsymbol{f}_F^{\mathfrak{R}} = F$. But now let us form a nonstandard structure,

[1] This section may be omitted without loss of continuity.

by using the compactness theorem. Let Γ be the set

$$\mathrm{Th}\,\mathfrak{R} \cup \{\mathbf{c}_r \mathbf{P}_< \mathbf{v}_1 : r \in \mathbb{R}\}.$$

(Here $\mathbf{c}_r \mathbf{P}_< \mathbf{v}_1$ formalizes "r is less than $\mathbf{v}_1$.") Any finite subset of Γ can be satisfied in $\mathfrak{R}$ by assigning to $\mathbf{v}_1$ some large real number. Hence by the compactness theorem there is a structure $\mathfrak{A}$ and an element $a \in |\mathfrak{A}|$ such that Γ is satisfied in $\mathfrak{A}$ when $\mathbf{v}_1$ is assigned a. Since $\mathfrak{A}$ is a model of $\mathrm{Th}\,\mathfrak{R}$, we have $\mathfrak{A} \equiv \mathfrak{R}$. There is also an isomorphism h of $\mathfrak{R}$ into (but not onto) $\mathfrak{A}$ defined by

$$h(r) = \mathbf{c}_r^{\mathfrak{A}}.$$

To check that this is indeed an isomorphism, we use the fact that $\mathfrak{A} \equiv \mathfrak{R}$. h is one-to-one, since for $r_1 \neq r_2$, the sentence $\mathbf{c}_{r_1} \not\approx \mathbf{c}_{r_2}$ holds in $\mathfrak{R}$ and hence in $\mathfrak{A}$. h preserves a binary relation $R(= \mathbf{P}_R^{\mathfrak{R}})$ since for any r and s in $\mathbb{R}$,

$$\begin{aligned}\langle r, s\rangle \in \mathbf{P}_R^{\mathfrak{R}} &\Leftrightarrow \models_{\mathfrak{R}} \mathbf{P}_R \mathbf{c}_r \mathbf{c}_s \\ &\Leftrightarrow \models_{\mathfrak{A}} \mathbf{P}_R \mathbf{c}_r \mathbf{c}_s \\ &\Leftrightarrow \langle \mathbf{c}_r^{\mathfrak{A}}, \mathbf{c}_s^{\mathfrak{A}}\rangle \in \mathbf{P}_R^{\mathfrak{A}} \\ &\Leftrightarrow \langle h(r), h(s)\rangle \in \mathbf{P}_R^{\mathfrak{A}}.\end{aligned}$$

A similar calculation applies to any n-ary relation. Next we must show that h preserves any function $F(= \mathbf{f}_F^{\mathfrak{R}})$. Again for notational ease, suppose that F is a binary operation. Consider any r and s in $\mathbb{R}$, and let $t = F(r, s)$. Then

$$\begin{aligned}h(\mathbf{f}_F^{\mathfrak{R}}(r, s)) &= h(F(r, s)) \\ &= h(t) \\ &= \mathbf{c}_t^{\mathfrak{A}}.\end{aligned}$$

Now the sentence $\mathbf{c}_t \approx \mathbf{f}_F \mathbf{c}_s \mathbf{c}_r$ holds in $\mathfrak{R}$ and hence in $\mathfrak{A}$. Hence

$$\begin{aligned}\mathbf{c}_t^{\mathfrak{A}} &= \mathbf{f}_F^{\mathfrak{A}}(\mathbf{c}_r^{\mathfrak{A}}, \mathbf{c}_s^{\mathfrak{A}}) \\ &= \mathbf{f}_F^{\mathfrak{A}}(h(r), h(s)).\end{aligned}$$

So h preserves $\mathbf{f}_F$. For the constant symbols we have by the definition of h,

$$\begin{aligned}h(\mathbf{c}_r^{\mathfrak{R}}) &= h(r) \\ &= \mathbf{c}_r^{\mathfrak{A}}.\end{aligned}$$

Since we have an isomorphic copy of $\mathfrak{R}$ inside $\mathfrak{A}$, we can find another structure $^*\mathfrak{R}$ isomorphic to $\mathfrak{A}$ such that $\mathfrak{R}$ is a substructure of $^*\mathfrak{R}$. The idea is simply to replace in $\mathfrak{A}$ the point $\mathbf{c}_r^{\mathfrak{A}}$ by the point r (provided that $|\mathfrak{A}| \cap \mathbb{R} = \varnothing$, as can always be arranged). For details, see Exercise 23 of Section 2.2. Since $^*\mathfrak{R}$ is isomorphic to $\mathfrak{A}$, there is a point $b \in |{}^*\mathfrak{R}|$ such that $^*\mathfrak{R}$ satisfies Γ when $\mathbf{v}_1$ is assigned b. In particular, $^*\mathfrak{R} \equiv \mathfrak{R}$.

To go much further, we need a less cumbersome notation. We will use an asterisk to indicate passage from $\mathfrak{R}$ to $^*\mathfrak{R}$.

1. For each n-ary relation R on $\mathbb{R}$, let $^*R$ be the relation $\mathbf{P}_R^{*\mathfrak{R}}$ assigned to the symbol $\mathbf{P}_R$ by $^*\mathfrak{R}$. In particular, $\mathbb{R}$ is a unary relation on $\mathbb{R}$. Its image $^*\mathbb{R}$ equals the universe of $^*\mathfrak{R}$, since the sentence $\forall x\, \mathbf{P}_{\mathbb{R}}x$ is true in $\mathfrak{R}$ and hence in $^*\mathfrak{R}$. Since $\mathfrak{R}$ is a substructure of $^*\mathfrak{R}$, we have that each relation R equals the restriction of $^*R$ to $\mathbb{R}$.

2. For each n-ary operation F on $\mathbb{R}$, let $^*F$ be the operation $\mathbf{f}_F^{*\mathfrak{R}}$ assigned to the symbol $\mathbf{f}_F$ by $^*\mathfrak{R}$. F is then the restriction to $\mathbb{R}$ of $^*F$.

Observe that $\mathbf{c}_r^{*\mathfrak{R}} = r$, so we need no special notation for this.

There is a general method (to be used heavily in the remainder of this section) for demonstrating properties of a relation $^*R$ or an operation $^*F$. One simply observes (1) that R or F has the property, (2) that the property can be expressed by a sentence of the language, and (3) that $\mathfrak{R} \equiv {}^*\mathfrak{R}$.

For example, the binary relation $^*<$ on $^*\mathbb{R}$ is transitive. This is because $<$ is transitive, and this property can be expressed by the sentence

$$\forall x\, \forall y\, \forall z (x\mathbf{P}_< y \rightarrow y\mathbf{P}_< z \rightarrow x\mathbf{P}_< z).$$

By similar reasoning, $^*<$ satisfies trichotomy on $^*\mathbb{R}$ and thus is an ordering relation on $^*<$.

For another example, we can prove that the binary operation $^*+$ on $^*\mathbb{R}$ is commutative, since $+$ is commutative and the commutative law can be expressed by a sentence. By applying this reasoning to each of the field axioms, we see that $(^*\mathbb{R}, 0, 1, {}^*+, {}^*\cdot)$ is a field.

This general method is used so much, we will shortly begin to take it for granted. If, for example, we assert that $^*|\, a \,{}^*+\, b \,| \;{}^*\!\leq\; {}^*|\, a \,| \,{}^*+\, {}^*|\, b \,|$ for a and b in $^*\mathbb{R}$, we will take for granted that the reader perceives that the general method yields this fact.

We have $\mathbb{R} \subseteq {}^*\mathbb{R}$, but $\mathbb{R} \neq {}^*\mathbb{R}$. For we have a point b such that $\models_{*\mathfrak{R}} \mathbf{c}_r \mathbf{P}_< \mathbf{v}_1 [\![b]\!]$; i.e., $r \,{}^*\!< b$. Thus b is infinitely large, being larger (in the ordering $^*<$) than any standard r, i.e., any $r \in \mathbb{R}$. Its reciprocal $1 \,{}^*\!/\, b$ will be a sample infinitesimal.

Properties of $\mathfrak{R}$ which *cannot* be expressed in the language are likely to fail in $^*\mathfrak{R}$. The least-upper-bound property is one such. There are nonempty bounded subsets S of $^*\mathbb{R}$ which have no least upper bound (with respect to the ordering $^*<$). For example, $\mathbb{R}$ is such a subset of $^*\mathbb{R}$. It is bounded by the infinite b of the preceding paragraph. But it has no least upper bound; see Exercise 7.

Define the set $\mathcal{F}$ of *finite* elements by the equation

$$\mathcal{F} = \{x \in {}^*\mathbb{R} : {}^*| x | \mathbin{{}^*<} y \text{ for some } y \in \mathbb{R}\}.$$

Similarly, define the set $\mathcal{I}$ of *infinitesimals* by the equation

$$\mathcal{I} = \{x \in {}^*\mathbb{R} : {}^*| x | \mathbin{{}^*<} y \text{ for all positive } y \in \mathbb{R}\}.$$

If $A \subseteq \mathbb{R}$ is unbounded, then $^*A$ contains infinite points. For the sentence "for any real r there is an element $a \in A$ larger than r" is true and formalizable. Take some infinite positive b; there must be a larger (and hence infinite) member of $^*A$. For example, $^*\mathbb{N}$ contains infinite numbers.

The only standard infinitesimal, i.e., the only member of $\mathbb{R} \cap \mathcal{I}$, is 0. But there are other infinitesimals. For by the usual (formalizable) rules for inequalities, the reciprocal of any infinite number is an infinitesimal.

Algebraic properties

In the next theorem we collect some algebraic facts about $\mathcal{F}$ and $\mathcal{I}$ which will be useful later.

Theorem 28A (a) $\mathcal{F}$ is closed under addition $^*+$, subtraction $^*-$, and multiplication $^*\cdot$.

(b) $\mathcal{I}$ is closed under addition $^*+$, subtraction $^*-$, and multiplication from $\mathcal{F}$:

$$x \in \mathcal{I} \quad \text{and} \quad y \in \mathcal{F} \Rightarrow x \mathbin{{}^*\cdot} y \in \mathcal{I}.$$

In algebraic terminology, part (a) says that $\mathcal{F}$ is a subring of the field $^*\mathbb{R}$, and part (b) says that $\mathcal{I}$ is an ideal in the ring $\mathcal{F}$. We will see a little later what the quotient ring $\mathcal{F}/\mathcal{I}$ is.

Proof (a) Let x and y be finite, so that $^*| x | \mathbin{{}^*<} a$, $^*| y | \mathbin{{}^*<} b$ for standard a and b in $\mathbb{R}$. Then

$${}^*| x \mathbin{{}^*\pm} y | \mathbin{{}^*\leq} {}^*| x | \mathbin{{}^*+} {}^*| y | \mathbin{{}^*<} a + b \in \mathbb{R},$$

whence $x \mathbin{*+} y$, $x \mathbin{*-} y$ are finite. Also

$$*| x \mathbin{*\cdot} y | \mathbin{*<} a \cdot b \in \mathbb{R},$$

whence $x \mathbin{*\cdot} y$ is finite, too.

(b) Let x and y be infinitesimals. Then for any positive standard a, $*| x | \mathbin{*<} a/2$ and $*| y | \mathbin{*<} a/2$. Hence

$$*| x \mathbin{*\pm} y | \mathbin{*<} a/2 + a/2 = a,$$

so that $x \mathbin{*+} y$ and $x \mathbin{*-} y$ are infinitesimal. If z is finite, then $*| z | \mathbin{*<} b$ for some standard b. Since x is infinitesimal, we have $*| x | \mathbin{*<} a/b$, whence

$$*| x \mathbin{*\cdot} z | \mathbin{*<} (a/b)b = a.$$

Thus $x \mathbin{*\cdot} z$ is also infinitesimal. ■

Definition x is *infinitely close* to $y (x \simeq y)$ iff $x \mathbin{*-} y$ is infinitesimal.

Theorem 28B (a) $\simeq$ is an equivalence relation on $*\mathbb{R}$.
(b) If $u \simeq v$ and $x \simeq y$, then $u \mathbin{*+} x \simeq v \mathbin{*+} y$ and $*{-}\, u \simeq *{-}\, v$.
(c) If $u \simeq v$ and $x \simeq y$ and x, y, u, v are finite, then $u \mathbin{*\cdot} x \simeq v \mathbin{*\cdot} y$.

Proof This is a consequence of part (b) of the preceding theorem ($\mathscr{I}$ is an ideal in $\mathscr{F}$).

(a) $\simeq$ is reflexive since 0 is infinitesimal. $\simeq$ is symmetric since the negative ($*-$) of an infinitesimal is infinitesimal. Finally, suppose that $x \simeq y$ and $y \simeq z$. Then

$$x \mathbin{*-} z = (x \mathbin{*-} y) \mathbin{*+} (y \mathbin{*-} z) \in \mathscr{I}$$

since $\mathscr{I}$ is closed under addition.

(b) If $u \simeq v$ and $x \simeq y$, then

$$(u \mathbin{*+} x) \mathbin{*-} (v \mathbin{*+} y) = (u \mathbin{*-} v) \mathbin{*+} (x \mathbin{*-} y) \in \mathscr{I}$$

since $\mathscr{I}$ is closed under addition. Also $*{-}u \simeq *{-}v$ since $\mathscr{I}$ is closed under negation.

(c)
$$\begin{aligned}(u \mathbin{*\cdot} x) \mathbin{*-} (v \mathbin{*\cdot} y) &= (u \mathbin{*\cdot} x) \mathbin{*-} (u \mathbin{*\cdot} y) \mathbin{*+} (u \mathbin{*\cdot} y) \mathbin{*-} (v \mathbin{*\cdot} y)\\ &= u \mathbin{*\cdot} (x \mathbin{*-} y) \mathbin{*+} (u \mathbin{*-} v) \mathbin{*\cdot} y \in \mathscr{I}\end{aligned}$$

since $\mathscr{I}$ is closed under multiplication from $\mathscr{F}$. ■

For standard r and s, we have $r \simeq s$ iff $r = s$, as 0 is the only standard infinitesimal.

Lemma 28C If $x \not\simeq y$ and at least one is finite, then there is a standard q strictly between x and y.

Proof We may suppose that $x \mathbin{*\!<} y$. In fact, we may further suppose that $0 \mathbin{*\!\leq} x \mathbin{*\!<} y$; the case $x \mathbin{*\!<} y \mathbin{*\!\leq} 0$ is similar and the case $x \mathbin{*\!<} 0 \mathbin{*\!<} y$ is trivial. Since $x \not\simeq y$ there is a standard b such that $0 < b \mathbin{*\!<} y \mathbin{*\!-} x$. Since x is finite we have $x \mathbin{*\!<} mb$ for some positive integer m; take the least such m. Then $x \mathbin{*\!<} mb \mathbin{*\!<} y$. (By the leastness of m, $(m-1)b \mathbin{*\!\leq} x$. So $mb \mathbin{*\!\leq} x \mathbin{*\!+} b \mathbin{*\!<} y$.) ■

Theorem 28D Every $x \in \mathscr{F}$ is infinitely close to a unique $r \in \mathbb{R}$.

Proof For $x \in \mathscr{F}$, the set

$$S = \{y \in \mathbb{R} : y \mathbin{*\!\leq} x\}$$

of standard points below x has an upper bound in $\mathbb{R}$. Let r be its least upper bound; we claim that $x \simeq r$.

If $x \not\simeq r$, then by the lemma there is a standard q between x and r. If $r < q \mathbin{*\!<} x$, then r fails to be an upper bound for S. If $x \mathbin{*\!<} q < r$, then q is also an upper bound for S, contradicting the leastness of r. Hence $x \simeq r$.

This establishes the existence of r. As for the uniqueness, note that if $x \simeq r$ and $x \simeq s$, then $r \simeq s$. For standard r and s this implies that $r = s$. ■

Corollary 28E Each finite x has a unique decomposition $x = s \mathbin{*\!+} i$, where s is standard and i is infinitesimal.

We call s the *standard part* of x, st(x). Of course for standard r, st(r) $= r$. In the next theorem we summarize some properties of the st function.

Theorem 28F (a) st maps $\mathscr{F}$ onto $\mathbb{R}$.
(b) $\operatorname{st}(x) = 0$ iff x is infinitesimal.
(c) $\operatorname{st}(x \mathbin{*\!+} y) = \operatorname{st}(x) + \operatorname{st}(y)$.
(d) $\operatorname{st}(x \mathbin{*\!\cdot} y) = \operatorname{st}(x) \cdot \operatorname{st}(y)$.

Proof (a) and (b) are clear. Since $\operatorname{st}(x) \simeq x$ and $\operatorname{st}(y) \simeq y$, we have by part (b) of Theorem 28B $\operatorname{st}(x) + \operatorname{st}(y) \simeq x \mathbin{*\!+} y$. Hence the left side equals $\operatorname{st}(x \mathbin{*\!+} y)$. Part (d) is similar and uses part (c) of Theorem 28B. ■

(In algebraic terminology, this theorem asserts that st is a homomorphism of the ring $\mathscr{F}$ onto the field $\mathbb{R}$, with kernel $\mathscr{I}$. Consequently, the quotient ring $\mathscr{F}/\mathscr{I}$ is isomorphic to the real field $\mathbb{R}$.)

Henceforth in this section we will streamline our notation by omitting the asterisks on the symbols for the arithmetic operations $*+$, $*-$, $*\cdot$, and $*/$.

Convergence

In calculus courses convergence is usually treated in terms of ε's and δ's and variables which come delicately close to certain values. We will give here the beginning of an alternative treatment of convergence, where variables come infinitely close to the limiting values.

Definition Let $F : \mathbb{R} \to \mathbb{R}$. Then *F converges at a to b* iff whenever x is infinitely close to (but different from) a, then $*F(x)$ is infinitely close to b.

Proof of equivalence with the ordinary definition: First suppose that F converges at a to b in the ordinary sense. That is, for any $\varepsilon > 0$ there is a $\delta > 0$ such that

$$0 \neq |x - a| < \delta \Rightarrow |b - F(x)| < \varepsilon \qquad \text{for any } x.$$

The displayed sentence (concerning the standard numbers ε and δ) is formalizable and thus holds in $*\mathfrak{R}$. Now if x in $*\mathbb{R}$ is infinitely close to (but different from) a, then certainly $0 \neq *|x - a| *< \delta$. Hence $*|b - *F(x)| *< \varepsilon$. Since ε was arbitrary, $b \simeq *F(x)$.

Conversely, assume that the condition stated in the definition is met. Then for any standard $\varepsilon > 0$, the sentence

There exists $\delta > 0$ such that for all x, $0 \neq |a - x| < \delta \Rightarrow |b - F(x)| < \varepsilon$

(when formalized) holds in $*\mathfrak{R}$, since we can take δ to be infinitesimal. Hence the sentence holds in $\mathfrak{R}$ also. ■

First remark: It is entirely possible that F does not converge at a to any number. On the other hand, F converges at a to at most one b. For if i is a nonzero infinitesimal, then $b = \text{st}(*F(a + i))$. It is traditional to denote this b by "$\lim_{x \to a} F(x)$." Thus

$$\lim_{x \to a} F(x) = \text{st}(*F(a + i)).$$

Second remark: It is not really necessary to have dom $F = \mathbb{R}$. It is enough for a to be an accumulation point of dom F. (a is an accumulation point of S iff a is infinitely close to, but different from, some member of $*S$.)

Corollary 28G F is continuous at a iff whenever $x \simeq a$, then $*F(x) \simeq F(a)$.

Now consider a function $F : \mathbb{R} \to \mathbb{R}$ and a standard $a \in \mathbb{R}$. Then the derivative $F'(a)$ is

$$\lim_{h \to a} \frac{F(a + h) - F(a)}{h}.$$

By our definition of limit, this can also be stated: $F'(a) = b$ iff for every nonzero infinitesimal dx we have $dF/dx \simeq b$, where $dF = *F(a + dx) - F(a)$. Thus if there is such a b (i.e., if $F'(a)$ exists), then

$$F'(a) = \text{st}(dF/dx)$$

for any nonzero infinitesimal dx. Here dF/dx is the result of *dividing* dF by dx. The fact that we just use division here greatly facilitates calculations.

Example Let $F(x) = x^2$. Then $F'(a) = 2a$, since

$$\frac{dF}{dx} = \frac{(a + dx)^2 - a^2}{dx} = \frac{2a(dx) + (dx)^2}{dx} = 2a + dx \simeq 2a.$$

Theorem 28H If $F'(a)$ exists, then F is continuous at a.

Proof For any nonzero infinitesimal dx we have

$$\frac{*F(a + dx) - F(a)}{dx} \simeq F'(a).$$

The right side is standard, so the left side is at least finite. Consequently, when we multiply the left side by the infinitesimal dx we are left with the fact that $*F(a + dx) - F(a) \in \mathscr{I}$; i.e., $*F(a + dx) \simeq F(a)$. ■

The reader should note that this result is not some nonstandard analog of a classical theorem. Nor even a generalization of a classical theorem. It *is* a classical theorem. It is only the proof which is nonstandard. The same remarks apply to the next theorem. Let $F \circ G$ be the function whose value at a is $F(G(a))$.

Chain Rule Assume that $G'(a)$ and $F'(G(a))$ exist. Then $(F \circ G)'(a)$ exists and equals $F'(G(a)) \cdot G'(a)$.

Proof First, notice that $*(F \circ G) = *F \circ *G$, since the sentence $\forall \boldsymbol{v}_1\, f_{F \circ G}\boldsymbol{v}_1 \approx f_F f_G \boldsymbol{v}_1$ holds in the structures. Now consider any nonzero infinitesimal dx. Let

$$\begin{aligned} dG &= *G(a + dx) - G(a), \\ dF &= *(F \circ G)(a + dx) - (F \circ G)(a) \\ &= *F(*G(a + dx)) - F(G(a)) \\ &= *F(G(a) + dG) - F(G(a)). \end{aligned}$$

Then $dG \simeq 0$ since G is continuous at a. If $dG \neq 0$, then by the last of these equations $dF/dG \simeq F'(G(a))$, whence

$$\frac{dF}{dx} = \frac{dF}{dG} \cdot \frac{dG}{dx} \simeq F'(G(a)) \cdot G'(a).$$

If $dG = 0$, then $dF = 0$ and $G'(a) \simeq dG/dx = 0$, so we again have

$$\frac{dF}{dx} \simeq F'(G(a)) \cdot G'(a). \qquad \blacksquare$$

These theorems are but samples of the treatment of convergence in terms of infinite proximity. The method is not at all limited to elementary topics. One can construct delta functions δ with the property that $\int_{-\infty}^{\infty} \delta = 1$ and yet $\delta(x) \simeq 0$ for $x \not\simeq 0$. Original results in analysis (e.g., in the theory of Hilbert spaces) have been obtained by the method of nonstandard analysis. Possibly the method will be more widely used in the future as more analysts become familiar with it.

EXERCISES

1. ($\mathbb{Q}$ is dense in $\mathbb{R}$.) Let $\mathbb{Q}$ be the set of rational numbers. Show that every member of $*\mathbb{R}$ is infinitely close to some member of $*\mathbb{Q}$.

2. (a) Let $A \subseteq \mathbb{R}$ and $F : A \to \mathbb{R}$. Then F is also a binary relation on $\mathbb{R}$; show that $*F : *A \to *\mathbb{R}$.

(b) Let $S : \mathbb{N} \to \mathbb{R}$. Recall that S is said to converge to b iff for every $\varepsilon > 0$ there is some k such that for all $n > k$, $| S(n) - b | < \varepsilon$. Show that this is equivalent to: $*S(x) \simeq b$ for every infinite $x \in *\mathbb{N}$.

(c) Assume that $S_i : \mathbb{N} \to \mathbb{R}$ and S_i converges to b_i for $i = 1, 2$. Show that $S_1 + S_2$ converges to $b_1 + b_2$ and $S_1 \cdot S_2$ converges to $b_1 \cdot b_2$.

3. Let $F : A \rightarrow \mathbb{R}$ be one-to-one, where $A \subseteq \mathbb{R}$. Show that if $x \in {}^*A$ but $x \notin A$, then ${}^*F(x) \notin \mathbb{R}$.

4. Let $A \subseteq \mathbb{R}$. Show that $A = {}^*A$ iff A is finite.

5. (Bolzano–Weierstrass theorem) Let $A \subseteq \mathbb{R}$ be bounded and infinite. Show that there is a point $p \in \mathbb{R}$ which is infinitely close to, but different from, some member of ${}^*A$. *Suggestion*: Let $S : \mathbb{N} \rightarrow A$ with S one-to-one; look at ${}^*S(x)$ for infinite $x \in {}^*\mathbb{N}$.

6. (a) Show that ${}^*\mathbb{Q}$ has cardinality at least $2^{\aleph_0}$, where $\mathbb{Q}$ is the set of rational numbers. *Suggestion*: Use Exercise 1.
(b) Show that ${}^*\mathbb{N}$ has cardinality at least $2^{\aleph_0}$.

7. Let A be a subset of $\mathbb{R}$ having no greatest member. Then as a subset of ${}^*\mathbb{R}$, A will have upper bounds (with respect to the ordering ${}^*<$) in ${}^*\mathbb{R}$. But show that A does not have a least such bound.

CHAPTER THREE

Undecidability

§ 3.0 NUMBER THEORY

In this chapter we will focus our attention on a specific language, the language of number theory. This will be the first-order language with equality and with the following parameters:

$\forall$, intended to mean "for all natural numbers." (Recall that the set N of natural numbers is the set $\{0, 1, 2, \ldots\}$.)

0, a constant symbol intended to denote the number 0.

S, a one-place function symbol intended to denote the successor function $S : N \to N$, i.e., the function for which $S(n) = n + 1$.

$<$, a two-place predicate symbol intended to denote the usual (strict) ordering relation on N.

$+$, $\cdot$, **E**, two-place function symbols intended to denote the operations $+$, $\cdot$, and E of addition, multiplication, and exponentiation, respectively.

We will let $\mathfrak{N}$ be the intended structure for this language. Thus we may informally write

$$\mathfrak{N} = (N, 0, S, <, +, \cdot, E).$$

(More precisely, $|\mathfrak{N}| = N$, $\mathbf{0}^{\mathfrak{N}} = 0$, etc.)

By *number theory* we mean the theory of this structure, Th $\mathfrak{N}$. As warming-up exercises we will study (in Sections 3.1 and 3.2) certain reducts of $\mathfrak{N}$, i.e., restrictions of $\mathfrak{N}$ to sublanguages:

$$\begin{aligned}\mathfrak{N}_S &= (N, 0, S),\\ \mathfrak{N}_L &= (N, 0, S, <),\\ \mathfrak{N}_A &= (N, 0, S, <, +).\end{aligned}$$

Finally, in Section 3.7 we will consider

$$\mathfrak{N}_M = (N, 0, S, <, +, \cdot).$$

For each of these structures we will raise the same questions:

(A) Is the theory of the structure decidable? If so, what is a nice set of axioms for the theory? Is there a finite set of axioms?

(B) What subsets of N are definable in the structure?

(C) What do the nonstandard models of the theory of the structure look like? (By "nonstandard" we mean "not isomorphic to the intended structure.")

Our reason for choosing number theory (rather than, say, group theory) for special study is this: We can show that a certain subtheory of number theory is an undecidable set of sentences. We will also be able to infer that any satisfiable theory which is at least as strong as this fragment of number theory (e.g., set theory) must be undecidable. In particular, such a theory cannot be both complete and axiomatizable.

In order to show that our subtheory of number theory is undecidable, we will show that it is strong enough to represent (in a sense to be made precise) facts about sequences of numbers, certain operations on numbers, and ultimately facts about decision procedures. This last feature then lets us perform a diagonal argument that demonstrates undecidability.

We could alternatively use, in place of a subtheory of number theory, some other theory (such as a fragment of the theory of finite sets) in which we could conveniently represent facts about decision procedures.

Before giving examples of the expressiveness of the language of number theory, it is convenient to introduce some notational conventions. As a concession to everyday usage, we will write

$$x < y, \quad x + y, \quad x \cdot y, \quad \text{and} \quad x \mathbf{E} y$$

in place of the official

$$<xy, \quad +xy, \quad \cdot xy, \quad \text{and} \quad \mathbf{E}xy.$$

For each natural number k we have a term $\mathbf{S}^k\mathbf{0}$ (the *numeral* for k) which denotes it

$$\mathbf{S}^0\mathbf{0} = \mathbf{0},\ \mathbf{S}^1\mathbf{0} = \mathbf{S0},\ \mathbf{S}^2\mathbf{0} = \mathbf{SS0},\ \text{etc.}$$

(The set of numerals is generated from $\{\mathbf{0}\}$ by the operation of prefixing $\mathbf{S}$.) The fact that every natural number can be named in the language will be a useful feature.

Even though only countably many relations on N are definable in $\mathfrak{N}$, almost all the familiar relations are definable. For example, the set of primes is defined in $\mathfrak{N}$ by

$$\boldsymbol{v}_1 \not\approx \mathbf{S}^1\mathbf{0} \wedge \forall \boldsymbol{v}_2 \forall \boldsymbol{v}_3 (\boldsymbol{v}_1 \approx \boldsymbol{v}_2 \cdot \boldsymbol{v}_3 \rightarrow \boldsymbol{v}_2 \approx \mathbf{S}^1\mathbf{0} \vee \boldsymbol{v}_3 \approx \mathbf{S}^1\mathbf{0}).$$

Later we will find it important to show that many other specific relations are definable in $\mathfrak{N}$.

One naturally expects the expressiveness of the language to be severely restricted when some of the parameters are omitted. For example, the set of primes, as we shall see, is not definable in $\mathfrak{N}_A$. On the other hand, in Section 3.7 we will show that any relation definable in $\mathfrak{N}$ is also definable in $\mathfrak{N}_M$.

Preview

The main theorems of this chapter are proved in Section 3.5. But we can already sketch here some of the ideas involved. We can assign to each formula α of the language of number theory an integer $\sharp\alpha$, called the Gödel number of α. Any straightforward way of assigning distinct integers to formulas would suffice for our purposes; a particular assignment is described at the beginning of Section 3.4. Similarly, to each finite sequence D of formulas (such as a deduction) we assign an integer $\mathscr{G}(D)$. We can now state the following result, which asserts that there is no definable (in $\mathfrak{N}$) complete axiomatization of Th $\mathfrak{N}$.

Theorem 30A Let $A \subseteq \text{Th}\,\mathfrak{N}$ be a set of sentences true in $\mathfrak{N}$, and assume that the set $\{\sharp\alpha : \alpha \in A\}$ of Gödel numbers of members of A is a set definable in $\mathfrak{N}$. Then we can find a sentence σ such that σ is true in $\mathfrak{N}$ but σ is not deducible from A.

Proof We will construct σ to express (in an indirect way) that σ itself is not a theorem of A. Then the argument will go roughly as follows: If $A \vdash \sigma$, then σ is false, contradicting the fact that A consists of true sentences. And so $A \nvdash \sigma$, whence σ is true.

To construct σ, we begin by considering the ternary relation R defined by

$$\langle a, b, c\rangle \in R \text{ iff } a \text{ is the Gödel number of some formula } \alpha \text{ and } c \text{ is the value of } \mathcal{G} \text{ at some deduction from } A \text{ of } \alpha(\mathbf{S}^b\mathbf{0}).$$

Then because $\{\#\alpha : \alpha \in A\}$ is definable in $\mathfrak{N}$, it follows that R is definable also. (The details of this step must wait until later sections.) Let ϱ be a formula which defines R in $\mathfrak{N}$. Let r be the Gödel number of

$$\forall \boldsymbol{v}_3 \neg\varrho(\boldsymbol{v}_1, \boldsymbol{v}_1, \boldsymbol{v}_3).$$

(We use here the notation of page 157.) Then let σ be

$$\forall \boldsymbol{v}_3 \neg\varrho(\mathbf{S}^r\mathbf{0}, \mathbf{S}^r\mathbf{0}, \boldsymbol{v}_3).$$

Thus σ says that no number is the value of $\mathcal{G}$ at a deduction from A of formula number r in which the numeral for r has replaced $\boldsymbol{v}_1$; i.e., no number is the value of $\mathcal{G}$ at a deduction of σ.

Suppose that, contrary to our expectations, there is a deduction of σ from A. Let k be the value of $\mathcal{G}$ at a deduction. Then $\langle r, r, k\rangle \in R$ and hence

$$\models_{\mathfrak{N}} \varrho(\mathbf{S}^r\mathbf{0}, \mathbf{S}^r\mathbf{0}, \mathbf{S}^k\mathbf{0}).$$

It is clear that

$$\sigma \vdash \neg\varrho(\mathbf{S}^r\mathbf{0}, \mathbf{S}^r\mathbf{0}, \mathbf{S}^k\mathbf{0})$$

and the two displayed lines tell us that σ is false in $\mathfrak{N}$. But $A \vdash \sigma$ and the members of A are true in $\mathfrak{N}$, so we have a contradiction.

Hence there is no deduction of σ from A. And so for every k, we have $\langle r, r, k\rangle \notin R$. Thus for every k

$$\models_{\mathfrak{N}} \neg\varrho(\mathbf{S}^r\mathbf{0}, \mathbf{S}^r\mathbf{0}, \mathbf{S}^k\mathbf{0}),$$

from which it follows (with the help of the substitution lemma) that

$$\models_{\mathfrak{N}} \forall \boldsymbol{v}_3 \neg\varrho(\mathbf{S}^r\mathbf{0}, \mathbf{S}^r\mathbf{0}, \boldsymbol{v}_3);$$

i.e., σ is true in $\mathfrak{N}$. ■

We will argue later that any decidable set of natural numbers must be definable in $\mathfrak{N}$. The conclusion will then be that Th $\mathfrak{N}$ is not axiomatizable.

Corollary 30B The set $\{\#\tau : \models_{\mathfrak{N}} \tau\}$ of Gödel numbers of sentences true in $\mathfrak{N}$ is a set which is not definable in $\mathfrak{N}$.

Proof If this were definable, we could take $A = \text{Th}\,\mathfrak{N}$ in the preceding theorem to obtain a contradiction. ■

§ 3.1 NATURAL NUMBERS WITH SUCCESSOR

We begin with a situation which is simple enough to let us give reasonably complete answers to our questions. We reduce the set of parameters to just $\forall$, $\mathbf{0}$, and $\mathbf{S}$, eliminating $<$, $+$, $\cdot$, and $\mathbf{E}$. The corresponding reduct of $\mathfrak{N}$ is

$$\mathfrak{N}_S = (N, 0, S).$$

In this restricted language we still have the numerals, naming each point in N. But the sentences we can express in the language are, from the viewpoint of arithmetic, uninteresting.

We want to ask about $\mathfrak{N}_S$ the same questions that interest us in the case of $\mathfrak{N}$. We want to know about the complexity of the set $\text{Th}\,\mathfrak{N}_S$; we want to study definability in $\mathfrak{N}_S$; and we want to survey the nonstandard models of $\mathfrak{N}_S$.

To study the theory of the natural numbers with successor ($\text{Th}\,\mathfrak{N}_S$), we begin by listing a few of its members, i.e., sentences true in $\mathfrak{N}_S$. (These sentences will ultimately provide an axiomatization for the theory.)

S1. $\forall x\, \mathbf{S}x \not\approx \mathbf{0}$, a sentence asserting that zero has no predecessor.

S2. $\forall x\, \forall y(\mathbf{S}x \approx \mathbf{S}y \rightarrow x \approx y)$. This asserts that the successor function is one-to-one.

S3. $\forall y(y \not\approx \mathbf{0} \rightarrow \exists x\, y \approx \mathbf{S}x)$. This asserts that any nonzero number is the successor of something.

S4.1 $\forall x\, \mathbf{S}x \not\approx x$.

S4.2 $\forall x\, \mathbf{SS}x \not\approx x$.

. . .

S4.n $\forall x\, \mathbf{S}^n x \not\approx x$, where the superscript n indicates that the symbol $\mathbf{S}$ occurs at n consecutive places.

Let A_S be the set consisting of the above sentences S1, S2, S3, S4.n ($n = 1, 2, \ldots$). Clearly these sentences are true in $\mathfrak{N}_S$; i.e., $\mathfrak{N}_S$ is a model of A_S. Hence

$$\text{Cn}\, A_S \subseteq \text{Th}\,\mathfrak{N}_S.$$

(Anything true in every model of A_S is true in this model.) What is not so obvious is that equality holds. We will prove this by considering arbitrary models of A_S.

What can be said of an arbitrary model

$$\mathfrak{A} = (|\mathfrak{A}|, \mathbf{0}^{\mathfrak{A}}, \mathbf{S}^{\mathfrak{A}})$$

of the axioms A_S? $\mathbf{S}^{\mathfrak{A}}$ must be a one-to-one map of $|\mathfrak{A}|$ onto $|\mathfrak{A}| - \{\mathbf{0}^{\mathfrak{A}}\}$, by S1, S2, and S3. And by S4.n, there can be no loops of size n. Thus $|\mathfrak{A}|$ must contain the "standard" points:

$$\mathbf{0}^{\mathfrak{A}} \to \mathbf{S}^{\mathfrak{A}}(\mathbf{0}^{\mathfrak{A}}) \to \mathbf{S}^{\mathfrak{A}}(\mathbf{S}^{\mathfrak{A}}(\mathbf{0}^{\mathfrak{A}})) \to \cdots,$$

which are all distinct. The arrow here indicates the action of $\mathbf{S}^{\mathfrak{A}}$. There may or may not be other points. If there is another point a in $|\mathfrak{A}|$, then there will be the successor of a, its successor, etc. Not only that, but since (by S3) each nonzero element has a predecessor (something of which it is the successor) which is (by S2) unique, $|\mathfrak{A}|$ must contain the predecessor of a, its predecessor, etc. These must all be distinct lest there be a finite loop. Thus a belongs to a "Z-chain":

$$\cdots * \to * \to a \to \mathbf{S}^{\mathfrak{A}}(a) \to \mathbf{S}^{\mathfrak{A}}(\mathbf{S}^{\mathfrak{A}}(a)) \to \cdots.$$

(We refer to these as Z-chains because they are arranged like the set $\mathbb{Z}$ of all integers $\{\ldots, -1, 0, 1, 2, \ldots\}$.) There can be any number of Z-chains. But any two Z-chains must be disjoint, as S2 prohibits merging. Similarly, any Z-chain must be disjoint from the standard part.

This can be restated in another way. Say that two points a and b in $|\mathfrak{A}|$ are *equivalent* if the function $\mathbf{S}^{\mathfrak{A}}$ can be applied a finite number of times to one point to yield the other point. This *is* an equivalence relation. (It is clearly reflexive and symmetric; the transitivity follows from the fact that $\mathbf{S}^{\mathfrak{A}}$ is one-to-one.) The standard part of $|\mathfrak{A}|$ is the equivalence class containing $\mathbf{0}^{\mathfrak{A}}$. For any other point (if any) a in $|\mathfrak{A}|$, the equivalence class of a is the set generated from $\{a\}$ by $\mathbf{S}^{\mathfrak{A}}$ and its inverse. This equivalence class is the Z-chain described above.

Conversely, any structure $\mathfrak{B}$ (for this language) which has a standard part

$$\mathbf{0}^{\mathfrak{B}} \to \mathbf{S}^{\mathfrak{B}}(\mathbf{0}^{\mathfrak{B}}) \to \mathbf{S}^{\mathfrak{B}}(\mathbf{S}^{\mathfrak{B}}(\mathbf{0}^{\mathfrak{B}})) \to \cdots$$

and a nonstandard part consisting of any number of separate Z-chains, is a model of A_S. (Check through the list of axioms in A_S, and note that each is true in $\mathfrak{B}$.) We thus have a complete characterization of what the models of A_S must look like.

If a model $\mathfrak{A}$ of A_S has only countably many Z-chains, then $|\mathfrak{A}|$ is countable. In general, if the set of Z-chains has cardinality λ, then altogether the number of points in $|\mathfrak{A}|$ is $\aleph_0 + \aleph_0 \cdot \lambda$. By facts of cardinal arithmetic (cf. Chapter 0) this number is the larger of $\aleph_0$ and λ. Hence

$$\text{card}\,|\mathfrak{A}| = \begin{cases} \aleph_0 & \text{if } \mathfrak{A} \text{ has countably many } Z\text{-chains,} \\ \lambda & \text{if } \mathfrak{A} \text{ has an uncountable number } \lambda \text{ of } Z\text{-chains.} \end{cases}$$

Lemma 31A If $\mathfrak{A}$ and $\mathfrak{A}'$ are models of A_S having the same number of Z-chains, then they are isomorphic.

Proof There is a unique isomorphism between the standard part of $\mathfrak{A}$ and the standard part of $\mathfrak{A}'$. By hypothesis we are given a one-to-one correspondence between the set of Z-chains of $\mathfrak{A}$ and the set of Z-chains of $\mathfrak{A}'$; thus each chain of $\mathfrak{A}$ is paired with a chain of $\mathfrak{A}'$. Clearly any two Z-chains are isomorphic. By combining all the pieces (which uses the axiom of choice) we have an isomorphism of $\mathfrak{A}$ onto $\mathfrak{A}'$. ■

Thus a model of $\mathfrak{A}_S$ is determined to within isomorphism by its number of Z-chains. For $\mathfrak{N}_S$ this number is zero, but any number is possible.

The reader should note that there is no sentence of the language which says, "There are no Z-chains." In fact, there is no set Σ of sentences such that a model $\mathfrak{A}$ of A_S satisfies Σ iff $\mathfrak{A}$ has no Z-chains. For by the LST theorem there is an uncountable structure $\mathfrak{A}$ with $\mathfrak{A} \equiv \mathfrak{N}_S$. But $\mathfrak{A}$ has uncountably many Z-chains and $\mathfrak{N}_S$ has none.

Theorem 31B Let $\mathfrak{A}$ and $\mathfrak{B}$ be uncountable models of A_S of the same cardinality. Then $\mathfrak{A}$ is isomorphic to $\mathfrak{B}$.

Proof By the above discussion, $\mathfrak{A}$ has card $\mathfrak{A}$ Z-chains, and $\mathfrak{B}$ has card $\mathfrak{B}$ Z-chains. Since card $\mathfrak{A}$ = card $\mathfrak{B}$, they have the same number of Z-chains and hence are isomorphic. ■

Theorem 31C Cn A_S is a complete theory.

Proof Apply the Łoś–Vaught theorem of Section 2.6. The preceding theorem asserts that the theory Cn A_S is categorical in any uncountable power. Furthermore, A_S has no finite models. Hence the Łoś–Vaught theorem applies. ■

Corollary 31D Cn A_S = Th $\mathfrak{N}_S$.

Proof We have Cn $A_S \subseteq$ Th $\mathfrak{N}_S$; the first theory is complete and the second is satisfiable. ■

★Corollary 31E Th $\mathfrak{N}_S$ is decidable.

Proof Any complete and axiomatizable theory is decidable (by Corollary 25G). A_S is a decidable set of axioms for this theory. ■

Elimination of quantifiers

Once one knows a theory to be decidable, it is tempting to try to find a realistically practical decision procedure. We will give such a procedure for Th $\mathfrak{N}_S$, based on "elimination of quantifiers."

Definition A theory T admits elimination of quantifiers iff for every formula φ there is a quantifier-free formula ψ such that

$$T \models (\varphi \leftrightarrow \psi).$$

Actually it is enough to consider only formulas φ of a rather special form:

Theorem 31F Assume that for every formula φ of the form

$$\exists x(\alpha_0 \wedge \cdots \wedge \alpha_n),$$

where each α_i is an atomic formula or the negation of an atomic formula, there is a quantifier-free formula ψ such that $T \models (\varphi \leftrightarrow \psi)$. Then T admits elimination of quantifiers.

Proof First we claim that we can find a quantifier-free equivalent for any formula of the form $\exists x\, \theta$ for quantifier-free θ. We begin by putting θ into disjunctive normal form (Corollary 15C). The resulting formula,

$$\exists x[(\alpha_0 \wedge \cdots \wedge \alpha_m) \vee (\beta_0 \wedge \cdots \wedge \beta_n) \vee \cdots \vee (\xi_0 \wedge \cdots \wedge \xi_t)],$$

is logically equivalent to

$$\exists x(\alpha_0 \wedge \cdots \wedge \alpha_m) \vee \exists x(\beta_0 \wedge \cdots \wedge \beta_n) \vee \cdots \vee \exists x(\xi_0 \wedge \cdots \wedge \xi_t).$$

By assumption, each disjunct of this formula can be replaced by a quantifier-free formula.

We leave it to the reader to show (in Exercise 2) that by using the above paragraph one can obtain a quantifier-free equivalent for an arbitrary formula. ■

In the special case where the theory in question is the theory Th $\mathfrak{A}$ of a structure $\mathfrak{A}$, the definition can be restated: Th $\mathfrak{A}$ admits elimination of quantifiers iff for every formula φ there is a quantifier-free formula ψ such that φ and ψ are "equivalent in $\mathfrak{A}$"; i.e.,

$$\models_{\mathfrak{A}} (\varphi \leftrightarrow \psi)\, [s]$$

for any map s of the variables into $|\mathfrak{A}|$.

Theorem 31G Th $\mathfrak{N}_S$ admits elimination of quantifiers.

Proof By the preceding theorem, it suffices to consider a formula

$$\exists x(\alpha_0 \wedge \cdots \wedge \alpha_q),$$

where each α_i is atomic or is the negation of an atomic formula. We will describe how to replace this formula by another which is quantifier-free. The equivalence of the new formula to the given one will, in fact, be a consequence of A_S; see Exercise 3.

In the language of $\mathfrak{N}_S$ the only terms are of the form $\mathbf{S}^k u$, where u is $\mathbf{0}$ or a variable. The only atomic formulas are equations. We may suppose that the variable x occurs in each α_i. For if x does not occur in α, then

$$\exists x(\alpha \wedge \beta) \models\!\dashv \alpha \wedge \exists x\, \beta.$$

Thus each α_i has the form

$$\mathbf{S}^m x \approx \mathbf{S}^n u$$

or the negation of this equation, where u is $\mathbf{0}$ or a variable. We may further suppose u is different from x, since $\mathbf{S}^m x \approx \mathbf{S}^n x$ could be replaced by $\mathbf{0} \approx \mathbf{0}$ if $m = n$, and by $\mathbf{0} \not\approx \mathbf{0}$ if $m \neq n$.

Case 1: Each α_i is the negation of an equation. Then the formula may be replaced by $\mathbf{0} \approx \mathbf{0}$.

Case 2: There is at least one α_i not negated; say α_0 is

$$\mathbf{S}^m x \approx t,$$

where the term t does not contain x. Since the solution for x must be non-negative, we replace α_0 by

$$t \not\approx \mathbf{0} \wedge \cdots \wedge t \not\approx \mathbf{S}^{m-1}\mathbf{0}$$

(or by $\mathbf{0} \approx \mathbf{0}$ if $m = 0$). Then in each other α_j we replace, say,

$$\mathbf{S}^k x \approx u$$

first by

$$\mathbf{S}^{k+m}x \approx \mathbf{S}^m u,$$

which in turn becomes

$$\mathbf{S}^k t \approx \mathbf{S}^m u.$$

We now have a formula in which x no longer occurs, so the quantifier may be omitted. ■

There are several interesting by-products of the quantifier elimination procedure. For one, we get an alternative proof of the completeness of Cn A_S. For suppose we begin with a sentence σ. The quantifier elimination procedure gives a quantifier-free *sentence* τ such that (by Exercise 3) $A_S \models (\sigma \leftrightarrow \tau)$. Now we claim that either $A_S \models \tau$ or $A_S \models \neg\tau$. For τ is built up from atomic sentences by means of $\neg$ and $\rightarrow$. An atomic sentence must be of the form $\mathbf{S}^k\mathbf{0} \approx \mathbf{S}^l\mathbf{0}$ and is deducible from A_S if $k = l$, but is refutable (i.e., its negation is deducible) from A_S if $k \neq l$. (In fact, just {S1, S2} suffices for this.) Since every atomic sentence can be deduced or refuted, so can every quantifier-free sentence. This establishes the claim. And so either $A_S \models \sigma$ or $A_S \models \neg\sigma$.

Another by-product concerns the problem of definability in $\mathfrak{N}_S$; see Exercises 4 and 5. For any formula φ in which just $\mathbf{v}_1$ and $\mathbf{v}_2$ occur free we now can find a quantifier-free ψ (with the same variables free) such that

$$\text{Th } \mathfrak{N}_S \models \forall \mathbf{v}_1 \forall \mathbf{v}_2(\varphi \leftrightarrow \psi);$$

i.e.,

$$\models_{\mathfrak{N}_S} \forall \mathbf{v}_1 \forall \mathbf{v}_2(\varphi \leftrightarrow \psi).$$

Thus the relation φ defined is also definable by a quantifier-free formula.

EXERCISES

1. Let A_S^* be the set of sentences consisting of S1, S2, and all sentences of the form

$$\varphi(\mathbf{0}) \rightarrow \forall \mathbf{v}_1(\varphi(\mathbf{v}_1) \rightarrow \varphi(\mathbf{S}\mathbf{v}_1)) \rightarrow \forall \mathbf{v}_1\, \varphi(\mathbf{v}_1),$$

where φ is a wff (in the language of $\mathfrak{N}_S$) in which no variable except $\mathbf{v}_1$ occurs free. Show that $A_S \subseteq \text{Cn } A_S^*$. Conclude that Cn $A_S^* = \text{Th } \mathfrak{N}_S$. (Here $\varphi(t)$ is by definition $\varphi_t^{\mathbf{v}_1}$. The sentence displayed above is called the *induction axiom* for φ.)

2. Complete the proof of Theorem 31F.

3. The proof of quantifier elimination for Th $\mathfrak{N}_S$ showed how, given a formula φ, to find a quantifier-free ψ. Show that

$$A_S \models (\varphi \leftrightarrow \psi)$$

without using the completeness of Cn A_S. (This yields an alternative proof of the completeness of Cn A_S, not involving Z-chains or the Łoś-Vaught theorem.)

4. Show that a subset of N is definable in $\mathfrak{N}_S$ iff either it is finite or its complement (in N) is finite.

5. Show that the ordering relation $\{\langle m, n\rangle : m < n \text{ in } N\}$ is not definable in $\mathfrak{N}_S$.

6. Show that Th $\mathfrak{N}_S$ is not finitely axiomatizable. *Suggestion*: Show that no finite subset of A_S suffices, and then apply Section 2.6.

§ 3.2 OTHER REDUCTS OF NUMBER THEORY[1]

First let us add the ordering symbol $<$ to the language. The intended structure is

$$\mathfrak{N}_L = (N, 0, S, <).$$

We want to show that the theory of this structure is (like Th $\mathfrak{N}_S$) decidable and also admits elimination of quantifiers. But unlike Th $\mathfrak{N}_S$, it is finitely axiomatizable and is not categorical in any infinite power.

As axioms of Th $\mathfrak{N}_S$ we will take the finite set A_L consisting of the six sentences listed below. Here $x \leq y$ is, of course, an abbreviation for $(x < y \vee x \approx y)$, and $x \not\leq y$ abbreviates the negation of this formula.

S3.	$\forall y$	$(y \not\approx \mathbf{0} \rightarrow \exists x\, y \approx \mathbf{S}x)$
L1.	$\forall x \forall y$	$(x < Sy \leftrightarrow x \leq y)$
L2.	$\forall x$	$x \not< \mathbf{0}$
L3.	$\forall x \forall y$	$(x < y \vee x \approx y \vee y < x)$
L4.	$\forall x \forall y$	$(x < y \rightarrow y \not< x)$
L5.	$\forall x \forall y \forall z$	$(x < y \rightarrow y < z \rightarrow x < z)$

[1] This section may be omitted without disastrous effects.

We begin by listing some consequence of these axioms.

(1) $A_L \vdash \forall x\, x < \mathbf{S}x$.

Proof In L1 take y to be x. ■

(2) $A_L \vdash \forall x\, x \nless x$.

Proof In L4 take y to x. ■

(3) $A_L \vdash \forall x\, \forall y(x \nless y \leftrightarrow y \leq x)$ (trichotomy).

Proof For "$\rightarrow$" use L3. For "$\leftarrow$" use L4 and (2). ■

(4) $A_L \vdash \forall x\, \forall y(x < y \leftrightarrow \mathbf{S}x < \mathbf{S}y)$.

Proof From A_L we can deduce the biconditionals:

$$\begin{aligned} x < y &\leftrightarrow y \nleq x && \text{by (3);} \\ &\leftrightarrow y \nless \mathbf{S}x && \text{by L1;} \\ &\leftrightarrow \mathbf{S}x \leq y && \text{by (3);} \\ &\leftrightarrow \mathbf{S}x < \mathbf{S}y && \text{by L1.} \end{aligned}$$ ■

(5) $A_L \vdash$ S1 and $A_L \vdash$ S2.

Proof S1 follows from L2 and (1). S2 comes from (4) by use of L3 and (2). ■

(6) $A_L \vdash$ S4.n for $n = 1, 2, \ldots$.

Proof This follows from (1) and (2), using L5. ■

Thus any model $\mathfrak{A}$ of A_L is (when we ignore $<^{\mathfrak{A}}$) also a model of A_S. So it must consist of a standard part plus zero or more Z-chains. In addition, it is ordered by $<^{\mathfrak{A}}$.

Theorem 32A The theory Cn A_L admits elimination of quantifiers.

Proof Again we consider a formula

$$\exists x(\beta_0 \wedge \cdots \wedge \beta_p),$$

where each β_i is atomic or the negation of an atomic formula. The terms are, as in Section 3.1, of the form $\mathbf{S}^k u$, where u is $\mathbf{0}$ or a variable. There are two possibilities for atomic formulas,

$$\mathbf{S}^k u \approx \mathbf{S}^l t \quad \text{and} \quad \mathbf{S}^k u < \mathbf{S}^l t.$$

1. We can eliminate the negation symbol. Replace $t_1 \not< t_2$ by $t_2 < t_1 \vee t_1 \approx t_2$ and replace $t_1 \not\approx t_2$ by $t_1 < t_2 \vee t_2 < t_1$. (This is justified by L3 and L4.) By regrouping the atomic formulas and noting that

$$\exists x(\varphi \vee \psi) \models\dashv \exists x\, \varphi \vee \exists x\, \psi,$$

we may again reach formulas of the form

$$\exists x(\alpha_0 \wedge \cdots \wedge \alpha_q),$$

where now each α_i is atomic.

2. We may suppose that the variable x occurs in each α_i. This is because if x does not occur in α, then

$$\exists x(\alpha \wedge \beta) \models\dashv \alpha \wedge \exists x\, \beta.$$

Furthermore, we may suppose that x occurs on only one side of the equality or inequality α_i. For $\mathbf{S}^k x \approx \mathbf{S}^l x$ can be dealt with as in Section 3.1. $\mathbf{S}^k x < \mathbf{S}^l x$ can be replaced by $\mathbf{0} \approx \mathbf{0}$ if $k < l$, and $\mathbf{0} \not\approx \mathbf{0}$ otherwise. (This is justified by (4) and L2, together with the observation that $A_L \vdash \mathbf{0} \leq x$.)

Case 1: Suppose that some α_i is an equality. Then we can proceed as in case 2 of the quantifier-elimination proof of Theorem 31G.

Case 2: Otherwise each α_i is an inequality. Then the formula can be rewritten

$$\exists x\Big(\bigwedge_i t_i < \mathbf{S}^{m_i} x \wedge \bigwedge_j \mathbf{S}^{n_j} x < u_j\Big).$$

(Here $\bigwedge_i$ indicates the conjunction of formulas indexed by i, so $\gamma_0 \wedge \gamma_1 \wedge \cdots \wedge \gamma_k$ can be abbreviated $\bigwedge_i \gamma_i$.) In the first conjunction, $\bigwedge_i t_i < \mathbf{S}^{m_i} x$, we have the lower bounds on x, in the second conjunction, $\bigwedge_j \mathbf{S}^{n_j} x < u_j$, we have the upper bounds. If the second conjunction is empty (i.e., if there are no upper bounds on x), then we can replace the formula by $\mathbf{0} \approx \mathbf{0}$. (Why?) If the first conjunction is empty (i.e., if there are no lower bounds on x), then we can replace the formula by

$$\bigwedge_j \mathbf{S}^{n_j}\mathbf{0} < u_j,$$

which asserts that zero satisfies the upper bounds. Otherwise, we rewrite the formula successively as

(1) $$\exists x \bigwedge_{i,j} (t_i < \mathbf{S}^{m_i}x \wedge \mathbf{S}^{n_j}x < u_j).$$

(2) $$\exists x \bigwedge_{i,j} (\mathbf{S}^{n_j}t_i < \mathbf{S}^{m_i+n_j}x < \mathbf{S}^{m_i}u_j).$$

(3) $$\left(\bigwedge_{i,j} \mathbf{S}^{n_j+1}t_i < \mathbf{S}^{m_i}u_j\right) \wedge \bigwedge_{j} \mathbf{S}^{n_j}\mathbf{0} < u_j.$$

This last formula says "any lower bound plus one satisfies any upper bound, and furthermore zero satisfies any upper bound." This implies that there is a gap between the greatest lower bound and the least upper bound, whence there is a solution for x. The second part guarantees that the solution for x is not forced to be negative.

In each case, we have arrived at a quantifier-free version of the given formula. ■

Corollary 32B (a) Cn A_L is complete.
(b) Cn A_L = Th $\mathfrak{N}_L$.
★(c) Th $\mathfrak{N}_L$ is decidable.

Proof (a) The argument which followed the proof of Theorem 31G is applicable here also. (b) This follows from (a), since Cn $A_L \subseteq$ Th $\mathfrak{N}_L$ and Th $\mathfrak{N}_L$ is satisfiable. For (c), we can use the fact that any complete axiomatizable theory is decidable. But the quantifier elimination proof yields a more efficient decision procedure. ■

Corollary 32C A subset of N is definable in $\mathfrak{N}_L$ iff it is either finite or has finite complement.

Proof Compare Exercise 4 of the preceding section. ■

On the other hand, $\mathfrak{N}_L$ has more definable binary relations than has $\mathfrak{N}_S$. For the ordering relation $\{\langle m, n\rangle : m < n\}$ is not definable in $\mathfrak{N}_S$, by Exercise 5 of the preceding section.

Corollary 32D The addition relation,

$$\{\langle m, n, p\rangle : m + n = p\},$$

is not definable in $\mathfrak{N}_L$.

Proof If we could define addition, we could then define the set of even natural numbers. But this set is neither finite nor has finite complement. ■

Now suppose we augment the language by the addition symbol $+$. The intended structure is

$$\mathfrak{N}_A = (N, 0, S, <, +).$$

The theory of this structure is also decidable, as we will prove shortly. But to keep matters from getting even more complicated, we will avoid listing any convenient set of axioms for the theory.

The nonstandard models of Th $\mathfrak{N}_A$ must also be models of Th $\mathfrak{N}_L$. So they have a standard part, followed by some Z-chains. But ordering among the Z-chains can no longer be arbitrary. Let $\mathfrak{A}$ be a nonstandard model of Th $\mathfrak{N}_A$. The ordering $<^{\mathfrak{A}}$ induces a well-defined ordering on the set of Z-chains. (See Exercise 3.) We claim that there is no largest Z-chain, there is no smallest Z-chain, and there is between any two Z-chains another one. The reasons, in outline, can be stated simply: If a belongs to some Z-chain (i.e., is an infinite element of $\mathfrak{A}$), then $a +^{\mathfrak{A}} a$ is in a larger Z-chain. There must be some b such that $b +^{\mathfrak{A}} b$ is either a or its successor; b must be in a smaller Z-chain. If a_1 and a_2 belong to different Z-chains, then there must be some b such that $b +^{\mathfrak{A}} b$ is either $a_1 +^{\mathfrak{A}} a_2$ or its successor. And b will lie in a Z-chain between that of a_1 and that of a_2. (These statements should seem quite plausible. The reader who enjoys working with infinite numbers might supply some details.)

***Theorem 32E (Presburger, 1929)** The theory of the structure $\mathfrak{N}_A = (N, 0, S, <, +)$ is decidable.

The proof is again based on a quantifier elimination procedure. The theory of $\mathfrak{N}_A$ itself does *not* admit elimination of quantifiers. For example, the formula defining the set of even numbers

$$\exists y\, \mathbf{v}_1 \approx y + y$$

is not equivalent to any quantifier-free formula. We can overcome this by adding a new symbol $\approxeq_2$ for congruence modulo 2. Similarly, we add symbols $\approxeq_3, \approxeq_4, \ldots$. The intended structure for this expanded language is

$$\mathfrak{N}^+ = (N, 0, S, <, +, \equiv_2, \equiv_3, \ldots),$$

where $\equiv_k$ is the binary relation of congruence modulo k. It turns out that the theory of this structure does admit elimination of quantifiers.

This by itself does not imply that the theory of either structure is decidable. After all, we can start with *any* structure, and expand it to a structure having additional relations until a structure is obtained that admits elimination of quantifiers. To obtain decidability, we must show that we can, given a sentence σ, (1) effectively find a quantifier-free equivalent σ', and then (2) effectively decide if σ' is true.

We will now give the quantifier elimination procedure for $\mathfrak{N}^+$. For a term t and a natural number n, let nt be the term $t + t + \cdots + t$, with n summands. $0t$ is $\mathbf{0}$. Then any term can be expanded to one of the form

$$\mathbf{S}^{n_0}\mathbf{0} + n_1 x_1 + \cdots + n_k x_k$$

for $k \geq 0$, $n_i \geq 0$. For example,

$$\mathbf{S}(x + \mathbf{S0})$$

becomes

$$\mathbf{S}^2\mathbf{0} + x.$$

As usual we begin with a formula $\exists y(\beta_1 \wedge \cdots \wedge \beta_n)$, where β_i is an atomic formula or the negation of one.

1. Eliminate negation. Replace $\neg(t_1 \approx t_2)$ by $(t_1 < t_2 \vee t_2 < t_1)$. Replace $\neg(t_1 < t_2)$ by $(t_1 \approx t_2 \vee t_2 < t_1)$. And replace $\neg(t_1 \approxeq_m t_2)$ by

$$t_1 \approxeq_m t_2 + \mathbf{S}^1\mathbf{0} \vee \cdots \vee t_1 \approxeq_m t_2 + \mathbf{S}^{m-1}\mathbf{0}.$$

Then regroup into a disjunction of formulas of the form

$$\exists y(\alpha_1 \wedge \cdots \wedge \alpha_m),$$

where each α_i is atomic. We may further suppose that y occurs in each α_i, and in fact that α_i has one of the four forms

$$\begin{aligned} ny + t &\approx u, \\ ny + t &\approxeq_m u, \\ ny + t &< u, \\ u &< ny + t, \end{aligned}$$

where u and t are terms not containing y. In what follows we will take the liberty of writing these formulas with a subtraction symbol:

$$\begin{aligned} ny &\approx u - t, \\ ny &\approxeq_m u - t, \\ ny &< u - t, \\ u - t &< ny. \end{aligned}$$

These are merely abbreviations for the formulas without subtraction obtained by transposing terms.

For example, we might have at this point the formula

$$\exists y(w < 4y \wedge 2y < u \wedge 3y < v \wedge y \approxeq_3 t),$$

where t, u, v, and w are terms not containing y.

2. Uniformize the coefficients of y. Let p be the least common multiple of the coefficients of y. Each atomic formula can be converted to one in which the coefficient of y is p, by "multiplying through" by the appropriate factor. This is obviously legitimate for equalities and inequalities. In the case of congruences one must remember to raise the modulus also:

$$a \equiv_m b \qquad \text{iff } ka \equiv_{km} kb.$$

In the example above p is 12, and we obtain

$$\exists y(3w < 12y \wedge 12y < 6u \wedge 12y < 4v \wedge 12y \approxeq_{36} 12t).$$

3. Eliminate the coefficient of y. Replace py by x and add the new conjunct $x \approxeq_p \mathbf{0}$. (In place of $\exists y \cdots 12y \cdots$ we can equally well have, "There exists a multiple x of 12 such that $\ldots x \ldots$.") Our example is now converted to

$$\exists x(3w < x \wedge x < 6u \wedge x < 4v \wedge x \approxeq_{36} 12t \wedge x \approxeq_{12} \mathbf{0}).$$

4. Special case. If one of the atomic formulas is an equality, $x + t \approx u$, then we can replace

$$\exists x\, \theta$$

by

$$\theta^x_{u-t} \wedge t \leq u.$$

Here replacement of x by "$u - t$" is the natural thing; we transpose terms to compensate for the absence of subtraction. For example,

$$(x \approxeq_m v)^x_{u-t} \quad \text{is} \quad u \approxeq_m v + t.$$

5. We may assume henceforth that $\approx$ does not occur. So we have a formula of the form

$$\begin{aligned}\exists x[r_0 - s_0 < x \wedge \cdots \wedge r_{l-1} - s_{l-1} < x \\ \wedge\, x < t_0 - u_0 \wedge \cdots \wedge x < t_{k-1} - u_{k-1} \\ \wedge\, x \approxeq_{m_0} v_0 - w_0 \wedge \cdots \wedge x \approxeq_{m_{n-1}} v_{n-1} - w_{n-1}],\end{aligned}$$

where r_i, s_i, t_i, u_i, v_i, and w_i are terms not containing x. This can be abbreviated

$$\exists x\left[\bigwedge_{j<l} r_j - s_j < x \wedge \bigwedge_{i<k} x < t_i - u_i \wedge \bigwedge_{i<n} x \approx_{m_i} v_i - w_i\right].$$

If there are no congruences (i.e., $n = 0$), then the formula asserts that there is a nonnegative space between the lower and upper bounds. We can replace the formula by the quantifier-free formula:

$$\bigwedge_{i<k}\bigwedge_{j<l} (r_j - s_j) + \mathbf{S0} < t_i - u_i \wedge \bigwedge_{i<k} \mathbf{0} < t_i - u_i.$$

Let M be the least common multiple of the moduli $m_0, \ldots, m_{n-1}$. Then $a + M \equiv_{m_i} a$. So as a increases, the pattern of residues of a modulo $m_0, \ldots, m_{n-1}$ has period M. Thus, in searching for a solution to the congruences, we need only search M consecutive integers.

We now have a formula which asserts the existence of a natural number which is not less than certain lower bounds $L_1, \ldots, L_l$ and which satisfies certain congruences and certain upper bounds. If there is such a solution, then one of the following is a solution:

$$\begin{aligned}
&L_1, L_1 + 1, \ldots, L_1 + M - 1,\\
&L_2, L_2 + 1, \ldots, L_2 + M - 1,\\
&\ldots\\
&L_l, L_l + 1, \ldots, L_l + M - 1,\\
&0, 1, \ldots, M - 1.
\end{aligned}$$

(The last line is needed to cover the case in which every L_j is negative. To avoid having to treat this line as special case, we will add a new lower bound of 0. That is, let $r_l = \mathbf{0}$ and $s_l = \mathbf{S0}$ so that

$$r_l - s_l < x$$

is a formula $\mathbf{0} < x + \mathbf{S0}$ asserting that x is nonnegative. We now have $l + 1$ lower bounds.)

Our formula (asserting the existence of a solution for x) can now be replaced by a quantifier-free disjunction which asserts that one of the numbers in the above matrix is a nonnegative solution:

$$\bigvee_{j\le l}\bigvee_{q<M}\left[\bigwedge_{i\le l} r_i - s_i < (r_j - s_j) + \mathbf{S}^q\mathbf{0} \wedge \bigwedge_{i<k} (r_j - s_j) + \mathbf{S}^q\mathbf{0} < t_i - u_i \right.$$
$$\left. \wedge \bigwedge_{i<n} (r_j - s_j) + \mathbf{S}^q\mathbf{0} \approx_{m_i} v_i - w_i\right].$$

In our continuing example we have, after adding the new lower bound on x,

$$\exists x(3w < x \wedge \mathbf{0} < x + \mathbf{S0} \wedge x < 6u \wedge x < 4v \wedge x \approx_{36} 12t \wedge x \approx_{12} \mathbf{0}).$$

The quantifier-free equivalent is a disjunction of seventy-two conjunctions. Each conjunction has six constituents.

This proves half of the theorem. If we are given a sentence σ, the above procedure tells us how to find effectively a quantifier-free sentence τ (in the language of $\mathfrak{N}^+$) which is true (in the intended structure) iff σ is. Now we must decide if τ is true.

But this is easy. It is enough to look at atomic sentences. Any variable-free term can be put in the form $\mathbf{S}^n\mathbf{0}$. Then, for example,

$$\mathbf{S}^n\mathbf{0} \approx_m \mathbf{S}^p\mathbf{0}$$

is true iff $n \equiv_m p$. ∎

A set D of natural numbers is *periodic* if for some positive p, a number n is in D iff $n + p$ is in D. D is *eventually periodic* iff there exist positive numbers M and p such that for all n greater than M, $n \in D$ iff $n + p \in D$.

Theorem 32F A set of natural numbers is definable in $(N, 0, S, <, +)$ iff it is eventually periodic.

Proof Exercise 1 asserts that every eventually periodic set is definable. Conversely, suppose D is definable. Then D is definable in $\mathfrak{N}^+$ by a quantifier-free formula (whose only variable is $\mathbf{v}_1$). Since the class of eventually periodic sets is closed under union, intersection, and complementation, it suffices to show that every atomic formula in the language of $\mathfrak{N}^+$ whose only variable is $\mathbf{v}_1$ defines an eventually periodic set. There are only four possibilities:

$$n\mathbf{v}_1 + t \approx u,$$

$$n\mathbf{v}_1 + t < u,$$

$$u < n\mathbf{v}_1 + t,$$

$$n\mathbf{v}_1 + t \approx_m u,$$

where u and t are numerals. The first two formulas define finite sets (which eventually have period 1), the third defines a set with finite complement, and the last formula defines a periodic set with period m. ∎

Corollary 32G The multiplication relation

$$\{\langle m, n, p\rangle : p = m \cdot n \text{ in } \mathbb{N}\}$$

is not definable in $(\mathbb{N}, 0, S, <, +)$.

Proof If we had a definition of multiplication, we could then use that to define the set of squares. But the set of squares is not eventually periodic. ■

EXERCISES

1. Show that any eventually periodic set of natural numbers is definable in the structure $\mathfrak{N}_A$.

2. Show that in the structure $(\mathbb{N}, +)$ the following relations are definable:

(a) Ordering, $\{\langle m, n\rangle : m < n\}$.
(b) Zero, $\{0\}$.
(c) Successor, $\{\langle m, n\rangle : n = S(m)\}$.

3. Let $\mathfrak{A}$ be a model of Th $\mathfrak{N}_L$ (or equivalently a model of A_L). For a and b in $|\mathfrak{A}|$ define the equivalence relation:

$a \sim b \Leftrightarrow \mathbf{S}^{\mathfrak{A}}$ can be applied a finite number of times to one of a, b to reach the other.

Let $[a]$ be the equivalence class to which a belongs. Order equivalence classes by

$$[a] < [b] \qquad \text{iff } a <^{\mathfrak{A}} b \quad \text{and} \quad a \nsim b.$$

Show that this is a well-defined ordering on the set of equivalence classes.

4. Show that the theory of the real numbers with its usual ordering, Th$(\mathbb{R}, <)$, admits elimination of quantifiers. (Assume that the language includes equality.)

§ 3.3 A SUBTHEORY OF NUMBER THEORY

We now return to the full language of number theory, as described in Section 3.0. The parameters of the language are $\forall, \mathbf{0}, \mathbf{S}, <, +, \cdot$, and $\mathbf{E}$. The intended structure for this language is

$$\mathfrak{N} = (\mathbb{N}, 0, S, <, +, \cdot, E).$$

Actually in $(N, \cdot, E)$ we can define $\{0\}$, S, $<$, and $+$. (See Exercise 1.) As we will show in Section 3.7, in $(N, +, \cdot)$ we can define E as well as 0, S, and $<$. So there are ways in which we could economize. The luxury of having all these parameters (particularly $\mathbf{E}$) will simplify some of the proofs.

As we shall see, Th $\mathfrak{N}$ is a very strong theory and is neither decidable nor axiomatizable. In order to prove this fact (and a number of related results) it will be strategically wise to select for study a finitely axiomatizable subtheory of Th $\mathfrak{N}$. As hinted at in Section 3.0, this subtheory should be strong enough to represent (in a sense to be made precise) facts about decidable sets. The subtheory we have selected is Cn A_E, where A_E is the set consisting of the eleven sentences listed below. (As in the preceding section, $x \leq y$ abbreviates $x < y \vee x \approx y$.)

SET A_E OF AXIOMS

S1.	$\forall x \quad \mathbf{S}x \not\approx \mathbf{0}$
S2.	$\forall x\, \forall y\, (\mathbf{S}x \approx \mathbf{S}y \rightarrow x \approx y)$
L1.	$\forall x\, \forall y\, (x < \mathbf{S}y \leftrightarrow x \leq y)$
L2.	$\forall x \quad x \not< 0$
L3.	$\forall x\, \forall y\, (x < y \vee x \approx y \vee y < x)$
A1.	$\forall x \quad x + \mathbf{0} \approx x$
A2.	$\forall x\, \forall y\; x + \mathbf{S}y \approx \mathbf{S}(x + y)$
M1.	$\forall x \quad x \cdot \mathbf{0} \approx \mathbf{0}$
M2.	$\forall x\, \forall y\; x \cdot \mathbf{S}y \approx x \cdot y + x$
E1.	$\forall x \quad x\, \mathbf{E}\, \mathbf{0} \approx \mathbf{S0}$
E2.	$\forall x\, \forall y\; x\, \mathbf{E}\, \mathbf{S}y \approx x\, \mathbf{E}\, y \cdot x$

Since $\mathfrak{N}$ is a model of A_E, we have Cn $A_E \subseteq$ Th $\mathfrak{N}$. But (as we will prove) equality does *not* hold here. In fact, it can be shown that $A_E \nvdash$ S3.

We first show that certain simple sentences in Th $\mathfrak{N}$ are deducible from A_E.

Lemma 33A (a) $A_E \vdash \forall x\, x \not< \mathbf{0}$.

(b) For any natural number k,

$$A_E \vdash \forall x(x < \mathbf{S}^{k+1}\mathbf{0} \leftrightarrow x \approx \mathbf{S}^0\mathbf{0} \vee \cdots \vee x \approx \mathbf{S}^k\mathbf{0}).$$

Notice that (a) can be thought of as the $k = -1$ case of (b), where the empty disjunction is $\perp$. The lemma tells us that the standard points are ordered in the natural way, and (by L3) the infinite points are all larger than any standard point.

Proof Part (a) is L2. For (b) we use induction (in English) on k. We have as a consequence of L1,

$$x < \mathbf{S0} \leftrightarrow x < \mathbf{0} \vee x \approx \mathbf{0},$$

which together with L2 gives

$$x < \mathbf{S0} \leftrightarrow x \approx \mathbf{0},$$

which is the $k = 0$ case of (b). For the inductive step we again apply L1:

$$x < \mathbf{S}^{k+1}\mathbf{0} \leftrightarrow x < \mathbf{S}^{k}\mathbf{0} \vee x \approx \mathbf{S}^{k}\mathbf{0}.$$

By the inductive hypothesis, $x < \mathbf{S}^{k}\mathbf{0}$ can be replaced by

$$x \approx \mathbf{S}^{0}\mathbf{0} \vee \cdots \vee x \approx \mathbf{S}^{k-1}\mathbf{0},$$

whereby we obtain (b). ■

Lemma 33B For any variable-free term t, there is a unique natural number n such that

$$A_E \vdash t \approx \mathbf{S}^{n}\mathbf{0}.$$

Proof The uniqueness is immediate. (Why?) For the existence half, we use induction on t. It t is $\mathbf{0}$, we take $n = 0$. If t is $\mathbf{S}u$, then by the inductive hypothesis $A_E \vdash u \approx \mathbf{S}^{m}\mathbf{0}$ for some m. Hence $A_E \vdash t \approx \mathbf{S}^{m+1}\mathbf{0}$.

Now suppose t is $u_1 + u_2$. By the inductive hypothesis $A_E \vdash t \approx \mathbf{S}^{m}\mathbf{0} + \mathbf{S}^{n}\mathbf{0}$ for some m and n. We now apply A2 n times and A1 once to obtain $A_E \vdash t \approx \mathbf{S}^{m+n}\mathbf{0}$. The arguments for multiplication and exponentiation are similar. ■

As a special case of this lemma we have "$2 + 2 = 4$" (i.e., $\mathbf{S}^{2}\mathbf{0} + \mathbf{S}^{2}\mathbf{0} \approx \mathbf{S}^{4}\mathbf{0}$) as a consequence of A_E.

Theorem 33C For any quantifier-free sentence τ true in $\mathfrak{N}$, $A_E \vdash \tau$.

Proof Exercise 2.

A simplified notation (used earlier in Section 2.7) for substitution will be helpful in the coming pages:

$$\varphi(t) = \varphi_t^{v_1},$$
$$\varphi(t_1, t_2) = (\varphi_{t_1}^{v_1})_{t_2}^{v_2},$$

and so forth. Thus $\varphi = \varphi(\mathbf{v}_1) = \varphi(\mathbf{v}_1, \mathbf{v}_2)$. Usually the term substituted will be a numeral, for example

$$\varphi(\mathbf{S}^a\mathbf{0},\ \mathbf{S}^b\mathbf{0}) = (\varphi_{\mathbf{S}^a\mathbf{0}}^{v_1})_{\mathbf{S}^b\mathbf{0}}^{v_2}.$$

But at times we will also substitute other terms, e.g., $\varphi(x) = \varphi_x^{v_1}$, where x is a variable. If, however, x is not substitutable for $\mathbf{v}_1$ in φ, then we must take $\varphi(x) = \psi_x^{v_1}$, where ψ is a suitable alphabetic variant of φ.

In the next proof (and elsewhere in this chapter) we make use of the following consequence of the substitution lemma of Section 2.5: For a formula φ in which at most $\mathbf{v}_1, \ldots, \mathbf{v}_n$ occur free and for natural numbers $a_1, \ldots, a_n$,

$$\models_{\mathfrak{N}} \varphi\, [\![a_1, \ldots, a_n]\!] \Leftrightarrow\ \models_{\mathfrak{N}} \varphi(\mathbf{S}^{a_1}\mathbf{0}, \ldots, \mathbf{S}^{a_n}\mathbf{0}).$$

An existential ($\exists_1$) formula is one of the form $\exists x_1 \cdots \exists x_k\, \theta$, where θ is quantifier-free. The following result improves Theorem 33C:

Corollary 33D If τ is an existential sentence true in $\mathfrak{N}$, then $A_E \vdash \tau$.

Proof If $\exists \mathbf{v}_1\, \exists \mathbf{v}_2\, \theta$ is true in $\mathfrak{N}$, then for some natural numbers m and n, $\theta(\mathbf{S}^m\mathbf{0}, \mathbf{S}^n\mathbf{0})$ is true in $\mathfrak{N}$. As this is a quantifier-free true sentence, it is deducible from A_E. But it in turn logically implies $\exists \mathbf{v}_1\, \exists \mathbf{v}_2\, \theta$. ■

On the other hand, it is known that there are true universal ($\forall_1$) sentences (i.e., of the form $\forall x_1 \cdots \forall x_k\, \theta$ for quantifier-free θ) which are *not* in $\mathrm{Cn}\ A_E$.

Representable relations

Let R be an m-ary relation on N; i.e., $R \subseteq N^m$. We know that a formula ϱ (in which only $\mathbf{v}_1, \ldots, \mathbf{v}_m$ occur free) defines R in $\mathfrak{N}$ iff for every $a_1, \ldots, a_m$ in N,

$$\langle a_1, \ldots, a_m \rangle \in R \Leftrightarrow\ \models_{\mathfrak{N}} \varrho\, [\![a_1, \ldots, a_m]\!]$$
$$\Leftrightarrow\ \models_{\mathfrak{N}} \varrho(\mathbf{S}^{a_1}\mathbf{0}, \ldots, \mathbf{S}^{a_m}\mathbf{0}).$$

(The last condition here is equivalent to the preceding one by the substitution lemma.) We can recast this into two implications:

$$\langle a_1, \ldots, a_m \rangle \in R \Rightarrow \models_{\mathfrak{N}} \varrho(\mathbf{S}^{a_1}\mathbf{0}, \ldots, \mathbf{S}^{a_m}\mathbf{0}),$$

$$\langle a_1, \ldots, a_m \rangle \notin R \Rightarrow \models_{\mathfrak{N}} \neg\varrho(\mathbf{S}^{a_1}\mathbf{0}, \ldots, \mathbf{S}^{a_m}\mathbf{0}).$$

We will say that ϱ also represents R in the theory Cn A_E if in these two implications the notion of truth in $\mathfrak{N}$ can be replaced by the stronger notion of deducibility from A_E.

More generally, let T be any theory in a language with $\mathbf{0}$ and $\mathbf{S}$. Then ϱ *represents* R in T iff for every $a_1, \ldots, a_m$ in N:

$$\langle a_1, \ldots, a_m \rangle \in R \Rightarrow \varrho(\mathbf{S}^{a_1}\mathbf{0}, \ldots, \mathbf{S}^{a_m}\mathbf{0}) \in T,$$

$$\langle a_1, \ldots, a_m \rangle \notin R \Rightarrow (\neg\varrho(\mathbf{S}^{a_1}\mathbf{0}, \ldots, \mathbf{S}^{a_m}\mathbf{0})) \in T.$$

For example, ϱ represents R in the theory Th $\mathfrak{N}$ iff ϱ defines R in $\mathfrak{N}$. But ϱ represents R in Cn A_E iff for all $a_1, \ldots, a_m$:

$$\langle a_1, \ldots, a_m \rangle \in R \Rightarrow A_E \vdash \varrho(\mathbf{S}^{a_1}\mathbf{0}, \ldots, \mathbf{S}^{a_m}\mathbf{0}),$$

$$\langle a_1, \ldots, a_m \rangle \notin R \Rightarrow A_E \vdash \neg\varrho(\mathbf{S}^{a_1}\mathbf{0}, \ldots, \mathbf{S}^{a_m}\mathbf{0}).$$

The equality relation on N, for example, is represented in Cn A_E by the formula $\boldsymbol{v}_1 \approx \boldsymbol{v}_2$. For

$$m = n \Rightarrow \vdash \mathbf{S}^m\mathbf{0} \approx \mathbf{S}^n\mathbf{0},$$

$$m \neq n \Rightarrow \{\text{S1, S2}\} \vdash \neg\mathbf{S}^m\mathbf{0} \approx \mathbf{S}^n\mathbf{0}.$$

A relation is *representable* in T iff there exists some formula which represents it in T.

The notion of representability should be compared with that of definability. In both cases we are somehow describing relations on the natural numbers by formulas. In the case of definability, we ask about the truth of sentences in the interpretation. In the case of representability in Cn A_E, we ask instead about the deducibility of sentences from the axioms.

Say that a formula φ, in which no variables other than $\boldsymbol{v}_1, \ldots, \boldsymbol{v}_m$ occur free, is *numeralwise determined* by A_E iff for any m-tuple $a_1, \ldots, a_m$ of natural numbers, either

$$A_E \vdash \varphi(\mathbf{S}^{a_1}\mathbf{0}, \ldots, \mathbf{S}^{a_m}\mathbf{0})$$

or

$$A_E \vdash \neg\varphi(\mathbf{S}^{a_1}\mathbf{0}, \ldots, \mathbf{S}^{a_m}\mathbf{0}).$$

Theorem 33E A formula ϱ represents a relation R in $\mathrm{Cn}\, A_E$ iff

(1) ϱ is numeralwise determined by A_E, and
(2) ϱ defines R in $\mathfrak{N}$.

Proof We use the fact that $\mathfrak{N}$ is a model of A_E. If ϱ represents R in $\mathrm{Cn}\, A_E$, then it is clear that (1) holds; (2) holds since "$A_E \vdash$" implies "$\models_{\mathfrak{N}}$." Conversely, if (1) and (2) hold, then we have

$$\begin{aligned}\langle a_1, \ldots, a_m\rangle \in R &\Rightarrow \models_{\mathfrak{N}} \varrho(\mathbf{S}^{a_1}\mathbf{0}, \ldots, \mathbf{S}^{a_m}\mathbf{0}) && \text{by (2)}\\ &\Rightarrow A_E \nvdash \neg\varrho(\mathbf{S}^{a_1}\mathbf{0}, \ldots, \mathbf{S}^{a_m}\mathbf{0}) && \text{since } \mathfrak{N} \text{ is a model of } A_E\\ &\Rightarrow A_E \vdash \varrho(\mathbf{S}^{a_1}\mathbf{0}, \ldots, \mathbf{S}^{a_m}\mathbf{0}) && \text{by (1).}\end{aligned}$$

Similarly for the complement of R and $\neg\varrho$. ■

Church's thesis

We now turn to the relationship of the concepts of representability and decidability.

★Theorem 33F Assume that R is a relation representable in a consistent axiomatizable theory. Then R is decidable.

Proof Say that ϱ represents R in the consistent axiomatizable theory T. Recall that T is effectively enumerable (Corollary 25F). The decision procedure is as follows:

Given $a_1, \ldots, a_m$, enumerate the members of T. If, in the enumeration, $\varrho(\mathbf{S}^{a_1}\mathbf{0}, \ldots, \mathbf{S}^{a_m}\mathbf{0})$ is found, then we are done and $\langle a_1, \ldots, a_m\rangle \in R$. If, in the enumeration, $\neg\varrho(\mathbf{S}^{a_1}\mathbf{0}, \ldots, \mathbf{S}^{a_m}\mathbf{0})$ is found, then we are done and $\langle a_1, \ldots, a_m\rangle \notin R$.

By the representability, one sentence or the other always appears eventually, whereupon the procedure terminates. Since T is consistent, the answer given by the procedure is correct. ■

★Corollary 33G Any relation representable in a consistent finitely axiomatizable theory is decidable.

What about the converse to the above corollary? We cannot really hope to prove the converse on the basis of our informal notion of decidability. For our informal approach is usable only for giving lower bounds on the class of decidable relations, and is unsuited to giving upper bounds.

It is nevertheless possible to make plausibility arguments in support of the converse. This will be easier to do at the end of Section 3.4 than here. Roughly, the idea is that in a finite number of axioms we could capture the (finitely long) instructions for the decision procedure.

The assertion that both the above corollary and its converse are correct is generally known as *Church's thesis.* This assertion is not really a mathematical statement susceptible of proof or disproof; rather it is a judgment that the correct formalization of the intuitive notion of decidability is by means of representability in consistent and finitely axiomatizable theories.

Definition A relation R on the natural numbers is *recursive* iff it is representable in some consistent finitely axiomatizable theory (in a language with $\mathbf{0}$ and $\mathbf{S}$).

Church's thesis now can be put more succinctly: A relation is decidable iff it is recursive. Or perhaps more accurately: The notion of recursiveness is the correct precise counterpart to the intuitive notion of decidability. The situation is analogous to one encountered in calculus. An intuitively continuous function (defined on an interval) is one whose graph you can draw without lifting your pencil off the paper. But to prove theorems, some formal counterpart of this notion is needed. And so one gives the usual definition of ε-δ-continuity. One should ask if the precise notion of ε-δ-continuity is an accurate formalization of intuitive continuity. If anything, the class of ε-δ-continuous functions is too broad. It includes nowhere differentiable functions, whose graphs cannot be drawn without lifting the pencil. But accurate or not, the class of ε-δ-continuous functions has been found to be a natural and important class in mathematical analysis.

Very much the same situation occurs with recursiveness. One should ask if the precise notion of recursiveness is an accurate formalization of the intuitive notion of decidability. Again, the precisely defined class (of recursive relations) appears to be, if anything, too broad. It includes relations for which any decision procedure will, for large inputs, require so much computing time and memory ("scratchpad") space as to make implementation absurd. Recursiveness corresponds to decidability in an idealized world, where length of computation and amount of memory are disregarded. But in any case, the class of recursive relations has been found to be a natural and imporant class in mathematical logic.

Empirical evidence that the class of recursive relations is not too narrow is provided by the following:

1. Any relation considered thus far which mathematicians have felt was decidable has been found to be recursive.

2. Several people have tried giving precise definitions of idealized computing agents (e.g., Turing machines, introduced by Turing in 1936, or the register machines described in Section 3.8.) The idea was to devise something that could carry out any effective procedure. In all cases, the class of relations having decision procedures executable by such a computing agent has been exactly the class of recursive relations.

The fact that so many different (yet equivalent) definitions for the class of recursive relations have been found is some indication of the naturalness and importance of the concept.

In this book we will continue to exclude the intuitive notion of decidability from nonstarred theorems. But in the remainder of the exposition we will accept Church's thesis. For example, we will speak of a set's being undecidable when we have a theorem stating it to be nonrecursive.

Obviously any relation representable in Cn A_E is recursive. We will prove later that the converse also holds; if a relation is representable in *any* consistent finitely axiomatizable theory, then it is representable in the one theory we have selected for special study. (This was, of course, a motivating factor in our selection.)

The functional analog of a decidable relation is a computable function.

★Definition A function $f : N^k \to N$ is *computable* iff there is an effective procedure which, given any k-tuple $\vec{a}$ of natural numbers, will produce $f(\vec{a})$.

For example, addition and multiplication are computable. Effective procedures for these functions are taught in the elementary schools. On the other hand, of the uncountably many functions from N^k into N, only countably many can be computable.

We want to give a mathematical counterpart to the intuitive concept of computability, just as in the case of decidable relations. The clue to the correct counterpart is provided by the next theorem. Recall that any function $f : N^k \to N$ is also a $(k + 1)$-ary relation on N:

$$\langle a_1, \ldots, a_k, b\rangle \in f \Leftrightarrow f(a_1, \ldots, a_k) = b.$$

At one time it was popular to distinguish between the function and the relation (which was called the *graph* of the function). Current set-theoretic usage takes a function to be the same thing as its graph. But we still have the two ways of looking at the function.

★Theorem 33H The following three conditions on a function $f : N^k \to N$ are equivalent:

(a) f is computable.
(b) When viewed as a relation, f is a decidable relation.
(c) When viewed as a relation, f is an effectively enumerable relation.

Proof (a) $\Rightarrow$ (b): Assume that f is computable; we will describe the decision procedure. Given $\langle a_1, \ldots, a_k, b\rangle$, first compute $f(a_1, \ldots, a_k)$. Then look to see if the result is equal to b. If it is, say "yes", otherwise say "no."

(b) $\Rightarrow$ (c): Any decidable relation is effectively enumerable. For we can enumerate the set of all $(k+1)$-tuples of numbers, and place on the output list those which meet the test of belonging to the relation.

(c) $\Rightarrow$ (a): Assume that we have an effective enumeration of (the graph of) f. To compute $f(a_1, \ldots, a_k)$ we examine the $(k+1)$-tuples in the enumeration until we find the one that begins with $a_1, \ldots, a_k$. Its last component is then the desired function value. ■

Thus by using Church's thesis, we can say that f is computable iff f (viewed as a relation) is recursive. The class of recursive functions is an interesting class even apart from its connection with incompleteness theorems of logic. It represents an upper bound to the class of functions which can actually be computed by programs for digital computers. The recursive functions are those which are calculable by digital computers, provided one ignores practical limitations on computing time and memory space.

We can now describe our plans for this section and the next. Our basic goal is to obtain the theorems of Section 3.5. But some groundwork is required before we can prove those theorems; we must verify that a number of relations (intuitively decidable) and a number of functions (intuitively computable) are representable in Cn A_E and hence are recursive. In the process we will show (Theorem 34A) that recursiveness is *equivalent* to representability in Cn A_E. In the remainder of the present section we will establish general facts about representability, and will show, for example, that certain functions for encoding finite sequences of numbers into single numbers are representable. Then in Section 3.4 we apply these results to particular relations and functions related to the syntactical features of the formal language.

The author is sufficiently realistic to know that many readers will be more interested in the theorems of Section 3.5 than in the preliminary spadework. If the reader is willing to believe that intuitively decidable relations are all representable in Cn A_E, and intuitively computable functions are functionally representable (a concept we will define shortly) there, then most if not all of the proofs in this spadework become unnecessary. But it is hoped that the definitions and the statements of the results will still receive some attention.

Numeralwise determined formulas

Theorem 33E tells us that we can show a relation to be representable in Cn A_E by finding a formula which defines it in $\mathfrak{N}$ and is numeralwise determined by A_E. The next theorem will be useful in establishing numeralwise determination.

Theorem 33I (a) Any atomic formula is numeralwise determined by A_E.

(b) If φ and ψ are numeralwise determined by A_E, then so are $\neg\varphi$ and $\varphi \rightarrow \psi$.

(c) If φ is numeralwise determined by A_E, then so are the following formulas (obtained from φ by "bounded quantification"):

$$\forall x(x < y \rightarrow \varphi),$$
$$\exists x(x < y \wedge \varphi).$$

Proof of (c) We will consider a formula

$$\exists x(x < y \wedge \varphi(x, y, z))$$

in which just the variables y and z occur free. Consider two natural numbers a and b; we must show that either

$$A_E \vdash \exists x(x < \mathbf{S}^a\mathbf{0} \wedge \varphi(x, \mathbf{S}^a\mathbf{0}, \mathbf{S}^b\mathbf{0}))$$

or

$$A_E \vdash \neg\exists x(x < \mathbf{S}^a\mathbf{0} \wedge \varphi(x, \mathbf{S}^a\mathbf{0}, \mathbf{S}^b\mathbf{0})).$$

Case 1: For some c less than a,

(1) $$A_E \vdash \varphi(\mathbf{S}^c\mathbf{0}, \mathbf{S}^a\mathbf{0}, \mathbf{S}^b\mathbf{0}).$$

(This case occurs iff $\exists x(x < \mathbf{S}^a\mathbf{0} \wedge \varphi(x, \mathbf{S}^a\mathbf{0}, \mathbf{S}^b\mathbf{0}))$ is true in $\mathfrak{N}$.) We also have

(2) $$A_E \vdash \mathbf{S}^c\mathbf{0} < \mathbf{S}^a\mathbf{0}$$

by (a). And the sentences in (1) and (2) imply the sentence

$$\exists x(x < \mathbf{S}^a\mathbf{0} \wedge \varphi(x, \mathbf{S}^a\mathbf{0}, \mathbf{S}^b\mathbf{0})).$$

Case 2: Otherwise for every c less than a,

$$A_E \vdash \neg\varphi(\mathbf{S}^c\mathbf{0}, \mathbf{S}^a\mathbf{0}, \mathbf{S}^b\mathbf{0}). \tag{3}$$

(This case occurs iff $\forall x(x < \mathbf{S}^a\mathbf{0} \rightarrow \neg\varphi(x, \mathbf{S}^a\mathbf{0}, \mathbf{S}^b\mathbf{0}))$ is true in $\mathfrak{N}$.) We know from Lemma 33A that

$$A_E \vdash \forall x(x < \mathbf{S}^a\mathbf{0} \rightarrow x \approx \mathbf{S}^0\mathbf{0} \vee \cdots \vee x \approx \mathbf{S}^{a-1}\mathbf{0}). \tag{4}$$

The sentence in (4) together with the sentences in (3) (for $c = 0, \ldots, a-1$) imply

$$\forall x(x < \mathbf{S}^a\mathbf{0} \rightarrow \neg\varphi(x, \mathbf{S}^a\mathbf{0}, \mathbf{S}^b\mathbf{0})).$$

And this is equivalent to

$$\neg\exists x(x < \mathbf{S}^a\mathbf{0} \wedge \varphi(x, \mathbf{S}^a\mathbf{0}, \mathbf{S}^b\mathbf{0})).$$

This shows that $\exists x(x < y \wedge \varphi(x, y, z))$ is numeralwise determined by A_E. By applying this result to $\neg\varphi$ we obtain the fact that the dual formula, $\forall x(x < y \rightarrow \varphi(x, y, z))$, is numeralwise determined by A_E as well. ■

The argument in case 2 relied on the fact that the x quantifier was bounded by $\mathbf{S}^a\mathbf{0}$. We will see that it is possible for

$$\neg\psi(\mathbf{S}^0\mathbf{0}), \quad \neg\psi(\mathbf{S}^1\mathbf{0}), \ldots$$

all to be consequences of A_E without having

$$\forall x \neg\psi(x)$$

be a consequence.

The preceding theorem is a useful tool for showing many relations to be representable in Cn A_E. For example, the set of primes is represented by

$$\mathbf{S}^1\mathbf{0} < \mathbf{v}_1 \wedge \forall x(x < \mathbf{v}_1 \rightarrow \forall y(y < \mathbf{v}_1 \rightarrow x \cdot y \not\approx \mathbf{v}_1)).$$

This formula defines the primes in $\mathfrak{N}$, and by the preceding theorem is numeralwise determined by A_E. It therefore represents the set of primes in Cn A_E.

Representable functions

Often it is more convenient to work with functions than with relations. Let $f : N^m \rightarrow N$ be an m-place function on the natural numbers. A formula φ in which only $\mathbf{v}_1, \ldots, \mathbf{v}_{m+1}$ occur free will be said to *functionally represent* f (in the theory Cn A_E) iff for any $a_1, \ldots, a_m$ in $\mathbb{N}$,

$$A_E \vdash \forall \mathbf{v}_{m+1}[\varphi(\mathbf{S}^{a_1}\mathbf{0}, \ldots, \mathbf{S}^{a_m}\mathbf{0}, \mathbf{v}_{m+1}) \leftrightarrow \mathbf{v}_{m+1} \approx \mathbf{S}^{f(a_1,\ldots,a_m)}\mathbf{0}].$$

(Observe that the "$\leftarrow$" half of this sentence is equivalent to $\varphi(\mathbf{S}^{a_1}\mathbf{0}, \ldots, \mathbf{S}^{a_m}\mathbf{0}, \mathbf{S}^{f(a_1,\ldots,a_m)}\mathbf{0})$. The "$\rightarrow$" half adds an assertion of uniqueness.)

Theorem 33J If φ functionally represents f in Cn A_E, then it also represents f (as a relation) in Cn A_E.

Proof, with $m = 1$ Since φ functionally represents f, we have for any b:

$$A_E \vdash \varphi(\mathbf{S}^a\mathbf{0}, \mathbf{S}^b\mathbf{0}) \leftrightarrow \mathbf{S}^b\mathbf{0} \approx \mathbf{S}^{f(a)}\mathbf{0}.$$

If $\langle a, b\rangle \in f$, i.e., if $f(a) = b$, then the right half of this biconditional is valid and we get

$$A_E \vdash \varphi(\mathbf{S}^a\mathbf{0}, \mathbf{S}^b\mathbf{0}).$$

But otherwise the right half is refutable from A_E (i.e., its negation is deducible), whence

$$A_E \vdash \neg\varphi(\mathbf{S}^a\mathbf{0}, \mathbf{S}^b\mathbf{0}). \qquad \blacksquare$$

The converse of this theorem is false. But we can change the formula:

Theorem 33K Let f be a function on $\mathbb{N}$ which is (as a relation) representable in Cn A_E. Then we can find a formula φ which functionally represents f in Cn A_E.

Proof To simplify the notation we will take f to be a one-place function on $\mathbb{N}$. The desired sentence,

$$\forall \mathbf{v}_2[\varphi(\mathbf{S}^a\mathbf{0}, \mathbf{v}_2) \leftrightarrow \mathbf{v}_2 \approx \mathbf{S}^{f(a)}\mathbf{0}],$$

is equivalent to the conjunction of the two sentences

(1) $$\varphi(\mathbf{S}^a\mathbf{0}, \mathbf{S}^{f(a)}\mathbf{0})$$

and

(2) $$\forall \mathbf{v}_2[\varphi(\mathbf{S}^a\mathbf{0}, \mathbf{v}_2) \rightarrow \mathbf{v}_2 \approx \mathbf{S}^{f(a)}\mathbf{0}].$$

The sentence (1) is a theorem of A_E whenever φ represents f. The sentence

(2) is an assertion of uniqueness; we must construct φ in such a way that this will also be a theorem of A_E.

Begin with a formula θ known to represent f (as a binary relation). Let φ be

$$\theta(\mathbf{v}_1, \mathbf{v}_2) \wedge \forall z(z < \mathbf{v}_2 \rightarrow \neg\theta(\mathbf{v}_1, z)).$$

We can then rewrite (2) as

$$\text{(2)} \qquad \forall \mathbf{v}_2[\theta(\mathbf{S}^a\mathbf{0}, \mathbf{v}_2) \wedge \forall z(z < \mathbf{v}_2 \rightarrow \neg\theta(\mathbf{S}^a\mathbf{0}, z)) \rightarrow \mathbf{v}_2 \approx \mathbf{S}^{f(a)}\mathbf{0}].$$

To show this to be a theorem of A_E it clearly suffices to show that

$$A_E \cup \{\theta(\mathbf{S}^a\mathbf{0}, \mathbf{v}_2), \forall z(z < \mathbf{v}_2 \rightarrow \neg\theta(\mathbf{S}^a\mathbf{0}, z))\} \vdash \mathbf{v}_2 \approx \mathbf{S}^{f(a)}\mathbf{0}.$$

Call this set of hypotheses (to the left of "$\vdash$") Γ. Since L3 $\in A_E$ it suffices to show that

$$\text{(3)} \qquad \Gamma \vdash \mathbf{v}_2 \not< \mathbf{S}^{f(a)}\mathbf{0}$$

and

$$\text{(4)} \qquad \Gamma \vdash \mathbf{S}^{f(a)}\mathbf{0} \not< \mathbf{v}_2.$$

It is easy to obtain (4), since from the last member of Γ we get

$$\mathbf{S}^{f(a)}\mathbf{0} < \mathbf{v}_2 \rightarrow \neg\theta(\mathbf{S}^a\mathbf{0}, \mathbf{S}^{f(a)}\mathbf{0})$$

and we know that

$$\text{(5)} \qquad A_E \vdash \theta(\mathbf{S}^a\mathbf{0}, \mathbf{S}^{f(a)}\mathbf{0}).$$

To obtain (3) we first note that we have as theorems of A_E,

$$\text{(6)} \qquad \mathbf{v}_2 < \mathbf{S}^{f(a)}\mathbf{0} \leftrightarrow \mathbf{v}_2 \approx \mathbf{S}^0\mathbf{0} \vee \cdots \vee \mathbf{v}_2 \approx \mathbf{S}^{f(a)-1}\mathbf{0}$$

and

$$\text{(7)} \qquad \neg\theta(\mathbf{S}^a\mathbf{0}, \mathbf{S}^b\mathbf{0}) \qquad \text{for } b = 0, \ldots, f(a) - 1.$$

The formulas (6) and (7) imply the formula

$$\text{(8)} \qquad \mathbf{v}_2 < \mathbf{S}^{f(a)}\mathbf{0} \rightarrow \neg\theta(\mathbf{S}^a\mathbf{0}, \mathbf{v}_2).$$

Since $\theta(\mathbf{S}^a\mathbf{0}, \mathbf{v}_2) \in \Gamma$, we have (3).

This shows (2) to be a theorem of A_E; (5) and (8) show (1) to be a theorem of A_E as well. ■

We next want to show that certain basic functions are representable (in Cn A_E) and that the class of representable functions has certain closure properties. Henceforth in this section, when we say that a function or relation is representable, we will mean that it is representable in the theory Cn A_E. But the phrase "in Cn A_E" will usually be omitted.

In simple cases, an m-place function might be represented by an equation

$$\mathbf{v}_{m+1} \approx t.$$

In fact, any such equation, when the variables in t are among $\mathbf{v}_1, \ldots, \mathbf{v}_m$, defines in $\mathfrak{N}$ an m-place function f. (The value of f at $\langle a_1, \ldots, a_m\rangle$ is the number assigned in $\mathfrak{N}$ to t when $\mathbf{v}_i$ is assigned a_i, $1 \leq i \leq m$.) Furthermore, we know that any equation is numeralwise determined by A_E, so the equation represents f as a relation. In fact, it even functionally represents f, for the sentence

$$\forall \mathbf{v}_{m+1}[\mathbf{v}_{m+1} \approx t(\mathbf{S}^{a_1}\mathbf{0}, \ldots, \mathbf{S}^{a_m}\mathbf{0}) \leftrightarrow \mathbf{v}_{m+1} \approx \mathbf{S}^{f(a_1,\ldots,a_m)}\mathbf{0}]$$

is logically equivalent to

$$t(\mathbf{S}^{a_1}\mathbf{0}, \ldots, \mathbf{S}^{a_m}\mathbf{0}) \approx \mathbf{S}^{f(a_1,\ldots,a_m)}\mathbf{0},$$

which is a quantifier-free sentence true in $\mathfrak{N}$. (Here $t(u_1, \ldots, u_m)$ is the term obtained by replacing $\mathbf{v}_1$ by u_1, then $\mathbf{v}_2$ by u_2, etc.) For example:

1. The successor function is represented (functionally) by the equation

$$\mathbf{v}_2 \approx \mathbf{S}\mathbf{v}_1.$$

2. Any constant function is representable. The m-place function which constantly assumes the value b is represented by the equation

$$\mathbf{v}_{m+1} \approx \mathbf{S}^b\mathbf{0}.$$

3. The projection function

$$I_i^m(a_1, \ldots, a_m) = a_i$$

is represented by the equation

$$\mathbf{v}_{m+1} \approx \mathbf{v}_i.$$

4. Addition, multiplication, and exponentiation are represented by the equations

$$\mathbf{v}_3 \approx \mathbf{v}_1 + \mathbf{v}_2,$$
$$\mathbf{v}_3 \approx \mathbf{v}_1 \cdot \mathbf{v}_2,$$
$$\mathbf{v}_3 \approx \mathbf{v}_1 \mathbf{E} \mathbf{v}_2,$$

respectively.

The reader should not be misled by these simple examples; not every representable function is representable by an equation.

We next want to show that the family of representable functions is closed under composition. To simplify the notation, we will consider a one-place function f on $\mathbb{N}$, where

$$f(a) = g(h_1(a), h_2(a)).$$

Suppose that g is functionally represented by ψ and h_i by θ_i. To represent f it would be reasonable to try either

$$\forall y_1 \forall y_2(\theta_1(\mathbf{v}_1, y_1) \to \theta_2(\mathbf{v}_1, y_2) \to \psi(y_1, y_2, \mathbf{v}_2))$$

or

$$\exists y_1 \exists y_2(\theta_1(\mathbf{v}_1, y_1) \wedge \theta_2(\mathbf{v}_1, y_2) \wedge \psi(y_1, y_2, \mathbf{v}_2)).$$

(Think of $\psi(y_1, y_2, \mathbf{v}_2)$ as saying "$g(y_1, y_2) = \mathbf{v}_2$" and think of $\theta_i(\mathbf{v}_1, y_i)$ as saying "$h_i(\mathbf{v}_1) = y_i$." Then the first formula translates, "For any y_1, y_2, if $h_1(\mathbf{v}_1) = y_1$ and $h_2(\mathbf{v}_1) = y_2$, then $g(y_1, y_2) = \mathbf{v}_2$." The second formula translates, "There exist y_1, y_2 such that $h_1(\mathbf{v}_1) = y_1$ and $h_2(\mathbf{v}_1) = y_2$ and $g(y_1, y_2) = \mathbf{v}_2$." Either one is a reasonable way of saying, "$g(h_1(\mathbf{v}_1), h_2(\mathbf{v}_1)) = \mathbf{v}_2$." There are two choices, because when something is unique, either quantifier can be used for it.)

Actually either formula would work; let φ be

$$\forall y_1 \forall y_2(\theta_1(\mathbf{v}_1, y_1) \to \theta_2(\mathbf{v}_1, y_2) \to \psi(y_1, y_2, \mathbf{v}_2)).$$

Consider any natural number a; we have at our disposal

(1) $$\forall \mathbf{v}_2[\psi(\mathbf{S}^{h_1(a)}\mathbf{0}, \mathbf{S}^{h_2(a)}\mathbf{0}, \mathbf{v}_2) \leftrightarrow \mathbf{v}_2 \approx \mathbf{S}^{f(a)}\mathbf{0}].$$

(2) $$\forall y_1[\theta_1(\mathbf{S}^a\mathbf{0}, y_1) \leftrightarrow y_1 \approx \mathbf{S}^{h_1(a)}\mathbf{0}].$$

(3) $$\forall y_2[\theta_2(\mathbf{S}^a\mathbf{0}, y_2) \leftrightarrow y_2 \approx \mathbf{S}^{h_2(a)}\mathbf{0}].$$

And we want

(4) $$\forall \mathbf{v}_2(\varphi(\mathbf{S}^a\mathbf{0}, \mathbf{v}_2) \leftrightarrow \mathbf{v}_2 \approx \mathbf{S}^{f(a)}\mathbf{0}),$$

i.e.,

$$\text{(4)}\quad \forall \boldsymbol{v}_2(\forall y_1 \forall y_2[\theta_1(\mathbf{S}^a\mathbf{0}, y_1) \rightarrow \theta_2(\mathbf{S}^a\mathbf{0}, y_2) \rightarrow \psi(y_1, y_2, \boldsymbol{v}_2)] \leftrightarrow \boldsymbol{v}_2 \approx \mathbf{S}^{f(a)}\mathbf{0}).$$

But (1), (2), and (3) imply (4), as the reader is asked to verify in Exercise 4.

More generally we have

Theorem 33L Let g be an n-place function, let $h_1, \ldots, h_n$ be m-place functions, and let f be defined by

$$f(a_1, \ldots, a_m) = g(h_1(a_1, \ldots, a_m), \ldots, h_n(a_1, \ldots, a_m)).$$

From formulas functionally representing g and $h_1, \ldots, h_n$ we can find a formula which functionally represents f.

In the above proof we have $m = 1$ and $n = 2$. But the general case is proved in exactly the same way.

In order to obtain a function such as

$$f(a, b) = g(h(a), b),$$

we note that

$$f(a, b) = g(h(I_1^2(a, b)), I_2^2(a, b)).$$

The above theorem then can be applied (twice) to show that f is representable (provided that g and h are).

To facilitate discussion of functions with an arbitrary number of variables, we will use vector notation. For example, the equation in the above theorem can be written

$$f(\vec{a}) = g(h_1(\vec{a}), \ldots, h_n(\vec{a})).$$

Another important closure property of the functions representable in Cn A_E is closure under the "least-zero" operator.

Theorem 33M Assume that the $(m + 1)$-place function g is representable and that for every $a_1, \ldots, a_m$ there is a b such that

$$g(a_1, \ldots, a_m, b) = 0.$$

Then we can find a formula which represents the m-place function f, where

$$f(a_1, \ldots, a_m) = \text{the least } b \text{ such that } g(a_1, \ldots, a_m, b) = 0.$$

(In vector notation we can rewrite this last equation:

$$f(\vec{a}) = \text{the least } b \text{ such that } g(\vec{a}, b) = 0.$$

The traditional notation for the least-zero operator is

$$f(\vec{a}) = \mu b[g(\vec{a}, b) = 0].)$$

Proof To simplify the notation we take $m = 1$; thus

$$f(a) = b \qquad \text{iff } g(a, b) = 0 \text{ and for all } c < b,\ g(a, c) \neq 0.$$

If ψ represents g, then we can obtain a formula representing f (as a relation) simply by formalizing the right side of this equivalence:

$$\psi(\mathbf{v}_1, \mathbf{v}_2, \mathbf{0}) \wedge \forall y(y < \mathbf{v}_2 \rightarrow \neg\psi(\mathbf{v}_1, y, \mathbf{0})).$$

This formula defines (the graph of) f and is numeralwise determined by A_E. ■

A catalog

We now construct a repertoire of representable (in Cn A_E) functions and relations, including functions for encoding and decoding sequences.

0. As a consequence of Theorem 33I any relation which has (in $\mathfrak{N}$) a quantifier-free definition is representable. And the class of representable relations is closed under unions, intersections, and complements. And if R is representable, then so are

$$\{\langle a_1, \ldots, a_m, b\rangle : \text{for all } c < b,\ \langle a_1, \ldots, a_m, c\rangle \in R\}$$

and

$$\{\langle a_1, \ldots, a_m, b\rangle : \text{for some } c < b,\ \langle a_1, \ldots, a_m, c\rangle \in R\}.$$

For example, any finite relation has a quantifier-free definition, as does the ordering relation.

1. A relation R is representable iff its characteristic function K_R is. (K_R is the function for which $K_R(\vec{a}) = 1$ when $\vec{a} \in R$, and $K_R(\vec{a}) = 0$ otherwise.)

Proof ($\Leftarrow$) Say that R is a unary relation (a subset of N) and that K_R is represented by $\psi(\mathbf{v}_1, \mathbf{v}_2)$. We claim that $\psi(\mathbf{v}_1, \mathbf{S0})$ represents R. For it defines R and is numeralwise determined by A_E.

($\Rightarrow$) Say that $\varphi(\boldsymbol{v}_1)$ represents R. Then

$$(\varphi(\boldsymbol{v}_1) \wedge \boldsymbol{v}_2 \approx \mathbf{S0}) \vee (\neg\varphi(\boldsymbol{v}_1) \wedge \boldsymbol{v}_2 \approx \mathbf{0})$$

represents (the graph of) K_R, for the same reason as in the last paragraph. (Actually this formula even functionally represents K_R, as the reader can verify.) ■

2. If R is a representable binary relation and f, g are representable functions, then

$$\{\vec{a} : \langle f(\vec{a}), g(\vec{a})\rangle \in R\}$$

is representable. Similarly for an m-ary relation R and functions $f_1, \ldots, f_m$.

Proof Its characteristic function at $\vec{a}$ has the value $K_R(f(\vec{a}), g(\vec{a}))$. Thus it is obtained from representable functions by composition. ■

For example, suppose that R is a representable ternary relation. Then

$$\{\langle x, y\rangle : \langle y, x, x\rangle \in R\}$$

is representable, being

$$\{\langle x, y\rangle : \langle I_2^2(x, y), I_1^2(x, y), I_1^2(x, y)\rangle \in R\}.$$

In this way we can rearrange and repeat variables in describing a representable relation.

3. If R is a representable binary relation, then so is

$$\{\langle a, b\rangle : \text{for some } c \leq b, \langle a, c\rangle \in R\}.$$

Proof We have from catalog item 0 that if

$$Q = \{\langle a, b\rangle : \text{for some } c < b, \langle a, c\rangle \in R\},$$

then Q is representable. And

$$\begin{aligned}\langle a, b\rangle \in R &\Leftrightarrow \langle a, S(b)\rangle \in Q \\ &\Leftrightarrow \langle I_1^2(a, b), S(I_2^2(a, b))\rangle \in Q.\end{aligned}$$

Hence by catalog item 2, R is representable. ■

More generally, if R is a representable $(m + 1)$-ary relation, then

$$\{\langle a_1, \ldots, a_m, b\rangle : \text{for some } c \leq b, \langle a_1, \ldots, a_m, c\rangle \in R\}$$

is also representable. In vector notation this relation becomes

$$\{\langle \vec{a}, b\rangle : \text{for some } c \leq b, \langle \vec{a}, c\rangle \in R\}.$$

Similarly,

$$\{\langle \vec{a}, b\rangle : \text{for all } c \leq b, \langle \vec{a}, c\rangle \in R\}$$

is representable.

4. The divisibility relation

$$\{\langle a, b\rangle : a \text{ divides } b \text{ in } \mathbb{N}\}$$

is representable.

Proof a divides b iff for some $q \leq b$, $a \cdot q = b$. We know that $\{\langle a, b, q\rangle : a \cdot q = b\}$ is representable, as it has a quantifier-free definition. Upon applying the above items, we get the divisibility relation. (In yet further detail, from catalog item 3 we get the representability of

$$R = \{\langle a, b, c\rangle : \text{for some } q \leq c,\ a \cdot q = b\}$$

and a divides b iff $\langle a, b, b\rangle \in R$.) ■

5. The set of primes is representable.

6. The set of pairs of adjacent primes is representable.

Proof $\langle a, b\rangle$ is a pair of adjacent primes iff a is prime and b is prime and $a < b$ and there does not exist any $c < b$ such that $a < c$ and c is prime. The right side of this equivalence is easily formalized by a numeralwise determined formula. ■

Note (for future use in Section 3.7) that we have not yet used the fact that exponentiation is representable.

7. The function whose value at a is p_a, the $(a + 1)$st prime, is representable. (Thus $p_0 = 2$, $p_1 = 3$, $p_2 = 5$, $p_3 = 7$, $p_4 = 11$, etc.)

Proof $p_a = b$ iff b is prime and there exists some $c \leq b^{a^2}$ such that

(i) 2 does not divide c.

(ii) For any $q < b$ and any $r \leq b$, if $\langle q, r\rangle$ is a pair of adjacent primes, then for all $j < c$,

$$q^j \text{ divides } c \Leftrightarrow r^{j+1} \text{ divides } c.$$

(iii) b^a divides c and b^{a+1} does not.

This equivalence is not obvious, but at least the relation defined by the right-hand side is representable. To verify the equivalence, first note that if $p_a = b$, then we can take

$$c = 2^0 \cdot 3^1 \cdot 5^2 \cdot \cdots \cdot p_a^a.$$

It is easy to check that this value of c meets all the conditions. Conversely, suppose c is a number meeting conditions (i)–(iii). We claim that c must be

$$2^0 \cdot 3^1 \cdot \cdots \cdot b^a \cdot \text{ powers of larger primes.}$$

Certainly the exponent of 2 in c is 0, by (i). We can use (ii) to work our way across to the prime b. But by (iii) the exponent of b is a, so b must be the $(a + 1)$st prime. ■

This function will be very useful in encoding finite sequences of numbers into single numbers. Let

$$\begin{aligned}\langle a_0, \ldots, a_m \rangle &= p_0^{a_0+1} \cdot \cdots \cdot p_m^{a_m+1} \\ &= \prod_{i \le m} p_i^{a_i+1}.\end{aligned}$$

This holds also for $m = -1$; we define $\langle\ \rangle = 1$. For example,

$$\langle 2, 1 \rangle = 2^3 \cdot 3^2 = 72.$$

The idea is that 72 safely encodes the pair $\langle 2, 1 \rangle$.

8. For each m, the function whose value at $a_0, \ldots, a_m$ is $\langle a_0, \ldots, a_m \rangle$ is representable.

9. There is a representable function (whose value at $\langle a, b \rangle$ is written: $(a)_b$) such that for $b \le m$,

$$(\langle a_0, \ldots, a_m \rangle)_b = a_b.$$

(This is our "decoding" function.)

Proof Let $(a)_b$ be the least n such that either $a = 0$ or p_b^{n+2} does not divide a. (There always *is* such an n.) Observe that $(0)_b = 0$, and for $a \ne 0$, $(a)_b$ is one less than the exponent of p_b in the prime factorization of a (but not less than 0). Hence for $b \le m$,

$$(\langle a_0, \ldots, a_m \rangle)_b = a_b.$$

To prove representability we use the least-zero operator. Let

$$R = \{\langle a, b, n\rangle : a = 0 \text{ or } p_b^{n+2} \text{ does not divide } a\}.$$

Then $(a)_b = \mu n[K_{\bar{R}}(a, b, n) = 0]$, where $\bar{R}$ is the complement of R. ■

Since the method used in the above proof will be useful elsewhere as well, we here state it separately:

Theorem 33N Assume that R is a representable relation such that for every $\vec{a}$ there is some n such that $\langle \vec{a}, n\rangle \in R$. Then the function f defined by

$$f(\vec{a}) = \text{the least } n \text{ such that } \langle \vec{a}, n\rangle \in R$$

is representable.

Proof $f(\vec{a}) = \mu n[K_{\bar{R}}(\vec{a}, n) = 0]$. ■

We will later use the notation

$$f(\vec{a}) = \mu n[\langle \vec{a}, n\rangle \in R].$$

10. Say that b is a *sequence number* iff for some $m \geq -1$ and some $a_0, \ldots, a_m$,

$$b = \langle a_0, \ldots, a_m\rangle.$$

(When $m = -1$ we get $\langle\ \rangle = 1$.) Then the set of sequence numbers is representable.

Proof Exercise 5.

11. There is a representable function lh such that

$$\text{lh}\langle a_0, \ldots, a_m\rangle = m + 1.$$

("lh" stands for "length.")

Proof Let lh a be the least n such that either $a = 0$ or p_n does not divide a. This works. ■

12. There is a representable function (whose value at $\langle a, b\rangle$ is called the *restriction* of a to b, written: $a \restriction b$) such that for any $b \leq m + 1$,

$$\langle a_0, \ldots, a_m\rangle \restriction b = \langle a_0, \ldots, a_{b-1}\rangle.$$

Proof Let $a \upharpoonright b$ be the least n such that either $a = 0$ or both $n \neq 0$ and for any $j < b$, any $k < a$

$$p_j^k \text{ divides } a \Rightarrow p_j^k \text{ divides } n.$$

This works. ■

13. (Primitive recursion) With a $(k+1)$-place function f we associate another function $\bar{f}$ such that $\bar{f}(a, b_1, \ldots, b_k)$ encodes the values of $f(j, b_1, \ldots, b_k)$ for all $j < a$. Specifically, let

$$\bar{f}(a, \vec{b}) = \langle f(0, \vec{b}), \ldots, f(a-1, \vec{b})\rangle.$$

For example, $\bar{f}(0, \vec{b}) = \langle\,\rangle = 1$, encoding the first zero values of f. $\bar{f}(1, \vec{b}) = \langle f(0, \vec{b})\rangle$. In any case, $\bar{f}(a, \vec{b})$ is a sequence number of length a, encoding the first a values of f.

Now suppose we are given a $(k+2)$-place function g. There exists a unique function f satisfying

$$f(a, \vec{b}) = g(\bar{f}(a, \vec{b}), a, \vec{b}).$$

For example,

$$f(0, \vec{b}) = g(\langle\,\rangle, 0, \vec{b}),$$
$$f(1, \vec{b}) = g(\langle f(0, \vec{b})\rangle, 1, \vec{b}).$$

(The existence and uniqueness of this f should be intuitively clear. For a proof, we can apply the recursion theorem of Section 1.2, obtaining first $\bar{f}$ and then extracting f.)

Theorem 33P Let g be a $(k+2)$-place function and let f be the unique $(k+1)$-place function such that for all a and (k-tuples) $\vec{b}$,

$$f(a, \vec{b}) = g(\bar{f}(a, \vec{b}), a, \vec{b}).$$

If g is representable, then so is f.

Proof First we claim that $\bar{f}$ is representable. This follows from the fact that

$$\bar{f}(a, b) = \text{the least } s \text{ such that } s \text{ is a sequence number of length } a \text{ and for } i \text{ less than } a,\ (s)_i = g(s \upharpoonright i, a, \vec{b}).$$

It then follows that f is representable, since

$$f(a, \vec{b}) = g(\bar{f}(a, \vec{b}), a, \vec{b})$$

and the functions on the right are representable. ■

Actually the phrase "primitive recursion" is more commonly applied to a simpler version of this, given in Exercise 8.

14. For a representable function F, the function whose value at $a, \vec{b}$ is

$$\prod_{i<a} F(i, \vec{b})$$

is also representable. Similarly with Σ in place of Π.

Proof Call this function G; then

$$G(0, \vec{b}) = 1,$$
$$G(a+1, \vec{b}) = F(a, \vec{b}) \cdot G(a, \vec{b}).$$

Apply Exercise 8. ■

15. Define the *concatenation* of a and b, $a * b$, by

$$a * b = a \cdot \prod_{i < \mathrm{lh}\, b} p_{i+\mathrm{lh}\, a}^{(b)_i + 1}.$$

This is a representable function of a and b, and

$$\langle a_1, \ldots, a_m \rangle * \langle b_1, \ldots, b_n \rangle = \langle a_1, \ldots, a_m, b_1, \ldots, b_n \rangle.$$

The concatenation operation has the further property of being associative on sequence numbers.

16. We will also want a "capital asterisk" operation. Let

$$\mathop{\ast}_{i<a} f(i) = f(0) * f(1) * \cdots * f(a-1).$$

For a representable function F, the function whose value at $a, \vec{b}$ is $\mathop{\ast}_{i<a} F(i, \vec{b})$ is representable.

Proof $\mathop{\ast}_{i<0} F(i, \vec{b}) = \langle\, \rangle = 1.$

$$\mathop{\ast}_{i<a+1} F(i, \vec{b}) = \mathop{\ast}_{i<a} F(i, \vec{b}) * F(a, \vec{b}).$$

So this is just like catalog item 14. ■

EXERCISES

1. Show that in the structure $(\mathbb{N}, \cdot, E)$ we can define $\{0\}$, $<$, the successor relation $\{\langle n, S(n)\rangle : n \in \mathbb{N}\}$, and the addition relation $\{\langle m, n, m+n\rangle : m, n \text{ in } \mathbb{N}\}$.

2. Prove Theorem 33C, stating that true (in $\mathfrak{N}$) quantifier-free sentences are theorems of A_E.

3. A theory T (in a language with **0** and **S**) is called *ω-complete* iff for any formula φ and variable x, if $\varphi^x_{\mathbf{S}^n\mathbf{0}}$ belongs to T for every natural number n, then $\forall x\, \varphi$ belongs to T. Show that if T is a consistent ω-complete theory in the language of $\mathfrak{N}$ and if $A_E \subseteq T$, then $T = \text{Th}\, \mathfrak{N}$.

4. Show that in the proof preceding Theorem 33L, formula (4) is logically implied by the set consisting of formulas (1), (2), and (3).

5. Show that the set of sequence numbers is representable (catalog item 10).

6. Is 3 a sequence number? What is lh 3? Find $(1 * 3) * 6$ and $1 * (3 * 6)$.

7. Establish the following facts:

(a) $a + 1 < p_a$.
(b) $(b)_k \leq b$; equality holds iff $b = 0$.
(c) $\text{lh}\, a \leq a$; equality holds iff $a = 0$.
(d) $a \restriction i \leq a$.
(e) $\text{lh}(a \restriction i)$ is the smaller of i and $\text{lh}\, a$.

8. Let g and h be representable functions, and assume that

$$f(0, b) = g(b),$$
$$f(a + 1, b) = h(f(a, b), a, b).$$

Show that f is representable.

9. Show that there is a representable function f such that for any n, $a_0, \ldots, a_n$,

$$f(\langle a_0, \ldots, a_n \rangle) = a_n.$$

10. Assume that R is a representable relation and that g and h are representable functions. Show that f is representable, where

$$f(\vec{a}) = \begin{cases} g(\vec{a}) & \text{if } \vec{a} \in R, \\ h(\vec{a}) & \text{if } \vec{a} \notin R. \end{cases}$$

11. (Monotone recursion) Assume that R is a representable binary relation on $\mathbb{N}$. Let C be the smallest subset of $\mathbb{N}$ (i.e., the intersection of all subsets) such that for all n, $a_0, \ldots, a_{n-1}$, b,

$$\langle \langle a_0, \ldots, a_{n-1} \rangle, b \rangle \in R \;\&\; a_i \in C \text{ (for all } i < n) \Rightarrow b \in C.$$

Further assume that (1) for all $n, a_0, \ldots, a_{n-1}, b$,

$$\langle\langle a_0, \ldots, a_{n-1}\rangle, b\rangle \in R \Rightarrow a_i < b \qquad \text{for all } i < n,$$

and (2) there is a representable function f such that for all $n, a_0, \ldots, a_{n-1}, b$,

$$\langle\langle a_0, \ldots, a_{n-1}\rangle, b\rangle \in R \Rightarrow n < f(b).$$

Show that C is representable. (C is, in a sense, generated by R. $C \neq \varnothing$ in general because if $\langle\langle\ \rangle, b\rangle \in R$, then $b \in C$.)

§ 3.4 ARITHMETIZATION OF SYNTAX

In this section we intend to develop two themes:

1. Certain assertions about wffs can be converted into assertions about natural numbers (by assigning numbers to expressions).
2. These (English) assertions about natural numbers can in some cases be translated into the formal language. And the theory Cn A_E is strong enough to prove some of the translations so obtained.

This will give us the ability to construct formulas which, by expressing facts about numbers, indirectly express facts about formulas (even about themselves). Such an ability will be exploited in Section 3.5 to obtain results of undefinability and undecidability.

Gödel numbers

We first want to assign numbers to expressions of the formal language. Recall that the symbols of our language are those listed in Table IX.

TABLE IX

Parameters	Logical symbols
0. $\forall$	1. (
2. **0**	3.)
4. **S**	5. $\neg$
6. $<$	7. $\rightarrow$
8. $+$	9. $\approx$
10. $\cdot$	11. v_1
12. **E**	13. v_2 etc.

There is a function h assigning to each symbol the integer listed to its left. Thus $h(\forall) = 0$, $h(\mathbf{0}) = 2$, and $h(\boldsymbol{v}_i) = 9 + 2i$. In order to make our subsequent work more widely applicable, we will assume only that we have some language with $\mathbf{0}$ and $\mathbf{S}$ which is *recursively numbered.* By this we mean that we have a one-to-one function h from the parameters of that language into the even numbers such that the two relations

$$\{\langle k, m\rangle : k \text{ is the value of } h \text{ at some } m\text{-place predicate symbol}\}$$

and

$$\{\langle k, m\rangle : k \text{ is the value of } h \text{ at some } m\text{-place function symbol}\}$$

are both representable in Cn A_E. (Of course in the case of the language of $\mathfrak{N}$ these sets are finite.) We define h on the logical symbols as before.

For an expression $\varepsilon = s_0 \cdots s_n$ of the language we define its *Gödel number*, $\sharp(\varepsilon)$, by

$$\sharp(s_0 \cdots s_n) = \langle h(s_0), \ldots, h(s_n)\rangle.$$

For example, using our original function h for the language of $\mathfrak{N}$, we obtain

$$\begin{aligned}&\sharp(\exists \boldsymbol{v}_3\, \boldsymbol{v}_3 \approx \mathbf{0})\\ &\quad = \sharp((\neg\forall \boldsymbol{v}_3(\neg{\approx}\boldsymbol{v}_3\mathbf{0})))\\ &\quad = 2^2 \cdot 3^6 \cdot 5^1 \cdot 7^{16} \cdot 11^2 \cdot 13^6 \cdot 17^{10} \cdot 19^{16} \cdot 23^3 \cdot 29^4 \cdot 31^4.\end{aligned}$$

This is a large number, being of the order of 1.3×10^{75}. To a set Φ of expressions we assign the set

$$\sharp\Phi = \{\sharp(\varepsilon) : \varepsilon \in \Phi\}$$

of Gödel numbers.

To a sequence $\langle \alpha_0, \ldots, \alpha_n\rangle$ of expressions (such as a deduction), we assign the number

$$\mathscr{G}(\langle \alpha_0, \ldots, \alpha_n\rangle) = \langle \sharp\alpha_0, \ldots, \sharp\alpha_n\rangle.$$

We now proceed to show that various relations and functions having to do with Gödel numbers are representable in Cn A_E (and hence are recursive). As in the preceding section, whenever we say that a relation or function is representable (without specifying a theory) we mean that it is representable in the theory Cn A_E.

We will make use of certain abbreviations in the language we use (i.e., English, although it is coming to differ more and more from what one ordinarily thinks of as English). For "there is a number a such that" we write "$\exists a$." In the same spirit, "$\exists a, b < c$" means "there are numbers a and b both of which are less than c such that." Similarly, we may employ "$\forall$." We would not have dared to employ such abbreviations in Chapter 2, for fear of creating confusion between the formal language and the metalanguage (English). But by now we trust the reader to avoid such erroneous ways.

1. The set of Gödel numbers of variables is representable.

Proof It is $\{a : (\exists b < a)a = \langle 11 + 2b \rangle\}$. It follows from results of the preceding section that this is a representable set. ■

2. The set of Gödel numbers of terms is representable.

Proof The set of terms was defined inductively. And terms were built up from constituents with smaller Gödel numbers. We will treat this case in some detail, since it is typical of the argument used for inductively defined relations.

Let f be the characteristic function of the set of Gödel numbers of terms. From the definition of "term" we obtain

$$f(a) = \begin{cases} 1 & \text{if } a \text{ is the Gödel number of a variable,} \\ 1 & \text{if } (\exists i, k < a)\ [i \text{ is a sequence number} \\ & \quad \& \ (\forall j < \operatorname{lh} i) f((i)_j) = 1 \ \& \ k \text{ is the value of} \\ & \quad h \text{ at some } (\operatorname{lh} i)\text{-place function symbol } \& \\ & \quad a = \langle k \rangle * \bigstar_{j < \operatorname{lh} i} (i)_j, \\ 0 & \text{otherwise.} \end{cases}$$

Although the right side of this equation refers to f, it refers only to $f((i)_j)$, where $(i)_j < a$. This feature permits us to apply primitive recursion. $f(a) = g(\bar{f}(a), a)$, where

$$g(s, a) = \begin{cases} 1 & \text{if } a \text{ is the Gödel number of a variable,} \\ 1 & \text{if } (\exists i, k < a)[i \text{ is a sequence number} \\ & \quad \& \ (\forall j < \operatorname{lh} i)(s)_{(i)_j} = 1 \ \& \ k \text{ is the} \\ & \quad \text{value of } h \text{ at some } (\operatorname{lh} i)\text{-place function} \\ & \quad \text{symbol } \& \ a = \langle k \rangle * \bigstar_{j < \operatorname{lh} i} (i)_j, \\ 0 & \text{otherwise.} \end{cases}$$

For if in this equation we set s equal to $\bar{f}(a)$, then $(s)_{(i)_j} = f((i)_j)$ for $(i)_j < a$. Hence by Theorem 33P, f is representable provided that g is.

It remains to show that g is representable. But this is straightforward, by using results of the preceding section. Briefly, the graph of g is the union of three relations, corresponding to the three clauses in the above equation. Each of the three is obtained from equality and other representable relations by bounded quantification and the substitution of representable functions. ■

3. The set of Gödel numbers of atomic formulas is representable.

Proof a is the Gödel number of an atomic formula iff $(\exists i, k < a)[i$ is a sequence number & $(\forall j < \mathrm{lh}\, i)(i)_j$ is the Gödel number of a term & k is the value of h at some $(\mathrm{lh}\, i)$-place predicate symbol & $a = \langle k \rangle * \mathop{\ast\!\!\!\ast}_{j<\mathrm{lh}\, i}(i)_j$. ■

4. The set of Gödel numbers of wffs is representable.

Proof The wffs were inductively defined. Let f be the characteristic function of the set, then

$$f(a) = \begin{cases} 1 & \text{if } a \text{ is the Gödel number of an atomic formula,} \\ 1 & \text{if } (\exists i < a)[a = \langle h(\,(\,),\ h(\neg)\rangle * i * \langle h(\,)\,)\rangle \\ & \quad \& f(i) = 1], \\ 1 & \text{if } (\exists i, j < a)[a = \langle h(\,(\,)\rangle * i * \langle h(\rightarrow)\rangle * j * \langle h(\,)\,)\rangle \\ & \quad \& f(i) = f(j) = 1], \\ 1 & \text{if } (\exists i, j < a)[a = \langle h(\forall)\rangle * i * j \,\&\, i \text{ is the Gödel} \\ & \quad \text{number of a variable and } f(j) = 1], \\ 0 & \quad \text{otherwise.} \end{cases}$$

By the same argument used for the set of Gödel numbers of terms, we have the representability of f. ■

5. There is a representable function Sb such that for a term or formula α, variable x, and term t,

$$\mathrm{Sb}(\sharp\alpha, \sharp x, \sharp t) = \sharp\alpha^x_t.$$

Proof We will define Sb to be the unique function such that $\mathrm{Sb}(a, b, c) = d$ iff one of the following holds:

(i) a is the Gödel number of a variable and $a = b$ and

$$d = c.$$

(ii) $(\exists i, k < a)[i$ is a sequence number & $(\forall j < \mathrm{lh}\, i)(i)_j$ is the Gödel number of a term & k is the value of h at some (lh i)-place function or predicate symbol and $a = \langle k \rangle * \ast_{j<\text{lh}\, i}(i)_j$ and

$$d = \langle k \rangle * \ast_{j<\text{lh}\, i}\, \text{Sb}((i)_j, b, c)].$$

(iii) $(\exists i < a)[i$ is the Gödel number of a wff & $a = \langle h(\,(\,),\ h(\neg)\rangle * i * \langle h(\,)\,)\rangle$ and

$$d = \langle h(\,(\,), h(\neg)\rangle * \text{Sb}(i, b, c) * \langle h()\,)\rangle].$$

(iv) $(\exists i, j < a)[i$ and j are Gödel numbers of wffs & $a = \langle h(\,(\,)\rangle * i * \langle h(\rightarrow)\rangle * j * \langle h(\,)\,)\rangle$ and

$$d = \langle h(\,(\,)\rangle * \text{Sb}(i, b, c) * \langle h(\rightarrow)\rangle * \text{Sb}(j, b, c) * \langle h(\,)\,)\rangle].$$

(v) $(\exists i, j < a)[i$ is the Gödel number of a variable & $i \neq b$ & j is the Gödel number of a wff & $a = \langle h(\forall)\rangle * i * j$ and

$$d = \langle h(\forall)\rangle * i * \text{Sb}(j, b, c)].$$

(vi) None of the above conditions on a and b are met (where we ignore the displayed equation for d) and $d = a$.

Observe that the above does uniquely determine the value of Sb(a, b, c). No two clauses could apply to one number a. And if, for example, clause (ii) applies to a, then we know from Section 2.3 that the numbers i and k are unique.

Finally, we apply the argument used for the set of Gödel numbers of terms, viewing Sb as a 4-ary relation. ■

6. The function whose value at n is $\#(\mathbf{S}^n\mathbf{0})$ is representable.

Proof Call this function f; then

$$f(0) = \langle h(\mathbf{0})\rangle,$$
$$f(n + 1) = \langle h(\mathbf{S})\rangle * f(n).$$

Apply Exercise 8 of the preceding section. ■

7. There is a representable relation Fr such that for a term or formula α and a variable x,

$$\langle \#\alpha, \#x\rangle \in \text{Fr} \Leftrightarrow x \text{ occurs free in } \alpha.$$

Proof $\langle a, b\rangle \in \text{Fr} \Leftrightarrow \text{Sb}(a, b, \#\mathbf{0}) \neq a.$ ■

8. The set of Gödel numbers of sentences is representable.

Proof a is the Gödel of a sentence iff a is the Gödel number of a formula and for any $b < a$, if b is the Gödel number of variable then $\langle a, b \rangle \notin \mathrm{Fr}$. ■

9. There is a representable relation Sbl such that for a formula α, variable x, and term t, $\langle \sharp\alpha, \sharp x, \sharp t \rangle \in \mathrm{Sbl}$ iff t is substitutable for x in α.

Proof Exercise 1. ■

10. The relation Gen, where $\langle a, b \rangle \in \mathrm{Gen}$ iff a is the Gödel number of a formula and b is the Gödel number of a generalization of that formula, is representable.

Proof $\langle a, b \rangle \in \mathrm{Gen}$ iff $a = b$ or $(\exists i, j < b)[i$ is the Gödel number of a variable and $\langle a, j \rangle \in \mathrm{Gen}$ and $b = \langle h(\forall) \rangle * i * j]$. Apply the usual argument to the characteristic function of Gen. ■

11. The set of Gödel numbers of tautologies is representable.

The set of tautologies is intuitively decidable since we can use the method of truth tables. To obtain representability, we recast truth tables in terms of Gödel numbers. There are several preliminary steps:

11.1 The relation R, such that $\langle a, b \rangle \in R$ iff a is the Gödel number of a formula α and b is the Gödel number of a prime constituent of α, is representable.

Proof $\langle a, b \rangle \in R \Leftrightarrow a$ is the Gödel number of a formula and one of

(i) $a = b \;\&\; (a)_0 \neq (\,()$.
(ii) $(\exists i < a)[a = \langle h(\,(\,), h(\neg) \rangle * i * \langle h(\,)\,) \rangle$ and $\langle i, b \rangle \in R]$.
(iii) The analog to (ii) but with $\rightarrow$.

Apply the usual argument to the characteristic function of R. ■

11.2 There is a representable function P such that for a formula α, $P(\sharp\alpha) = \langle \sharp\beta_1, \ldots, \sharp\beta_n \rangle$, the list of Gödel numbers of prime constituents of α, in numerical order.

Proof First define a function g for locating the next prime constituent in $\natural a$ after $\natural y$ (where $\natural a$ is the formula α for which $a = \sharp\alpha$).

$$g(a, y) = \text{the least } n \text{ such that either } n = a + 1 \text{ or both } y < n \text{ and } \langle a, n \rangle \in R.$$

Next define a function h such that $h(a, n)$ lists the first n prime constituents of $\natural a$ (if there are that many):

$$h(a, 0) = \begin{cases} \langle\ \rangle & \text{if } g(a, 0) > a, \\ \langle g(a, 0)\rangle & \text{otherwise.} \end{cases}$$

$$h(a, n+1) = \begin{cases} h(a, n) & \text{if } g(a, t) > a, \\ h(a, n) * \langle g(a, t)\rangle & \text{otherwise.} \end{cases}$$

Here t is the last component of $h(a, n)$, obtained from $h(a, n)$ by the representable function of Exercise 9 of the preceding section. Finally, let $P(a) = h(a, a)$. ■

11.3 Say that the integer v *encodes a truth assignment* for α iff v is a sequence number and $\operatorname{lh} v = \operatorname{lh} P(\sharp\alpha)$ and $(\forall i < \operatorname{lh} v)(\exists e < 2)(v)_i = \langle(P(\sharp\alpha))_i, e\rangle$. This is a representable condition on v and $\sharp\alpha$.

For example, if $P(\alpha) = \langle \sharp\beta_0, \ldots, \sharp\beta_n\rangle$, then

$$v = \langle\langle\sharp\beta_0, e_0\rangle, \ldots, \langle\sharp\beta_n, e_n\rangle\rangle,$$

where each e_i is 0 or 1. We will later need an upper bound for v in terms of $\sharp\alpha$. The largest v is obtained when each e_i is 1. Also $\sharp\beta_i \leq \sharp\alpha$, so that

$$\begin{aligned} v &\leq \langle\langle\sharp\alpha, 1\rangle, \ldots, \langle\sharp\alpha, 1\rangle\rangle \\ &= \ast_{i<\operatorname{lh} P(\sharp\alpha)}\langle\langle\sharp\alpha, 1\rangle\rangle. \end{aligned}$$

11.4 There is a representable relation Tr such that for a formula α and a v which encodes a truth assignment for α (or more), $\langle\sharp\alpha, v\rangle \in \text{Tr}$ iff that truth assignment satisfies α.

Proof Exercise 2. ■

Finally, α is a tautology iff α is a formula and for every v encoding a truth assignment for α, $\langle\sharp\alpha, v\rangle \in \text{Tr}$. The (English) quantifier on v can be bounded by a representable function of $\sharp\alpha$, as explained in 11.3.

12. The set of Gödel numbers of formulas of the form $\forall x\, \varphi \to \varphi^x_t$, where t is a term substitutable for the variable x in φ, is representable.

Proof α is of this form iff $(\exists \text{ wff } \varphi < \alpha)(\exists \text{ variable } x < \alpha)(\exists \text{ term } t < \alpha)[t$ is substitutable for x in φ and $\alpha = \forall x\, \varphi \to \varphi^x_t]$. Here "$\varphi < \alpha$" means that $\sharp\varphi < \sharp\alpha$. This is easily rewritten in terms of Gödel numbers:

a belongs to the set iff $(\exists f < a)(\exists x < a)(\exists t < a)[f$ is the Gödel number of a formula & x is the Gödel number of a variable & t is the Gödel number of term and $\langle f, x, t\rangle \in \text{Sbl}$ &

$$a = \langle h(\,(\,),\ h(\forall)\rangle * x * f * \langle h(\rightarrow)\rangle * \text{Sb}(f, x, t) * \langle h(\,)\,)\rangle].$$

13. The set of Gödel numbers of formulas of the form $\forall x(\alpha \rightarrow \beta) \rightarrow \forall x\,\alpha \rightarrow \forall x\,\beta$ is representable.

Proof γ is of this form iff ($\exists$ variable $x < \gamma$) ($\exists$ formulas $\alpha, \beta < \gamma$)[$\gamma = \forall x(\alpha \rightarrow \beta) \rightarrow \forall x\,\alpha \rightarrow \forall x\,\beta$]. This is easily rewritten in terms of Gödel numbers, as in 12. ■

14. The set of Gödel numbers of formulas of the form $\alpha \rightarrow \forall x\,\alpha$, where x does not occur free in α, is representable.

Proof Similar to 13. ■

15. The set of Gödel numbers of formulas of the form $x \approx x$ is representable.

Proof Similar to 13. ■

16. The set of Gödel numbers of formulas of the form $x \approx y \rightarrow \alpha \rightarrow \alpha'$, where α is atomic and α' is obtained from α by replacing x at zero or more places by y, is representable.

Proof This is similar to 13, except for the relation of "partial substitution." Let $\langle a, b, x, y\rangle \in \text{Psb}$ iff x and y are Gödel numbers of variables, a is the Gödel number of an atomic formula, b is a sequence number of length lh a, and for all $j < \text{lh}\ a$, either $(a)_j = (b)_j$ or $(a)_j = x$ and $(b)_j = y$. This relation is representable. ■

17. The set of Gödel numbers of logical axioms is representable.

Proof α is a logical axiom iff $\exists\beta \leq \alpha$ such that α is a generalization of β and β is in one of the sets in items 11–16. ■

18. For a finite set A of formulas,

$$\{\mathscr{G}(D) : D \text{ is a deduction from } A\}$$

is representable.

Proof A number d belongs to this set iff d is a sequence number of positive length and for every i less than lh d, either

1. $(d)_i \in \sharp A$,
2. $(d)_i$ is the Gödel number of a logical axiom, or
3. $(\exists j, k < i)\ [(d)_j = \langle h(\,(\,)\rangle * (d)_k * \langle h(\rightarrow)\rangle * (d)_i * \langle h(\,)\,)\rangle]$.

This is representable whenever $\sharp A$ is, as is certainly the case for finite A. ■

19. Any recursive relation is representable in Cn A_E.

Proof Recall that the relation R is recursive iff there is *some* finite consistent set A of sentences such that some formula ϱ represents R in Cn A. (There is no loss of generality in assuming that the language has only finitely many parameters: those in the finite set A, those in ϱ, and **0**, **S**, and $\forall$.) In the case of a unary relation R, we have that $a \in R$ iff the least D which is a deduction from A of either $\varrho(\mathbf{S}^a\mathbf{0})$ or $\neg\varrho(\mathbf{S}^a\mathbf{0})$ is, in fact, a deduction of the former.

More formally, $a \in R$ iff the last component of $f(a)$ is $\sharp\varrho(\mathbf{S}^a\mathbf{0})$, where

$f(a) =$ the least d such that d is in the set of item 18 and the last component of d is either $\sharp\varrho(\mathbf{S}^a\mathbf{0})$ or $\sharp\neg\varrho(\mathbf{S}^a\mathbf{0})$.

For this (fixed) ϱ, there always is such a d. ■

Since the converse to item 19 is immediate, we have

Theorem 34A A relation is recursive iff it is representable in the theory Cn A_E.

Henceforth we will usually use the word "recursive" in preference to "representable."

Corollary 34B Any recursive relation is definable in $\mathfrak{N}$.

20. Now suppose we have a set A of sentences such that $\sharp A$ is recursive. Then $\sharp$Cn A need *not* be recursive (as we will show in the next section). But we do have a way of defining Cn A from A:

$a \in \sharp$Cn A iff $\exists d$[d is the number of a deduction from A and the last component of d is a and a is the Gödel number of a sentence]

The part in square brackets is recursive, by the proof to item 18. But we cannot in general put any bound on the number d. The best we can say is that $\sharp\text{Cn } A$ is the domain of a recursive relation (or, as we will say later, is *recursively enumerable*).

21. If $\sharp A$ is recursive and Cn A is a complete theory, then $\sharp\text{Cn } A$ is recursive.

In other words, a complete recursively axiomatizable theory is recursive. This is the analog to Corollary 25G, which asserts that a complete axiomatizable theory is decidable.

The proof is essentially unchanged. Let

$g(s) =$ the least d such that s is not a sentence, or d is in the set of item 18 and the last component of d is either s or is $\langle h(\,(\,), h(\neg)\rangle * s * \langle h(\,)\,)\rangle$.

Thus $g(\sharp\sigma)$ is $\mathcal{G}$ of the least deduction of σ or $(\neg\sigma)$ from A. And $s \in \sharp\text{Cn } A$ iff $s > 0$ and the last component of $g(s)$ is s. ■

At this point we might reconsider the plausibility of Church's thesis. Suppose that the relation R is decidable. Then there is a finite list of explicit instructions for the decision procedure. The procedure itself will presumably consist of certain atomic steps, which are then performed repeatedly. (The reader familiar with computer programming will know that a short program can still require much time for its execution, but some commands will be utilized over and over.) Any one atomic step is presumably very simple.

By devices akin to Gödel numbering, we can mirror the decision procedure in the integers. The characteristic function of R can then be put in the form

$K_R(\vec{a}) = U$[the least s such that

(i) $(s)_0$ encodes the input $\vec{a}$;

(ii) for all positive $i < \text{lh } s$, $(s)_i$ is obtained from $(s)_{i-1}$ by performance of the applicable atomic step;

(iii) the last component of s describes a terminal situation, at which the computation is completed],

where U (the upshot function) is some simple function that extracts from the last component of s the answer (affirmative or negative). The recursiveness of R is now reduced to the recursiveness of U and of the relations indicated in (i), (ii), and (iii). In special cases, such as decision procedures provided by the register machines of Section 3.8, the recursiveness of these components is easily verified. It seems most improbable that any decision procedure will ever be regarded as effective and yet will have components which are nonrecursive. For example in (ii), it seems that it ought to be possible to make each atomic step extremely simple, and in particular to make each one recursive.

EXERCISES

1. Supply a proof for item 9 of this section.

2. Supply a proof for item 11.4 of this section.

3. Use Exercise 11 of Section 3.3 to give a new proof that the set of Gödel numbers of terms is representable (item 2).

4. Let T be a consistent recursively axiomatizable theory in a recursively numbered language with $\mathbf{0}$ and $\mathbf{S}$. Show that any relation representable in T must be recursive.

§ 3.5 INCOMPLETENESS AND UNDECIDABILITY

In this section we reap the rewards of our work in Sections 3.3 and 3.4. We have assigned Gödel numbers to expressions, and we have shown that certain intuitively decidable relations on N (related to syntactical notions about expressions) are representable in Cn A_E.

Throughout this section we assume that the language in question is the language of $\mathfrak{N}$. (This affects the meaning of "Cn" and "theory.")

Fixed-Point Lemma Given any formula β in which only $\boldsymbol{v}_1$ occurs free, we can find a sentence σ such that

$$A_E \vdash [\sigma \leftrightarrow \beta(\mathbf{S}^{\#\sigma}\mathbf{0})].$$

We can think of σ as *indirectly* saying, "β is true of me." Actually σ doesn't say anything of course, it's just a string of symbols. And even when translated into English according to the intended structure $\mathfrak{N}$, it then talks of numbers and their successors and products and so forth. It is only by virtue of our having associated numbers with expressions that we can think of σ as referring to a formula, in this case to σ itself.

Proof Let $\theta(\mathbf{v}_1, \mathbf{v}_2, \mathbf{v}_3)$ functionally represent in Cn A_E a function whose value at $\langle \sharp\alpha, n \rangle$ is $\sharp(\alpha(\mathbf{S}^n\mathbf{0}))$. (See items 5 and 6 in Section 3.4.) First consider the formula

$$\text{(1)} \qquad \forall \mathbf{v}_3[\theta(\mathbf{v}_1, \mathbf{v}_1, \mathbf{v}_3) \rightarrow \beta(\mathbf{v}_3)].$$

(This formula has only $\mathbf{v}_1$ free. It defines in $\mathfrak{N}$ a set to which $\sharp\alpha$ belongs iff $\sharp(\alpha(\mathbf{S}^{\sharp\alpha}\mathbf{0}))$ is in the set defined by β.) Let q be the Gödel number of (1). Let σ be

$$\forall \mathbf{v}_3[\theta(\mathbf{S}^q\mathbf{0}, \mathbf{S}^q\mathbf{0}, \mathbf{v}_3) \rightarrow \beta(\mathbf{v}_3)].$$

Thus σ is obtained from (1) by replacing $\mathbf{v}_1$ by $\mathbf{S}^q\mathbf{0}$. Notice that σ does assert (under $\mathfrak{N}$) that $\sharp\sigma$ is in the set defined by β. But we must check that

$$\text{(2)} \qquad \sigma \leftrightarrow \beta(\mathbf{S}^{\sharp\sigma}\mathbf{0})$$

is a consequence of A_E. We have, by our choice of θ,

$$\text{(3)} \qquad A_E \vdash \forall \mathbf{v}_3[\theta(\mathbf{S}^q\mathbf{0}, \mathbf{S}^q\mathbf{0}, \mathbf{v}_3) \leftrightarrow \mathbf{v}_3 \approx \mathbf{S}^{\sharp\sigma}\mathbf{0}].$$

We can obtain (2) as follows:

($\rightarrow$): It is clear (by looking at σ) that

$$\sigma \vdash \theta(\mathbf{S}^q\mathbf{0}, \mathbf{S}^q\mathbf{0}, \mathbf{S}^{\sharp\sigma}\mathbf{0}) \rightarrow \beta(\mathbf{S}^{\sharp\sigma}\mathbf{0}).$$

And, by (3),

$$A_E \vdash \theta(\mathbf{S}^q\mathbf{0}, \mathbf{S}^q\mathbf{0}, \mathbf{S}^{\sharp\sigma}\mathbf{0}).$$

Hence

$$A_E ; \sigma \vdash \beta(\mathbf{S}^{\sharp\sigma}\mathbf{0}),$$

which gives half of (2).

($\leftarrow$): The sentence in (3) implies

$$\beta(\mathbf{S}^{\sharp\sigma}\mathbf{0}) \rightarrow [\forall \mathbf{v}_3(\theta(\mathbf{S}^q\mathbf{0}, \mathbf{S}^q\mathbf{0}, \mathbf{v}_3) \rightarrow \beta(\mathbf{v}_3))].$$

But the part in square brackets is just σ. ■

Our first application of this lemma does not concern the subtheory Cn A_E, and requires only the weaker fact that

$$\models_{\mathfrak{N}}[\sigma \leftrightarrow \beta(\mathbf{S}^{\sharp\sigma}\mathbf{0})].$$

Tarski Undefinability Theorem The set $\sharp$Th $\mathfrak{N}$ is not definable in $\mathfrak{N}$.

Proof Consider any formula β (which you suspect *might* define $\sharp\mathrm{Th}\,\mathfrak{N}$). By the fixed-point lemma (applied to $\neg\beta$) we have a sentence σ such that

$$\models_{\mathfrak{N}}[\sigma \leftrightarrow \neg\beta(\mathbf{S}^{\sharp\sigma}\mathbf{0})].$$

(If β did define $\sharp\mathrm{Th}\,\mathfrak{N}$, then σ would indirectly say "I am false.") Then

$$\models_{\mathfrak{N}} \sigma \Leftrightarrow \not\models_{\mathfrak{N}} \beta(\mathbf{S}^{\sharp\sigma}\mathbf{0}),$$

so either σ is true but not in the set β defines, or it is false and in that set. Either way σ shows that β cannot define $\sharp\mathrm{Th}\,\mathfrak{N}$. ∎

The above theorem gives at once the undecidability of the theory of $\mathfrak{N}$:

Corollary 35A $\sharp\mathrm{Th}\,\mathfrak{N}$ is not recursive.

Proof Any recursive set is (by Corollary 34B) definable in $\mathfrak{N}$. ∎

Gödel Incompleteness Theorem (1931) If $A \subseteq \mathrm{Th}\,\mathfrak{N}$ and $\sharp A$ is recursive, then Cn A is not a complete theory.

Thus there is no complete recursive axiomatization of Th $\mathfrak{N}$.

Proof Since $A \subseteq \mathrm{Th}\,\mathfrak{N}$, we have $\mathrm{Cn}\,A \subseteq \mathrm{Th}\,\mathfrak{N}$. If Cn A is a complete theory, then equality holds. But if Cn A is a complete theory, then $\sharp\mathrm{Cn}\,A$ is recursive (item 21 of the preceding section). And by the above corollary, $\sharp\mathrm{Th}\,\mathfrak{N}$ is not recursive. ∎

In particular, Cn A_E is not a complete theory and so is not equal to Th $\mathfrak{N}$. And the incompleteness would not be eliminated by the addition of any recursive set of true axioms. (By a recursive set of sentences we mean of course a set Σ for which $\sharp\Sigma$ is recursive.)

We can extract more information from the proof of Gödel's theorem. Suppose we have a particular recursive $A \subseteq \mathrm{Th}\,\mathfrak{N}$ in mind. Then by Theorem 34A we can find a formula β which defines $\sharp\mathrm{Cn}\,A$ in $\mathfrak{N}$. The sentence σ produced by the proof to Tarski's theorem is (as we noted there) a true sentence *not* in Cn A. This sentence asserts that $\sharp\sigma$ does not belong to the set defined by β, i.e., it indirectly says, "I am not a theorem of A." Thus $A \nvdash \sigma$, and of course $A \nvdash \neg\sigma$ as well. This way of viewing the proof is closer to Gödel's original proof, which did not involve a detour through Tarski's theorem. For that matter, Gödel's statement of the theorem did not involve Th $\mathfrak{N}$; we have taken some liberties in the labeling of theorems.

Next we need a lemma which says (roughly) that one can add one new axiom (and hence finitely many new axioms) to a recursive theory without losing the property of recursiveness.

Lemma 35B If $\sharp\mathrm{Cn}\,\Sigma$ is recursive, then $\sharp\mathrm{Cn}(\Sigma\,;\tau)$ is recursive.

Proof $\alpha \in \mathrm{Cn}(\Sigma\,;\tau) \Leftrightarrow (\tau \rightarrow \alpha) \in \mathrm{Cn}\,\Sigma$. Thus

$$a \in \sharp\mathrm{Cn}(\Sigma\,;\tau) \Leftrightarrow a \text{ is the Gödel number of a sentence and } \langle h(\,(\,)\rangle * \sharp\tau * \langle h(\rightarrow)\rangle * a * \langle h(\,)\,)\rangle \text{ is in } \sharp\mathrm{Cn}\,\Sigma.$$

This is recursive by the results of the preceding sections. ■

Theorem 35C (strong undecidability of Cn A_E) Let T be a theory such that $T \cup A_E$ is consistent. Then $\sharp T$ is not recursive.

(Notice that because throughout this section the language in question is the language of $\mathfrak{N}$, the word "theory" here means "theory in the language of $\mathfrak{N}$.")

Proof Let T' be the theory $\mathrm{Cn}(T \cup A_E)$. If $\sharp T$ is recursive, then since A_E is finite we can conclude by the above lemma that $\sharp T'$ is also recursive.

Suppose, then, that $\sharp T'$ is recursive and so is represented in Cn A_E by some formula β. From the fixed-point lemma we get a sentence σ such that

$$(*) \qquad A_E \vdash [\sigma \leftrightarrow \neg\beta(\mathbf{S}^{\sharp\sigma}\mathbf{0})].$$

(Indirectly σ asserts, "I am not in T'.")

$$\begin{aligned} \sigma \notin T' &\Rightarrow \sharp\sigma \notin \sharp T' \\ &\Rightarrow A_E \vdash \neg\beta(\mathbf{S}^{\sharp\sigma}\mathbf{0}) \\ &\Rightarrow A_E \vdash \sigma \qquad \text{by } (*) \\ &\Rightarrow \sigma \in T'. \end{aligned}$$

So we get $\sigma \in T'$. But this, too, is untenable:

$$\begin{aligned} \sigma \in T' &\Rightarrow \sharp\sigma \in \sharp T' \\ &\Rightarrow A_E \vdash \beta(\mathbf{S}^{\sharp\sigma}\mathbf{0}) \\ &\Rightarrow A_E \vdash \neg\sigma \qquad \text{by } (*) \\ &\Rightarrow (\neg\sigma) \in T', \end{aligned}$$

which contradicts the consistency of T'. ■

Corollary 35D Assume that $\sharp\Sigma$ is recursive and $\Sigma \cup A_E$ is consistent. Then $\mathrm{Cn}\,\Sigma$ is not a complete theory.

Proof A complete recursively axiomatizable theory is recursive (item 21 of Section 3.4). But $\sharp\mathrm{Cn}\,\Sigma$ is not recursive, by the above theorem. ■

This corollary is Gödel's incompleteness theorem again but with truth in $\mathfrak{N}$ replaced by consistency with A_E.

Church's Theorem (1936) The set of Gödel numbers of valid sentences (in the language of $\mathfrak{N}$) is not recursive.

Proof In the strong undecidability of $\mathrm{Cn}\,A_E$, take T to be the smallest theory in the language, the set of valid sentences. ■

The set of Gödel numbers of valid wffs is not recursive either, lest the set of valid sentences be recursive.

This proof applies to the language of $\mathfrak{N}$. For a language with more parameters, the set of valid sentences is still nonrecursive (lest its intersection with the language of $\mathfrak{N}$ be recursive). Actually it is enough for the language to contain at least one two-place predicate symbol. (See Corollary 36D.) On the other hand, *some* restriction on the language is needed. If the language has $\forall$ as its only parameter (the language of equality), then the set of valid formulas is decidable. (See Exercise 6.) More generally, it is known that if the only parameters are $\forall$ and one-place predicate symbols, then the set of valid formulas is decidable.

Recursive enumerability

A relation on the natural numbers is said to be *recursively enumerable* iff it is of the form

$$\{\vec{a} : \exists b\, \langle \vec{a}, b\rangle \in Q\}$$

with Q recursive. Recursively enumerable relations play an important role in logic. They constitute the formal counterpart to the effectively enumerable relations (as will be explained presently).

Theorem 35E The following conditions on an m-ary relation R are equivalent:

1. R is recursively enumerable.
2. R is the domain of some recursive relation Q.

3. For some recursive $(m+1)$-ary relation Q,

$$R = \{\langle a_1, \ldots, a_m\rangle : \exists b\langle a_1, \ldots, a_m, b\rangle \in Q\}.$$

4. For some recursive $(m+n)$-ary relation Q,

$$R = \{\langle a_1, \ldots, a_m\rangle : \exists b_1, \ldots, b_n\langle a_1, \ldots, a_m, b_1, \ldots, b_n\rangle \in Q\}.$$

Proof By definition 1 and 3 are equivalent. Also 2 and 3 are equivalent by our definition (in Chapter 0) of domain and $(m+1)$-tuple. Clearly 3 implies 4. So we have only to show that 4 implies 3. This is because

$$\exists b_1, \ldots, b_n \langle a_1, \ldots, a_m, b_1, \ldots, b_n\rangle \in Q$$
$$\text{iff } \exists c\langle a_1, \ldots, a_m, (c)_0, \ldots, (c)_{n-1}\rangle \in Q$$

and

$$\{\langle a_1, \ldots, a_m, c\rangle : \langle a_1, \ldots, a_m, (c)_0, \ldots, (c)_{n-1}\rangle \in Q\}$$

is recursive whenever Q is recursive. (Here we haver used our sequence decoding function to compress a string of quantifiers into a single one.) ■

By part 4 of this theorem, R is recursively enumerable iff it is definable in $\mathfrak{N}$ by a formula $\exists x_1 \cdots \exists x_n \varphi$, where φ is numeralwise determined by A_E. In fact, we can require here that φ be quantifier-free; this result was proved in 1961 (with exponentiation) and 1970 (without exponentiation). The proofs involve some number theory; we will omit them here.

Notice that any recursive relation is also recursively enumerable. For if R is recursive, then it is defined in $\mathfrak{N}$ by a formula $\exists x_1 \cdots \exists x_n \varphi$, where φ is numeralwise determined by A_E and $x_1, \ldots, x_n$ do not occur in φ.

Theorem 35F A relation is recursive iff both it and its complement are recursively enumerable.

This is the formal counterpart to the fact (cf. Theorem 17F) that a relation is decidable iff both it and its complement can be effectively enumerated.

Proof If a relation is recursive, then so is its complement, whence both are recursively enumerable.

Conversely, suppose that both P and its complement are recursively enumerable; thus for any $\vec{a}$,

$$\vec{a} \in P \Leftrightarrow \exists b \langle \vec{a}, b\rangle \in Q,$$
$$\vec{a} \notin P \Leftrightarrow \exists b \langle \vec{a}, b\rangle \in R$$

for some recursive Q and R. Let

$$f(\vec{a}) = \text{the least } b \text{ such that either } \langle \vec{a}, b \rangle \in Q \text{ or } \langle \vec{a}, b \rangle \in R.$$

Such a number b always exists, and f is recursive. Finally,

$$\vec{a} \in P \Leftrightarrow \langle \vec{a}, f(\vec{a}) \rangle \in Q,$$

so P is recursive. ■

The recursively enumerable relations constitute the formal counterpart of the effectively enumerable relations. For we have the following informal result, parallelling a characterization of recursive enumerability given by Theorem 35E.

⋆Lemma 35G A relation is effectively enumerable iff it is the domain of a decidable relation.

Proof Assume that Q is effectively enumerated by some procedure. Then $\vec{a} \in Q$ iff $\exists n[\vec{a}$ appears in the enumeration in n steps]. The relation defined in square brackets is decidable and has domain Q.

Conversely, to enumerate $\{\langle a, b \rangle : \exists n \; \langle a, b, n \rangle \in R\}$ for decidable R, we check to see if $\langle (m)_0, (m)_1, (m)_2 \rangle \in R$ for $m = 0, 1, 2, \ldots$. Whenever the answer is affirmative, we place $\langle (m)_0, (m)_1 \rangle$ on the output list. ■

⋆Corollary 35H (Church's thesis, second form) A relation is effectively enumerable iff it is recursively enumerable.

Proof By identifying the class of decidable relations with the class of recursive relations, we automatically identify the domains of decidable relations with domains of recursive relations. ■

The second form of Church's thesis is, in fact, equivalent to the first form. To prove the first form from the second, we use Theorems 35F and 17F.

We have already shown that a recursively axiomatizable theory is recursively enumerable, but using different words. We restate the result here, as it indicated the role recursive enumerability plays in logic.

Theorem 35I If A is a set of sentences such that $\sharp A$ is recursive, then $\sharp \mathrm{Cn}\, A$ is recursively enumerable.

Proof Item 20 of Section 3.4. ■

This theorem is the precise counterpart of the intuitive fact that a theory with a decidable set of axioms is effectively enumerable (Corollaries 25F and 26F). It indicates the gap between what is *provable* in an axiomatic theory and what is *true* in the intended structure. With a recursive set of axioms, all we can possibly obtain is a recursively enumerable set of consequences. But by Tarski's theorem, Th $\mathfrak{N}$ is not even definable in $\mathfrak{N}$, much less recursively enumerable.

Even if we expand the language or add new axioms, the same phenomenon is present. As long as we can recursively distinguish deductions from nondeductions, the set of theorems can be only recursively enumerable. For example, the set of sentences of number theory provable in your favorite system of axiomatic set theory is recursively enumerable. Furthermore, this set includes A_E and is consistent (unless you have very strange favorites). It follows that this set theory is nonrecursive and incomplete. (This topic is discussed more carefully in Section 3.6.)

Weak representability

Consider a recursively enumerable set Q, where

$$a \in Q \Leftrightarrow \exists b \, \langle a, b \rangle \in R$$

for recursive R. We know there is a formula ϱ which represents R in Cn A_E. Consequently, the formula $\exists \mathbf{v}_2 \, \varrho$ defines Q in $\mathfrak{N}$. This formula cannot represent Q in Cn A_E unless Q is recursive. But it can come halfway.

$$\begin{aligned} a \in Q &\Rightarrow \langle a, b \rangle \in R && \text{for some } b \\ &\Rightarrow A_E \vdash \varrho(\mathbf{S}^a\mathbf{0}, \mathbf{S}^b\mathbf{0}) && \text{for some } b \\ &\Rightarrow A_E \vdash \exists \mathbf{v}_2 \, \varrho(\mathbf{S}^a\mathbf{0}, \mathbf{v}_2). \\ a \notin Q &\Rightarrow \langle a, b \rangle \notin R && \text{for all } b \\ &\Rightarrow A_E \vdash \neg\varrho(\mathbf{S}^a\mathbf{0}, \mathbf{S}^b\mathbf{0}) && \text{for all } b \\ &\Rightarrow A_E \nvdash \exists \mathbf{v}_2 \, \varrho(\mathbf{S}^a\mathbf{0}, \mathbf{v}_2). \end{aligned}$$

The last step is justified by the fact that if $A_E \vdash \neg\varphi(\mathbf{S}^b\mathbf{0})$ for all b, then $A_E \nvdash \exists x \, \varphi(x)$. (The term *ω-consistency* is given to this property.) For it is impossible for $\exists x \, \varphi(x)$, $\neg\varphi(\mathbf{S}^0\mathbf{0})$, $\neg\varphi(\mathbf{S}^1\mathbf{0})$, ... all to be true in $\mathfrak{N}$.

Thus we have

$$a \in Q \Leftrightarrow A_E \vdash \exists \mathbf{v}_2 \, \varphi(\mathbf{S}^a\mathbf{0}, \mathbf{v}_2).$$

It will be convenient to formulate a definition of this half of representability.

Definition Let Q be an n-ary relation on $\mathbb{N}$, ψ a formula in which only $\mathbf{v}_1, \ldots, \mathbf{v}_n$ occur free. Then ψ *weakly represents* Q in a theory T iff for every $a_1, \ldots, a_n$ in $\mathbb{N}$,

$$\langle a_1, \ldots, a_n \rangle \in Q \Leftrightarrow \psi(\mathbf{S}^{a_1}\mathbf{0}, \ldots, \mathbf{S}^{a_n}\mathbf{0}) \in T.$$

Observe that if Q is representable in a consistent theory T, then Q is also weakly representable in T.

Theorem 35J A relation is weakly representable in Cn A_E iff it is recursively enumerable.

Proof We just showed that a recursively enumerable unary relation Q is weakly representable in Cn A_E; the same proof applies to n-ary Q with only notational changes. Conversely, let Q be weakly represented by ψ in Cn A_E. Then

$$\begin{aligned} \langle a_1, \ldots, a_n \rangle \in R &\Leftrightarrow \exists D[D \text{ is a deduction of } \psi(\mathbf{S}^{a_1}\mathbf{0}, \ldots, \mathbf{S}^{a_n}\mathbf{0}) \\ &\qquad \text{from the axioms } A_E] \\ &\Leftrightarrow \exists d \langle d, f(a_1, \ldots, a_n) \rangle \in P \end{aligned}$$

for a certain recursive function f and recursive relation P. ■

Arithmetical hierarchy

Define a relation on the natural numbers to be *arithmetical* iff it is definable in $\mathfrak{N}$. But some arithmetical relations are, in a sense, more definable than others. We can organize the arithmetical relations into a hierarchy according to how definable the relations are.

Let Δ_1 be the class of recursive relations, and let Σ_1 be the class of recursively enumerable relations. Then we proceed to define the classes Π_k and Σ_k by recursion on k. Let a relation belong to Π_k iff its complement is in Σ_k. And let Σ_{k+1} be the class of domains of Π_k relations. (If we let $\Sigma_0 = \Pi_0 = \Delta_1$, then this will hold also for $k = 0$.) For example, the first few classes consist of relations of the form shown in the second column:

Σ_1: $\{\vec{a} : \exists b \langle \vec{a}, b \rangle \in R\}$, R recursive.

Π_1: $\{\vec{a} : \forall b \langle \vec{a}, b \rangle \in R\}$, R recursive.

Σ_2: $\{\vec{a} : \exists c\, \forall b \langle \vec{a}, b, c \rangle \in R\}$, R recursive.

Π_2: $\{\vec{a} : \forall c\, \exists b \langle \vec{a}, b, c \rangle \in R\}$, R recursive.

In general, a relation Q is in Π_k iff it is of then form

$$\{\vec{a} : \forall b_1\, \exists b_2 \cdots \square\, b_k \langle \vec{a}, \vec{b}\rangle \in R\}$$

for a recursive relation R. Here "$\square$" is to be replaced by "$\forall$" if k is odd and by "$\exists$" if k is even. Similarly, Q is in Σ_k iff it has the form

$$\{\vec{a} : \exists b_1\, \forall b_2 \cdots \square\, b_k \langle \vec{a}, \vec{b}\rangle \in R\}$$

for recursive R, where now "$\square$" is replaced by "$\exists$" if k is odd and by "$\forall$" if k is even.

EXAMPLE The set of Gödel numbers of formulas numeralwise determined by A_E is in Π_2.

Proof a belongs to this set iff [a is the Gödel number of a formula α] and $\forall b\, \exists d$[d is $\mathscr{G}$ of a deduction from A_E either of $\alpha(\mathbf{S}^{(b)_0}\mathbf{0}, \mathbf{S}^{(b)_1}\mathbf{0}, \ldots)$ or of the negation of this sentence]. By the technique of Section 3.4, we can show that the phrases in square brackets define recursive relations. By using the English counterpart to prenex form, we obtain the desired form,

$$\{a : \forall b\, \exists d \langle a, b, d\rangle \in R\},$$

with R recursive. ■

Our earlier result (Theorem 35F) stating that a relation is recursive iff both it and its complement are recursively enumerable can now be stated by the equation

$$\Delta_1 = \Sigma_1 \cap \Pi_1.$$

Since this equation holds, we proceed to define Δ_n for $n > 1$ by the analogous equation,

$$\Delta_n = \Sigma_n \cap \Pi_n.$$

The following inclusions hold:

$$\begin{array}{ccccccccccc}
 & & \Sigma_1 & & & & \Sigma_2 & & & & \Sigma_3 \\
 & \rotatebox{45}{$\subseteq$} & & \rotatebox{-45}{$\subseteq$} & & \rotatebox{45}{$\subseteq$} & & \rotatebox{-45}{$\subseteq$} & & \rotatebox{45}{$\subseteq$} & \\
\Delta_1 & & & & \Delta_2 & & & & \Delta_3 & & & \cdots \\
 & \rotatebox{-45}{$\subseteq$} & & \rotatebox{45}{$\subseteq$} & & \rotatebox{-45}{$\subseteq$} & & \rotatebox{45}{$\subseteq$} & & \rotatebox{-45}{$\subseteq$} & \\
 & & \Pi_1 & & & & \Pi_2 & & & & \Pi_3
\end{array}$$

The case $\Delta_1 \subseteq \Sigma_1$ was mentioned previously (cf. Theorem 35F); its proof

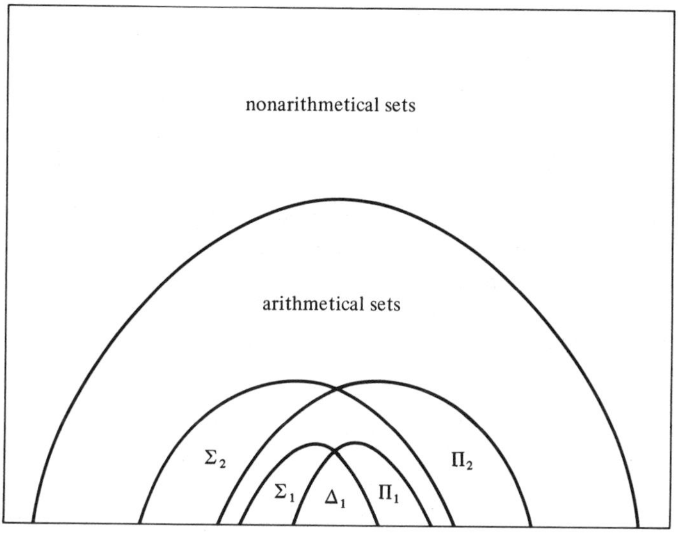

Figure 9. Picture of $\mathscr{P}N$.

hinged on the possibility of "vacuous quantification." The proofs of the other cases are conceptually the same. If x does not occur in φ, then φ, $\forall x\, \varphi$, and $\exists x\, \varphi$ are all equivalent. For example, a relation in Σ_1 is defined by a formula $\exists y\, \varphi$, where φ is numeralwise determined by A_E. But the same relation is defined by $\exists y\, \forall x\, \varphi$ and $\forall x\, \exists y\, \varphi$ (where x does not occur in φ). Hence the relation is also in Σ_2 and Π_2.

It is also true that all the inclusion shown are proper inclusions, i.e., equality does not hold. But we will not prove this fact here. The inclusions are shown pictorially in Fig. 9.

The class of arithmetical relations equals $\bigcup_k \Sigma_k$ and also $\bigcup_k \Pi_k$. For example, any relation in Σ_2 is arithmetical, being defined in $\mathfrak{N}$ by a formula $\exists x\, \forall y\, \varphi$, where φ is numeralwise determined by A_E. Conversely, any arithmetical relation is defined in $\mathfrak{N}$ by some prenex formula. The quantifier-free part of this prenex formula defines a recursive relation (since quantifier-free formulas are numeralwise determined by A_E). Consequently, the defined relation falls somewhere in the hierarchy.

There are certain tricks which are useful in locating specific arithmetical relations in the hierarchy. For example, let A be the set of Gödel numbers of formulas α such that for some n,

$$A_E \vdash \alpha(\mathbf{S}^n\mathbf{0}) \quad \text{and} \quad (\forall i < n)\, A_E \vdash \neg\alpha(\mathbf{S}^i\mathbf{0}).$$

Then $a \in A$ iff [a is the Gödel of a wff α] and $\exists n\{\exists D[D$ is a deduction from

A_E of $\alpha(\mathbf{S}^n\mathbf{0})$] and $(\forall i < n)(\exists D_i)[D_i$ is a deduction from A_E of $\neg\alpha(\mathbf{S}^i\mathbf{0})]\}$. The parts in square brackets are recursive so we count the remaining quantifiers. The bounded quantifier "$\forall i < n$" need not be counted. For we have

$$(\forall i < n)(\exists d)\,\langle d, i\rangle \in P \Leftrightarrow (\exists d)(\forall i < n)\,\langle (d)_i, i\rangle \in P.$$

Use of this fact lets us push the bounded quantifier inward until it merges with the recursive part. Consequently, $A \in \Sigma_1$.

The following theorem generalizes Theorem 35I.

Theorem 35K Let A be a set of sentences such that $\sharp A$ is in Σ_k. Then $\sharp\mathrm{Cn}\,A$ is also in Σ_k.

Proof Return to the proofs of items 18 and 20 of Section 3.4. We had there:

$a \in \sharp\mathrm{Cn}\,A \Leftrightarrow a$ is the Gödel number of a sentence and $\exists d[d$ is a sequence number and the last component of d is a and for every i less than $\mathrm{lh}\,d$, either (1) $(d)_i \in \sharp A$. (2) $(d)_i$ is the Gödel number of a logical axiom, or (3) for some j and k less than i, $(d)_j = \langle h(()\rangle * (d)_k * \langle h(\rightarrow)\rangle * (d)_i * \langle h())\rangle]$.

Since $\sharp A \in \Sigma_k$, in (1) we must replace "$(d)_i \in \sharp A$" by something of the form

$$\exists b_1 \forall b_2 \cdots \Box\, b_k \,\langle (d)_i, \vec{b}\rangle \in Q$$

for recursive Q. It remains to convert the result into an English prenex expression in Σ_k form. We suggest that the reader set $k = 2$ and write out this expression; the device used in the preceding example will help. ■

EXERCISES

1. Show that there is no recursive set R such that $\sharp\mathrm{Cn}\,A_E \subseteq R$ and $\sharp\{\sigma : (\neg\sigma) \in \mathrm{Cn}\,A_E\} \subseteq \bar{R}$. (This result can be stated: The theorems of A_E cannot be recursively separated from the refutable sentences.)

2. Let A be a recursive set of sentences in a recursively numbered language with $\mathbf{0}$ and $\mathbf{S}$. Assume that every recursive set is representable in the theory $\mathrm{Cn}\,A$. Further assume that A is ω-consistent; i.e., there is no formula φ such that $A \vdash \exists x\,\varphi(x)$ and for all $a \in N$, $A \vdash \neg\varphi(\mathbf{S}^a\mathbf{0})$. Construct a sentence σ indirectly asserting that it is not a theorem of A, and show that neither $A \vdash \sigma$ nor $A \vdash \neg\sigma$.

3. Let T be a theory in a recursively numbered language (with $\mathbf{0}$ and $\mathbf{S}$). Assume that all recursive subsets of $\mathbb{N}$ are weakly representable in T. Show that $\sharp T$ is not recursive. *Suggestion*: Construct a binary relation P such that any weakly representable subset of $\mathbb{N}$ equals $\{b : \langle a, b\rangle \in P\}$ for some a, and such that P is recursive if $\sharp T$ is. Consider $\{b : \langle b, b\rangle \notin P\}$.

4. Show that there exist $2^{\aleph_0}$ nonisomorphic countable models of Th $\mathfrak{N}$. *Suggestion*: For each set A of primes, make a model having an element divisible by exactly the primes in A.

5. (Lindenbaum) Let T be a decidable consistent theory (in a reasonable language). Show that T can be extended to a complete decidable consistent theory T'.

6. Consider the language of equality, having $\forall$ as its only parameter. Let λ_n be the translation of "There are at least n things," cf. the proof of Theorem 26B. Call a formula *simple* iff it can be built up from atomic formulas and the λ_n's by use of connective symbols (but no quantifiers). Show how, given any formula in the language of equality, we can find a logically equivalent simple formula. *Suggestion*: View this as an elimination-of-quantifers result (where the quantifers in λ_n do not count). Use Theorem 31F.

7. (a) Assume that A and B are subsets of $\mathbb{N}^m$ belonging to Σ_k (or Π_k). Show that $A \cup B$ and $A \cap B$ also belong to Σ_k (or Π_k, respectively).
(b) Assume that A is in Σ_k (or Π_k) and that the functions $f_1, \ldots, f_m$ are recursive. Show that

$$\{\vec{a} : \langle f_1(\vec{a}), \ldots, f_m(\vec{a})\rangle \in A\}$$

is also in Σ_k (or Π_k, respectively).

8. Let T be a theory in a recursively numbered language (with $\mathbf{0}$ and $\mathbf{S}$). Let n be fixed, $n \geq 0$. Assume that all subsets of $\mathbb{N}$ in Σ_n are weakly representable in T. Show that $\sharp T$ is not in Π_n. (Observe that Exercise 3 is a special case of this, wherein $n = 0$. The suggestions given there carry over to the present case.)

§ 3.6 APPLICATIONS TO SET THEORY

We know that in the language of number theory, Cn A_E is incomplete and nonrecursive, as is any compatible recursively axiomatizable theory in the language.

But now suppose we leave arithmetic for a while and look at set theory. Here we have a language (with the parameters ∀ and ∈) and a set of axioms. In all presently accepted cases the set of axioms is recursive. Or more precisely, the set of Gödel numbers of the axioms is recursive. And so the theory (set theory) obtained is recursively enumerable. We claim that this theory, if consistent, is not recursive and hence not complete. We will presently prove this, but we can already sketch the argument in rough form. We can, in a very real sense, embed the language of number theory in set theory. We can then look at that fragment of set theory which deals with the natural numbers and their arithmetic (the shaded area in Fig. 10). That is a theory compatible with A_E. And so it is nonrecursive. Now if set theory were recursive, then its arithmetical part would also be re-

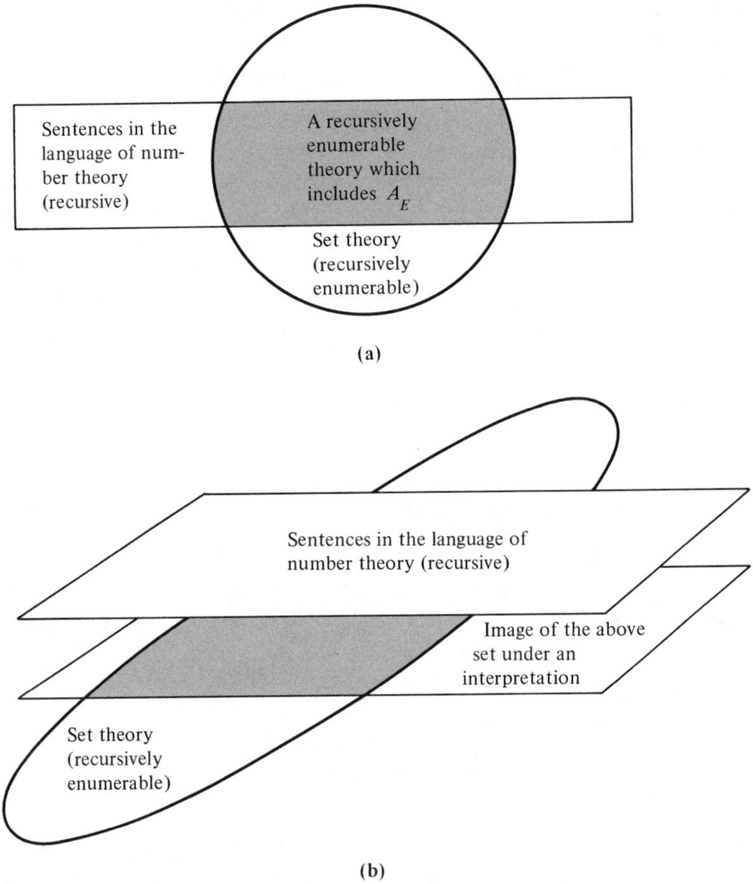

Figure 10. Set theory and number theory. (a) Flat picture. (b) A more accurate picture.

cursive, which it isn't. As a bonus, we will come across a remarkable theorem regarding the difficulty of proving set theory to be consistent.

Henceforth by set theory (*ST*) we mean that theory (in the language with equality having the two parameters $\forall$ and $\in$) which is the set of consequences of the reader's favorite set-theoretic axioms. (The standard Zermelo–Fraenkel axioms will do nicely, if the reader has no favorite. We ask only that the set of axioms be recursive, and that it be strong enough to yield certain everyday facts about sets.) We need an interpretation π of Cn A_E into *ST*. (For the remainder of this section we assume a familiarity with Section 2.7.) But the existence of such a π is a standard result of set theory, although it is not usually stated in these words. We need formulas of the language of *ST* which adequately express the concept of being a natural number, being the sum of two given numbers, and so forth.

The formula $\pi_\forall$ is the result of eliminating the defined symbol ω from the formula $\mathbf{v}_1 \in \omega$. The formula $\pi_\mathbf{0}$ is similarly obtained from the set-theoretic formula $\mathbf{v}_1 \approx \varnothing$, and the formula $\pi_\mathbf{S}$ is obtained from $\mathbf{v}_2 \approx \mathbf{v}_1 \cup \{\mathbf{v}_1\}$. The formula $\pi_<$ is simply $\mathbf{v}_1 \in \mathbf{v}_2$. For π_+ we use the translation into the language of *ST* of

For any f, if $f : \omega \times \omega \to \omega$ and for all a and b in ω we have $f(a, \varnothing) = a$ and $f(a, b \cup \{b\}) = f(a, b) \cup \{f(a, b)\}$, then $f(\mathbf{v}_1, \mathbf{v}_2) = \mathbf{v}_3$.

(The manner of translation is partially indicated in Chapter 0.) The formulas $\pi_\cdot$ and $\pi_\mathbf{E}$ are obtained in much the same fashion.

The claim that this π is an interpretation of Cn A_E into *ST* makes a number (and the number is seventeen) of demands on *ST*.

(i) $\exists \mathbf{v}_1 \pi_\forall$ must be in *ST*. It is, since we can prove in set theory that ω is nonempty.

(ii) For each of the five function symbols f in the language of A_E, *ST* must contain a sentence asserting, roughly, that π_f defines a function on the set defined by $\pi_\forall$. (The exact sentence is set forth in the definition of interpretation in Section 2.7.) In the case of $\mathbf{0}$, we have in *ST* the result that there is a unique empty set and that it belongs to ω. The case for $\mathbf{S}$ is simple, since $\pi_\mathbf{S}$ defines a unary operation on the universe of all sets, and ω is closed under this operation. For $+$ we must use the recursion theorem on ω. That is, we can prove in *ST* (as sketched in Section 1.2) that there is a unique $f : \omega \times \omega \to \omega$ such that $f(a, \varnothing) = a$ and $f(a, b \cup \{b\}) = f(a,b) \cup \{f(a, b)\}$ for a, b in ω. The required property of π_+ then follows. Similar arguments apply to $\cdot$ and $\mathbf{E}$.

(iii) For each of the eleven sentences σ in A_E, the sentence σ^π must be in ST. For example, in the case of L3, we have in ST the fact that for any m and n in ω, either $m \in n$, $m = n$, or $n \in m$.

Since these demands are finite in number, there is a finite $\Phi \subseteq ST$ such that π is also an interpretation of Cn A_E into Cn Φ.

Theorem 36A (strong undecidability of set theory) Let T be a theory in the language of set theory such that $T \cup ST$ (or at least $T \cup \Phi$) is consistent. Then $\sharp T$ is not recursive.

Proof Let Δ be the consistent theory $\text{Cn}(T \cup \Phi)$. Let Δ_0 be the corresponding theory $\pi^{-1}[\Delta]$ in the language of number theory. From Section 2.7 we know that Δ_0 is a consistent theory (since Δ is). Also $A_E \subseteq \Delta_0$, since if $\sigma \in A_E$, then $\sigma^\pi \in \text{Cn}\,\Phi \subseteq \Delta$. Hence by the strong undecidability of Cn A_E (Theorem 35C), $\sharp\Delta_0$ is not recursive.

Now we must derive the nonrecursiveness of T from that of Δ_0. We have

$$\sigma \in \Delta_0 \qquad \text{iff } \sigma^\pi \in \Delta$$

and by the lemma below, $\sharp\sigma^\pi$ depends recursively on $\sharp\sigma$. Hence $\sharp\Delta$ cannot be recursive, lest $\sharp\Delta_0$ be. Similarly, we have

$$\tau \in \Delta \qquad \text{iff } (\varphi \to \tau) \in T,$$

where φ is the conjunction of the members of Φ. Since $\sharp(\varphi \to \tau)$ depends recursively on $\sharp\tau$, we have that $\sharp T$ cannot be recursive lest $\sharp\Delta$ be. ■

Lemma 36B There is a recursive function p such that for any formula α of the language of number theory, $p(\sharp\alpha) = \sharp(\alpha^\pi)$.

Proof In Section 2.7 we gave explicit instructions for constructing α^π. The construction in some cases utilized formulas β^π for formulas β simpler than α. The methods of the Section 3.3 and 3.4 can be applied to the Gödel numbers of these formulas to show that p is recursive. But the details are not particularly attractive, and we omit them here. ■

Corollary 36C If set theory is consistent, then it is not complete.

Proof Set theory has a recursive set of axioms. If complete, the theory is then recursive (by item 21 of Section 3.4). By the foregoing theorem, this cannot happen if ST is consistent. ■

Corollary 36D In the language with equality and a two-place predicate symbol, the set of (Gödel numbers of) valid sentences is not recursive.

Partial proof In the foregoing theorem take $T = \text{Cn}\,\varnothing$, the set of valid sentences. The theorem then assures us that $\sharp T$ is nonrecursive, provided that Φ is consistent. We have not given the finite set Φ explicitly. But we assure the reader that Φ can be chosen in such a way as to be provably consistent. ∎

It should be noted that π is *not* an interpretation of Th $\mathfrak{N}$ into ST (unless ST is inconsistent). For $\pi^{-1}[ST]$ is a recursively enumerable theory in the language of $\mathfrak{N}$, as a consequence of Lemma 36B. Hence it cannot coincide with Th $\mathfrak{N}$, and it can include the complete theory Th $\mathfrak{N}$ only if it is inconsistent.

Gödel's second incompleteness theorem

We can employ our usual tricks to find a sentence σ of number theory which indirectly asserts that its own interpretation σ^π is not a theorem of set theory. For let D be the ternary relation on N such that

$\langle a, b, c\rangle \in D$ iff a is the Gödel number of a formula α of number theory and c is the Gödel number of a deduction from the axioms of ST of $\alpha(\mathbf{S}^b\mathbf{0})^\pi$.

The relation D is recursive (by the usual arguments); let $\delta(\boldsymbol{v}_1, \boldsymbol{v}_2, \boldsymbol{v}_3)$ represent D in Cn A_E. Let r be the Gödel number of

$$\forall \boldsymbol{v}_3 \neg\delta(\boldsymbol{v}_1, \boldsymbol{v}_1, \boldsymbol{v}_3)$$

and let σ be

$$\forall \boldsymbol{v}_3 \neg\delta(\mathbf{S}^r\mathbf{0}, \mathbf{S}^r\mathbf{0}, \boldsymbol{v}_3).$$

Observe that σ *does* indirectly assert that $\sigma^\pi \notin ST$. We will now prove that the assertion is correct:

Lemma 36E If ST is consistent, then $\sigma^\pi \notin ST$.

Proof Suppose to the contrary that σ^π is deducible from the axioms of ST; let k be $\mathscr{G}$ of such a deduction. Then $\langle r, r, k\rangle \in D$.

$$\therefore A_E \vdash \delta(\mathbf{S}^r\mathbf{0}, \mathbf{S}^r\mathbf{0}, \mathbf{S}^k\mathbf{0});$$

$$\therefore A_E \vdash \exists \boldsymbol{v}_3\, \delta(\mathbf{S}^r\mathbf{0}, \mathbf{S}^r\mathbf{0}, \boldsymbol{v}_3);$$

i.e.,

$$A_E \vdash \neg\sigma.$$

Applying our interpretation π, we conclude that $\neg\sigma^\pi$ is in *ST*, whence *ST* is inconsistent. Thus

$$ST \text{ is consistent} \Rightarrow \sigma^\pi \notin ST. \qquad \blacksquare$$

Now the above proof, like all those in this book, is carried out in intuitive mathematics. But all of our work in the book could have been carried out within *ST*. Indeed it is common knowledge that essentially all work in mathematics can be carried out in *ST*. Imagine actually doing so. Then instead of a proof of an English sentence, "*ST* is consistent $\Rightarrow \sigma^\pi \notin ST$," we have a deduction from the axioms of *ST* of certain sentence in the formal language of set theory:

$$(Cons(ST) \rightarrow \square).$$

Here *Cons*(*ST*) is the result of translating (in a nice way) "*ST* is consistent" into the language of set theory. Similarly, $\square$ is the result of translating "$\sigma^\pi \notin ST$." But we already *have* a sentence in the language of set theory asserting that $\sigma^\pi \notin ST$. It is σ^π. This strongly suggests that $\square$ is (or is provably equivalent in *ST* to) σ^π, from which we get

$$(Cons(ST) \rightarrow \sigma^\pi)$$

as a theorem of *ST*.

Now this *can* actually be carried out in such a way as to have $\square$ be σ^π. We have given above an argument which we hope will convince the reader that this is at least probable. And from it we now have

Gödel's Second Incompleteness Theorem The sentence *Cons*(*ST*) is not a theorem of *ST*, unless *ST* is inconsistent.

Proof By the above (plausibility) argument

$$(Cons(ST) \rightarrow \sigma^\pi)$$

is a theorem of *ST*. So if *Cons*(*ST*) is also a theorem of *ST*, then σ^π is too. But by Lemma 36E, if $\sigma^\pi \in ST$, then *ST* is inconsistent. $\blacksquare$

Of course if *ST* is inconsistent, then every sentence is a theorem, including *Cons*(*ST*). Because of this, a proof of *Cons*(*ST*) within *ST* would not convince anyone that *ST* was consistent. (And by Gödel's second theorem, it would convince him of the opposite.) But prior to Gödel's work it was possible to hope that *Cons*(*ST*) might be provable from assumptions weaker than the axioms of set theory, ideally assumptions already known to be consistent. But we now see that *Cons*(*ST*) is not in any subtheory of *ST*, unless of course *ST* is inconsistent.

EXERCISE

Let T be a theory in a recursively numbered language, and assume that there is an interpretation of Cn A_E into T. Show that T is strongly undecidable; i.e., whenever T' is a theory in the language for which $T \cup T'$ is consistent, then $\sharp T'$ is not recursive.

§ 3.7 REPRESENTING EXPONENTIATION[1]

In Sections 3.1 and 3.2 we studied the theory of certain reducts of $\mathfrak{N}$ and found them to be decidable. Then in Section 3.3 we added *both* multiplication and exponentiation. The resulting theory was found (in Section 3.5) to be undecidable. Actually it would have been enough to add only multiplication (and forego exponentiation); we would still have undecidability.

Let $\mathfrak{N}_M$ be the reduct of $\mathfrak{N}$ obtained by dropping exponentiation:

$$\mathfrak{N}_M = (N, 0, S, <, +, \cdot).$$

Thus the symbol **E** does not appear in the language of $\mathfrak{N}_M$. Let A_M be the set obtained from A_E by dropping E1 and E2. The purpose of this section is to show that all the theorems of Sections 3.3–3.5 continue to hold when "A_E" and "$\mathfrak{N}$" are replaced by "A_M" and "$\mathfrak{N}_M$". The key fact needed to establish this claim is that exponentiation is representable in Cn A_M. That is, there is a formula ε in the language of $\mathfrak{N}_M$ such that for any a and b,

$$A_M \vdash \forall z[\varepsilon(\mathbf{S}^a\mathbf{0}, \mathbf{S}^b\mathbf{0}, z) \leftrightarrow z \approx \mathbf{S}^{(a^b)}\mathbf{0}].$$

Thus ε can be used to simulate the formula $x \mathbf{E} y \approx z$ without actual use of the symbol **E**.

If we look to see what relations and functions are representable in Cn A_M, we find at first that everything (except for exponentiation itself) that was shown to be representable in Cn A_E is (by the same proof) representable in Cn A_M. Until, that is, we reach item 7 in the catalog listing of Section 3.3. To go further, we must show that exponentiation itself is representable in Cn A_M.

[1] This section may be omitted without loss of continuity.

We know that exponentiation can be characterized by the equations

$$a^0 = 1,$$
$$a^{b+1} = a^b \cdot a.$$

From what we know about primitive recursion (catalog item 13 in Section 3.3 plus Exercise 8 there), we might think of defining

$$E^*(a, b) = \text{the least } s \text{ such that } [(s)_0 = 1 \text{ and for all } i < b,\ (s)_{i+1} = (s)_i \cdot a].$$

For then $a^b = (E^*(a, b))_b$. This fails to yield a proof of representability, because we do not yet know that the decomposition function $(a)_b$ is representable in Cn A_M. But we do not really need that particular decomposition function (which corresponded to a particular way of encoding sequences). All we need is any function δ which acts like a decomposition function; the properties we need are summarized in the following lemma.

Lemma 37A There is a function δ representable in Cn A_M such that for every $n, a_0, \ldots, a_n$, there is an s for which $\delta(s, i) = a_i$ for all $i \leq n$.

Once the lemma has been established, we can define

$$E^{**}(a, b) = \text{the least } s \text{ such that } [\delta(s, 0) = 1 \text{ and for all } i < b,\ \delta(s, i + 1) = \delta(s, i) \cdot a].$$

The lemma assures us that such an s exists. E^{**} is then representable in Cn A_M, as is exponentiation, since

$$a^b = \delta(E^{**}(a, b), b).$$

A function δ which establishes the lemma will be provided by some facts of number theory.

A pairing function

As a first step toward proving the foregoing lemma, we will construct functions for encoding and decoding pairs of numbers. It is well known that there exist functions mapping $N \times N$ one-to-one onto N. In particular, the function J does this, where in the diagram shown, $J(a, b)$ has been written at the point with coordinates $\langle a, b\rangle$.

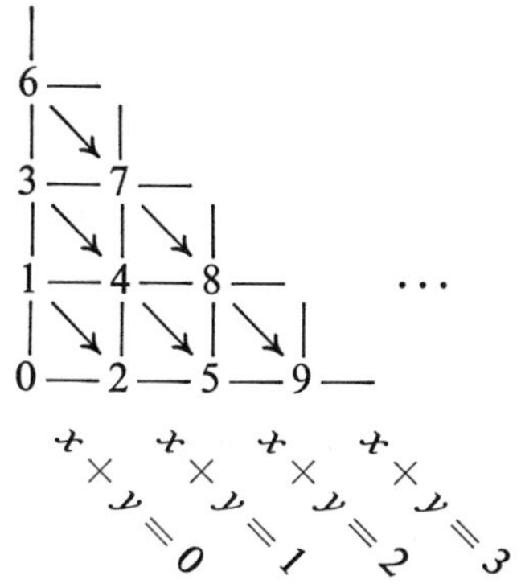

For example, $J(2, 1) = 8$ and $J(0, 2) = 3$. To obtain an equation for $J(a, b)$, we note that along the line $x + y = n$ there are $n + 1$ points (with coordinates in $\mathbb{N}$). Thus

$$\begin{aligned} J(a, b) &= \text{the number of points in the plane to which } J \text{ assigns smaller values} \\ &= [\text{the number of points on lines } x + y = n \text{ for } n = 0, 1, \ldots, (a + b - 1)] + [\text{the number of points on the line } x + y = a + b \text{ for which } x < a] \\ &= [1 + 2 + \cdots + (a + b)] + a \\ &= \tfrac{1}{2}(a + b)(a + b + 1) + a \\ &= \tfrac{1}{2}[(a + b)^2 + 3a + b]. \end{aligned}$$

Let K and L be the corresponding projection functions onto the axes, i.e., the unique functions such that

$$K(J(a, b)) = a, \qquad L(J(a, b)) = b.$$

For example, $K(7) = 1$, the x-coordinate of the point $\langle 1, 2\rangle$ in the plane to which J assigned the number 7. Similarly, $L(7) = 2$, the y-coordinate of that point.

We claim that J, K, and L are representable in Cn A_M. The function

$$H(a) = \text{the least } b \text{ such that } a \leq 2b$$

has the property that $H(a) = \frac{1}{2}a$ for even a. Then we can write

$$\begin{aligned} J(a, b) &= H((a + b) \cdot (a + b + 1)) + a, \\ K(p) &= \text{the least } a \text{ such that [for some } b \leq p, J(a, b) = p], \\ L(p) &= \text{the least } b \text{ such that [for some } a \leq p, J(a, b) = p]. \end{aligned}$$

From the form of the four preceding equations we conclude that H, J, K, and L are representable in Cn A_M.

The Gödel β-function

Let β be the function defined as follows:

$$\begin{aligned}\beta(c, d, i) &= \text{the remainder in } c \div [1 + (i + 1) \cdot d] \\ &= \text{the least } r \text{ such that for some } q \leq c,\ c = q \cdot [1 + (i + 1) \cdot d] + r.\end{aligned}$$

This unlikely-looking function produces a satisfactory decomposition function for Lemma 37A. Let

$$\delta(s, i) = \beta(K(s), L(s), i).$$

It is clear that δ is representable in Cn A_M. What is not so obvious is that it meets the conditions of Lemma 37A. We want to show:

(∗) For any n and any $a_0, \ldots, a_n$, there are numbers c and d such that $\beta(c, d, i) = a_i$ for all $i \leq n$.

For then it follows that $\delta(J(c, d), i) = \beta(c, d, i) = a_i$ for $i \leq n$.

Now (∗) is a statement of number theory, not logic. The proof of (∗) is based on the Chinese remainder theorem. Numbers $d_0, \ldots, d_n$ are said to be *relatively prime in pairs* iff no prime divides both d_i and d_j for $i \neq j$.

Chinese Remainder Theorem Let $d_0, \ldots, d_n$ be relatively prime in pairs; let $a_0, \ldots, a_n$ be natural numbers with each $a_i < d_i$. Then we can find a number c such that for all $i \leq n$,

$$a_i = \text{the remainder in } c \div d_i.$$

Proof Let $p = \Pi_{i \leq n} d_i$, and for any c let $F(c)$ be the $(n + 1)$-tuple of remainders when c is divided by $d_0, \ldots, d_n$. Notice that there are p possible values for this $(n + 1)$-tuple.

We claim that F is one-to-one on $\{k : 0 \leq k < p\}$. For suppose that $F(c_1) = F(c_2)$. Then each d_i divides $| c_1 - c_2 |$. Since the d_i's are relatively prime, p must divide $| c_1 - c_2 |$. For c_1, c_2 less than p, this implies that $c_1 = c_2$.

Hence the restriction of F to $\{k : 0 \leq k < p\}$ takes on all p possible values. In particular, it assumes (at some point c) the value $\langle a_0, \ldots, a_n \rangle$. And that is the c we want. ■

Lemma 37B For any $s \geq 0$, the $s + 1$ numbers

$$1 + 1 \cdot s!,\ 1 + 2 \cdot s!,\ \ldots,\ 1 + (s + 1) \cdot s!$$

are relatively prime in pairs.

Proof All these numbers have the property that any prime factor q cannot divide $s!$, whence $q > s$. If the prime q divides both $1 + j \cdot s!$ and $1 + k \cdot s!$, then it divides their difference, $|j - k| \cdot s!$. Since q does not divide $s!$, it divides $|j - k|$. But $|j - k| \leq s < q$. This is possible only if $|j - k| = 0$. ∎

Proof of (∗) *page* 248 Assume we are given $a_0, \ldots, a_n$; we need numbers c and d such that the remainder when c is divided by $1 + (i + 1) \cdot d$ is a_i, for $i \leq n$.

Let s be the largest of $\{n, a_0, \ldots, a_n\}$ and let $d = s!$. Then by Lemma 37B, the numbers $1 + (i + 1) \cdot d$ are relatively prime in pairs for $i \leq n$. So by the Chinese remainder theorem there is a c such that the remainder in $c \div [1 + (i + 1) \cdot d]$ is a_i for $i \leq n$. ∎

This completes the proof of Lemma 37A. And by the argument which followed that lemma, we can conclude:

Theorem 37C Exponentiation is representable in Cn A_M.

Armed with this theorem, we can now return to catalog item 7 of Section 3.3. The proof given there now establishes that the function in question (whose value at n is p_n) is representable in Cn A_M. For it was formed by allowable methods from relations and functions (including exponentiation) known to be representable in Cn A_M.

The same phenomenon persists throughout Sections 3.3 and 3.4. The representability proofs given there now establish representability in Cn A_M. Thus any recursive relation is representable in Cn A_M, and if the relation happens to be a function, then it is functionally representable. The proofs given in Section 3.5 then apply to $\mathfrak{N}_M$ and A_M as well as to $\mathfrak{N}$ and A_E. In particular, we have the strong undecidability of Cn A_M: Any theory T in the language of $\mathfrak{N}_M$ for which $T \cup A_M$ is consistent cannot be recursive.

Notice that any relation definable in $\mathfrak{N}$ (i.e., any arithmetical relation) is also definable in $\mathfrak{N}_M$. For exponentiation, being representable in a subtheory of Th $\mathfrak{N}_M$, is *a fortiori* definable in $\mathfrak{N}_M$. By the new version of Tarski's theorem, $\sharp$Th $\mathfrak{N}_M$ is not definable in $\mathfrak{N}_M$, and consequently $\sharp$Th $\mathfrak{N}_M$ cannot be arithmetical.

TABLE X

Structure	Theory	Models of the theory	Definable sets	Comments
(N)	Decidable. Not finitely axiomatizable. Admits elimination of quantifiers.	Any infinite set.	$\varnothing$ and N. $\{0\}$ is not definable.	
$(N, 0)$	As above.	Any infinite set with distinguished element.	$\varnothing$, $\{0\}$, $N - \{0\}$, N. S is not definable.	
$(N, 0, S)$	As above.	Standard part plus any number of Z-chains.	Finite and cofinite sets. $<$ is not definable.	$\{0\}$ is definable in (N, S).
$(N, 0, S, <)$	Decidable. Finitely axiomatizable. Admits elimination of quantifiers.	As above, with any ordering of the Z-chains.	Finite and cofinite sets. $+$ is not definable.	$\{0\}$ and S are definable in $(N, <)$.
$(N, 0, S, <, +)$	Decidable (Presburger).	The Z-chains are densely ordered without endpoints. Also there is a suitable addition operation.	Eventually periodic sets. $\cdot$ is not definable.	$\{0\}$, S, and $<$ are definable in $(N, +)$.
$(N, 0, S, <, +, \cdot)$	Not arithmetical. $\therefore$ not recursively axiomatizable.	As above, but with a suitable multiplication operation.	All arithmetical relations are definable.	The arithmetical relations are definable in $(N, S, \cdot)$, $(N, +, \cdot)$, and $(N, <, D)$, where $D(x, y) = (x)_y$.

In the terminology of Section 2.7, we can say that there is a faithful interpretation of Th $\mathfrak{N}$ into Th $\mathfrak{N}_M$. It equals the identity interpretation on all parameters except **E**, and to **E** it assigns a formula defining exponentiation in $\mathfrak{N}_M$.

In Table X we summarize some of the results of Chapter 3 on number theory and its reducts.

EXERCISES

1. Let $D(a, b) = (a)_b$. Show that any arithmetical relation is definable in the structure $(N, <, D)$.

2. Show that the addition relation $\{\langle a, b, c\rangle : a + b = c\}$ is definable in the structure $(N, S, \cdot)$. *Suggestion*: Under what conditions does the equation $S(ac) \cdot S(bc) = S(c \cdot c \cdot S(ab))$ hold?

3. (a) Show that $\text{Th}(\mathbb{Z}, +, \cdot)$ is strongly undecidable. (See the exercise of Section 3.6.)

(b) (This part assumes a background in algebra.) Show that the theory of rings is undecidable and that the theory of commutative rings is undecidable.

§ 3.8 RECURSIVE FUNCTIONS

We have used recursive functions (i.e., functions which, when viewed as relations, are recursive) to obtain theorems of incompleteness and undecidability of theories. But the class of recursive functions is also an interesting class in its own right, and in this section we will indicate a few of its properties.

Recall that by Church's thesis, a function is recursive iff it is computable by an effective procedure (page 201). This fact is responsible for much of the interest in recursive functions. At the same time, this fact makes possible an intuitive understanding of recursiveness, which greatly facilitates the study of the subject. Suppose, for example, that you are suddenly asked whether or not the inverse of a recursive permutation of N is recursive. Before trying to prove this, you should first ask yourself the intuitive counterpart: Is the inverse of a computable permutation f also computable? You then (hopefully) perceive that the answer is affirmative. To compute $f^{-1}(3)$, you can compute $f(0)$, $f(1)$, ... until for some k it is found that $f(k) = 3$. Then $f^{-1}(3) = k$. Having done this, you have gained two advantages. For one, you feel certain that the answer to the question regarding recursive permutations must also be affirmative. And second, you have a good outline of how to prove this; the proof is found by making rigorous the intuitive proof. This strategy for approaching problems involving recursiveness will be very useful in this section.

Before proceeding, it might be wise to summarize here some of the facts about recursive functions we have already established. We know that a function f is recursive iff it (as a relation) is representable in Cn A_E, by Theorem 34A. Consequently, every recursive function is weakly representable in this theory.

In Section 3.3 a repertoire of recursive functions was developed. In addition, it was shown that the class of recursive functions is closed under certain operations, such as composition (Theorem 33L) and the "least-zero" operator (Theorem 33M).

We also know of a few functions which are not recursive. There are $2^{\aleph_0}$ functions from $\mathbb{N}^m$ into $\mathbb{N}$ altogether, but only $\aleph_0$ of them can be recursive. So an abundance of nonrecursive functions exists, despite the fact that the most commonly met functions (like the polynomials) were shown in Section 3.3 to be recursive. By catalog item 1 of Section 3.3, the characteristic function of a nonrecursive set is nonrecursive. For example, if $f(a) = 1$ whenever a is the Gödel number of a member of Cn A_E and $f(a) = 0$ otherwise, then f is not recursive.

Normal form

For any computable function, such as the polynomial function $a^2 + 3a + 5$, one can in principle design a digital computer into which one feeds a and out of which comes $a^2 + 3a + 5$ (Fig. 11). But if you then want a different function, you must build a different computer. (Or change the wiring in the one you have.) It was recognized long ago that it is usually more desirable to build a single general-purpose stored-program computer. Into this you feed both a and the program for computing your polynomial

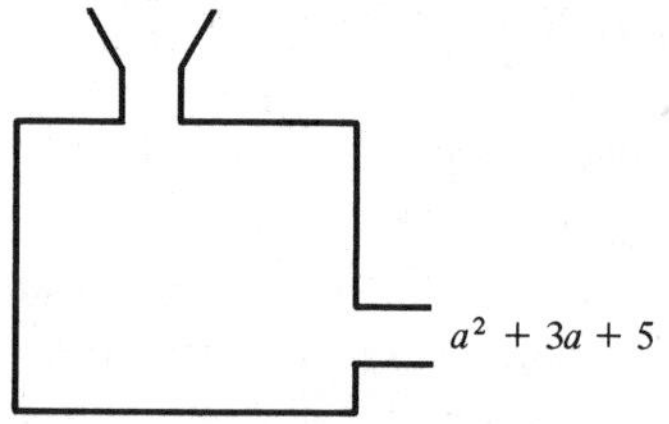

Figure 11. Special-purpose computer.

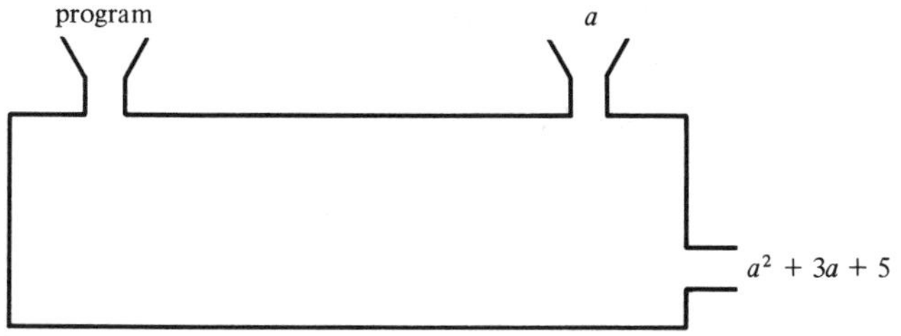

Figure 12. General-purpose computer.

(Fig. 12). This "universal" computer requires two inputs, and it will compute any one-place computable function (if supplied with enough memory space), provided that the right program is fed into it. Of course, there are some programs which do not produce any function on $\mathbb{N}$, as many a programmer has, to his sorrow, discovered.

In this subsection and the next, we will repeat what has just been said, but with recursive functions and with proofs. For our universal computer we will have a recursive relation T_1 and a recursive function U. Then for any recursive $f : \mathbb{N} \to \mathbb{N}$ there will exist an e (analogous to the program) such that

$$\begin{aligned} f(a) &= U(\text{the least } k \text{ such that } \langle e, a, k\rangle \in T_1) \\ &= U(\mu k\, \langle e, a, k\rangle \in T_1), \end{aligned}$$

where the second equation is to be understood as being an abbreviation for the first. Actually e will here be the Gödel number of a formula φ which represents (or at least weakly represents) f in Cn A_E. And the numbers k for which $\langle e, a, k\rangle \in T_1$ will encode both $f(a)$ and $\mathcal{G}$ of a deduction from A_E of $\varphi(\mathbf{S}^a\mathbf{0}, \mathbf{S}^{f(a)}\mathbf{0})$.

Definition For each positive integer m, let T_m be the $(m + 2)$-ary relation to which an $(m + 2)$-tuple $\langle e, a_1, \ldots, a_m, k\rangle$ belongs iff

(i) e is the Gödel number of a formula φ in which only $\mathbf{v}_1, \ldots, \mathbf{v}_m$, $\mathbf{v}_{m+1}$ occur free;

(ii) k is a sequence number of length 2, and $(k)_0$ is $\mathcal{G}$ of a deduction from A_E of $\varphi(\mathbf{S}^{a_1}\mathbf{0}, \ldots, \mathbf{S}^{a_m}\mathbf{0}, \mathbf{S}^{(k)_1}\mathbf{0})$.

The idea here is that for any one-place recursive function f we can first of all take e to be the Gödel number of a formula φ weakly representing f (as a relation). Then we know that for any a and b,

$$A_E \vdash \varphi(\mathbf{S}^a\mathbf{0}, \mathbf{S}^b\mathbf{0}) \qquad \text{iff } b = f(a).$$

So any number k meeting clause (ii) of the definition must equal $\langle (k)_0, f(a)\rangle$, where $(k)_0$ is $\mathscr{G}$ of a deduction of $\varphi(\mathbf{S}^a\mathbf{0}, \mathbf{S}^{f(a)}\mathbf{0})$ from A_E. (We have departed from the usual definition of T_m here by not requiring that k be as small as possible.)

Take for the "upshot" function U the function

$$U(k) = (k)_1.$$

This U is recursive and in the situation described in the preceding paragraph we have $U(k) = f(a)$.

Lemma 38A For each m, the relation T_m is recursive.

Proof, for $m = 2$ $\langle e, a_1, a_2, k\rangle \in T_2$ iff e is the Gödel number of a formula, $\sharp(\forall \mathbf{v}_1 \forall \mathbf{v}_2 \forall \mathbf{v}_3) * e$ is the Gödel number of a sentence, k is a sequence number of length 2, $(k)_0$ is $\mathscr{G}$ of a deduction from A_E of

$$\mathrm{Sb}(\mathrm{Sb}(\mathrm{Sb}(e, \sharp\mathbf{v}_1, g(a_1)), \sharp\mathbf{v}_2, g(a_2)), \sharp\mathbf{v}_3, g((k)_1)),$$

where $g(n) = \sharp\mathbf{S}^n\mathbf{0}$. From Section 3.4 we know all this to be recursive. ■

Theorem 38B (a) For any recursive function $f : \mathbb{N}^m \to \mathbb{N}$, there is an e such that for all $a_1, \ldots, a_m$,

$$f(a_1, \ldots, a_m) = U(\mu k\langle e, a_1, \ldots, a_m, k\rangle \in T_m).$$

(In particular, such a number k exists.)

(b) Conversely, for any e such that $\forall a_1 \cdots a_m \exists k \langle e, a, \ldots, a_m, k\rangle \in T_m$, the function whose value at $a_1, \ldots, a_m$ is $U(\mu k\langle e, a_1, \ldots, a_m, k\rangle \in T_m)$ is recursive.

Proof Part (*b*) follows immediately the fact that U and T_m are recursive. As for part (*a*), we take for e the Gödel number of a formula φ weakly representing f in Cn A_E. Given any $\vec{a}$, we know that $A_E \vdash \varphi(\mathbf{S}^{a_1}\mathbf{0}, \ldots, \mathbf{S}^{a_m}\mathbf{0}, \mathbf{S}^{f(\vec{a})}\mathbf{0})$. If we let d be $\mathscr{G}$ of a deduction from A_E of this sentence, then $\langle e, \vec{a}, \langle d, f(\vec{a})\rangle\rangle \in T_m$. Hence there is some k for which $\langle e, \vec{a}, k\rangle \in T_m$. And for any such k, we know that $A_E \vdash \varphi(\mathbf{S}^{a_1}\mathbf{0}, \ldots, \mathbf{S}^{a_m}\mathbf{0}, \mathbf{S}^{(k)_1}\mathbf{0})$, since $(k)_0$ is $\mathscr{G}$ of a deduction. Consequently, $U(k) = (k)_1 = f(\vec{a})$ by our choice of φ. Thus we have $U(\mu k\langle e, \vec{a}, k\rangle \in T_m) = f(\vec{a})$. ■

This theorem, due to Kleene in 1936, shows that every recursive function is representable in the normal form

$$f(\vec{a}) = U(\mu k\langle e, \vec{a}, k\rangle \in T_m).$$

Thus a computing machine able to calculate U and the characteristic function of T_1 is a "universal" computer for one-place recursive functions. The input e corresponds to the program, and it must be chosen with care if any output is to result (i.e., if there is to be any k such that $\langle e, a, k\rangle \in T_1$).

Recursive partial functions

The theory of recursive functions becomes more natural if we consider the broader context of partial functions.

Definition A m-place *partial function* is a function f with $\operatorname{dom} f \subseteq \mathbb{N}^m$ and $\operatorname{ran} f \subseteq \mathbb{N}$. If $\vec{a} \notin \operatorname{dom} f$, then $f(\vec{a})$ is said to be *undefined*. If $\operatorname{dom} f = \mathbb{N}^m$, then f is said to be *total.*

The reader is hereby cautioned against reading too much into our choice of the words "partial" and "total" (or the word "undefined," for that matter). A partial function f may or may not be total; the words "partial" and "total" are not antonyms.

We will begin by looking at those partial functions which are intuitively computable.

***Definition** An m-place partial function f is *computable* iff there is an effective precedure such that (a) given an m-tuple $\vec{a}$ in $\operatorname{dom} f$, the procedure produces $f(\vec{a})$; and (b) given an m-tuple $\vec{a}$ not in $\operatorname{dom} f$, the procedure produces no output at all.

This definition extends the one previously given for total functions. At that time we proved a result (Theorem 33H), part of which generalizes to partial functions.

***Theorem 38C** An m-place partial function f is computable iff f (as an $(m+1)$-ary relation) is effectively enumerable.

Proof The proof is reminiscent of the proof of another result, Theorem 17E. First suppose we have a way of effectively enumerating f. Given an m-tuple $\vec{a}$, we examine the listing of the relation as the procedure churns it out. If and when an $(m+1)$-tuple beginning with $\vec{a}$ appears, we print out its last component as $f(\vec{a})$.

Conversely, assume that f is computable, and first suppose that f is a one-place partial function. We can enumerate f as a relation by the following procedure:

1. Spend one minute calculating $f(0)$.
2. Spend two minutes calculating $f(0)$, then two minutes calculating $f(1)$.
3. Spend three minutes calculating $f(0)$, three minutes calculating $f(1)$, and three minutes calculating $f(2)$.

And so forth. Of course, whenever one of these calculations produces any output, we place the corresponding pair on the list of members of the relation f.

For a computable m-place partial function, instead of calculating the value of f at 0, 1, 2, ... we calculate its value at $\langle (0)_0, \ldots, (0)_{m-1} \rangle$, $\langle (1)_0, \ldots, (1)_{m-1} \rangle$, $\langle (2)_0, \ldots, (2)_{m-1} \rangle$, etc. ■

In the case of a computable total function f, we were also able to conclude that f was a decidable relation. But this may fail for a nontotal f. For example, let

$$f(a) = \begin{cases} 0 & \text{if } a \in \sharp\text{Cn } A_E, \\ \text{undefined} & \text{otherwise.} \end{cases}$$

Then f is computable. (We compute $f(a)$ by enumerating $\sharp\text{Cn } A_E$ and looking for a.) But f is not a decidable relation, lest $\sharp\text{Cn } A_E$ be decidable. On the basis of this example and the foregoing theorem, we select our definition for the precise counterpart of the concept of computable partial function.

Definition A *recursive partial function* is a partial function which, as a relation, is recursively enumerable.

The reader should be warned that "recursive partial function" is an indivisible phrase; a recursive partial function need *not* be (as relation) recursive. But at least for a total function our terminology is consistent with past practice.

Theorem 38D Let $f : N^m \to N$ be a total function. Then f is a recursive partial function iff f is recursive (as a relation).

Proof If f is recursive (as a relation), then *a fortiori* f is recursively enumerable. Conversely, suppose that f is recursively enumerable. Since f is total,

$$f(\vec{a}) \neq b \Leftrightarrow \exists c[f(\vec{a}) = c \;\&\; b \neq c].$$

The form of the right-hand side shows that the complement of f is also recursively enumerable. Thus by Theorem 35F, f is recursive. ■

In first discussing normal form results, we pictured a two-input device (Fig. 13). For any computable partial function, there is some program which computes it. But now the converse holds: Any program will produce some computable partial function. Of course many programs will produce the empty function, but that is a computable partial function.

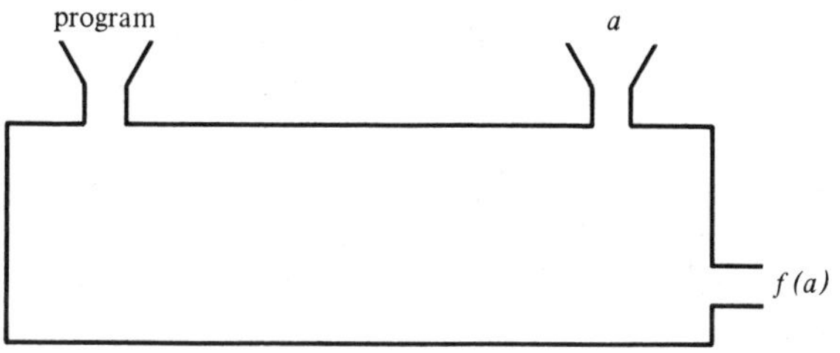

Figure 13. Computer with program for f.

For the recursive partial functions the same considerations apply. Define, for each $e \in N$, the m-place partial function $[\![e]\!]_m$ by

$$[\![e]\!]_m(a_1, \ldots, a_m) = U(\mu k \langle e, a_1, \ldots, a_m, k\rangle \in T_m).$$

The right-hand side is to be understood as undefined if there is no such k. In other words,

$$\vec{a} \in \text{dom } [\![e]\!]_m \qquad \text{iff } \exists k \langle e, a_1, \ldots, a_m, k\rangle \in T_m,$$

in which case the value $[\![e]\!]_m(\vec{a})$ is given by the above equation.

Normal Form Theorem (Kleene, 1943) (a) The $(m + 1)$-place partial function whose value at $\langle e, a_1, \ldots, a_m\rangle$ is $[\![e]\!]_m(a_1, \ldots, a_m)$ is a recursive partial function.

(b) For each $e \geq 0$, $[\![e]\!]_m$ is a m-place recursive partial function.

(c) Any m-place recursive partial function equals $[\![e]\!]_m$ for some e.

Proof (a) We have $[\![e]\!]_m(\vec{a}) = b \Leftrightarrow \exists k[\langle e, \vec{a}, k\rangle \in T_m \ \& \ U(k) = b \ \& \ (\forall k' < k) \langle e, \vec{a}, k'\rangle \notin T_m]$. The part in square brackets is recursive, so the function (as a relation) is recursively enumerable.

(b) The above proof still applies, e now being held fixed.

(c) Let f be an m-place recursive partial function, so that $\{\langle \vec{a}, b\rangle : f(\vec{a}) = b\}$ is recursively enumerable. Hence there is a formula φ which weakly represents this relation in Cn A_E. We claim that $f = [\![\sharp\varphi]\!]_m$. For if $f(\vec{a}) = b$, then $A_E \vdash \varphi(\mathbf{S}^{a_1}\mathbf{0}, \ldots, \mathbf{S}^{a_m}\mathbf{0}, \mathbf{S}^b\mathbf{0})$. Hence there is a k such that $\langle \sharp\varphi, \vec{a}, k\rangle$

$\in T_m$. For any such k, $U(k) = b$, since $A_E \nvdash \varphi(\mathbf{S}^{a_1}\mathbf{0}, \ldots, \mathbf{S}^{a_m}\mathbf{0}, \mathbf{S}^c\mathbf{0})$ for $c \neq b$. Similarly, if $f(\vec{a})$ is undefined, then $A_E \nvdash \varphi(\mathbf{S}^{a_1}\mathbf{0}, \ldots, \mathbf{S}^{a_m}\mathbf{0}, \mathbf{S}^c\mathbf{0})$ for any c, whence $[\![\sharp\varphi]\!]_m$ is undefined here also. ■

The function $[\![e]\!]_m$ is said to be the m-place recursive partial function with *index e*. Part (c) of the normal form theorem tells us that every recursive partial function has an index. The proof shows that the Gödel number of a formula weakly representing a function is always an index of the function.

We now have a convenient indexing $[\![0]\!]_1, [\![1]\!]_1, \ldots$ of the one-place recursive partial functions. Function $[\![e]\!]_1$ is produced by the "instructions" encoded by e. Of course, that function will be empty unless e is the Gödel number of a formula and certain other conditions are met.

All the recursive total functions are included in our enumeration of recursive partial functions. But we cannot tell effectively by looking at a number e whether or not it is the index of a total function:

Theorem 38E $\{e : [\![e]\!]_1 \text{ is total}\}$ is not recursive.

Proof Call this set A. Consider the function defined by

$$f(a) = \begin{cases} [\![a]\!]_1(a) + 1 & \text{if } a \in A, \\ 0 & \text{if } a \notin A. \end{cases}$$

Then f, by its construction, is total. Is it recursive? We have

$$f(a) = b \Leftrightarrow [(a \notin A \;\&\; b = 0) \quad \text{or} \quad (a \in A \;\&\; \exists k(\langle a, a, k\rangle \in T_1 \\ \&\; b = U(k) + 1 \;\&\; (\forall j < k)\langle a, a, j\rangle \notin T_1))].$$

Thus if A is recursive, then f (as a relation) is recursively enumerable. But then f is a total recursive function, and so equals $[\![e]\!]_1$ for some $e \in A$. But $f(e) = [\![e]\!]_1(e) + 1$, so we cannot have $f = [\![e]\!]_1$. This contradiction shows that A cannot be recursive. ■

It is not hard to show that A is in Π_2. This classification is the best possible, as it can be shown that A is not in Σ_2.

Theorem 38F The set

$$K = \{a : [\![a]\!]_1(a) \text{ is defined}\}$$

is recursively enumerable but not recursive.

Proof K is recursively enumerable, since $a \in K \Leftrightarrow \exists k \langle a, a, k \rangle \in T_1$. To see that K cannot be recursive, consider the function defined by

$$g(a) = \begin{cases} [\![a]\!]_1(a) + 1 & \text{if } a \in K, \\ 0 & \text{if } a \notin K. \end{cases}$$

This is a total function. Exactly as in the preceding theorem, we have that K cannot be recursive. ■

Corollary 38G (unsolvability of the halting problem) The set

$$\{\langle a, b \rangle : [\![b]\!]_1(a) \text{ is defined}\}$$

is not recursive.

This corollary tells us that there is no effective way to tell, given a program b for a recursive partial function and an input a, whether or not the function $[\![b]\!]_1$ is defined at a.

We can obtain an indexing of the recursively enumerable relations by using the following characterization.

Theorem 38H A relation on N is recursively enumerable iff it is the domain of some recursive partial function.

Proof The domain of any recursively enumerable relation is also recursively enumerable; cf. part 4 of Theorem 35E. In particular, the domain of any recursive partial function is recursively enumerable.

Conversely, let Q be any recursively enumerable relation, where

$$\vec{a} \in Q \Leftrightarrow \exists b \langle \vec{a}, b \rangle \in R$$

with R recursive. Let

$$f(\vec{a}) = \mu b \langle \vec{a}, b \rangle \in R;$$

i.e.,

$$f(\vec{a}) = b \Leftrightarrow \langle \vec{a}, b \rangle \in R \;\&\; (\forall c < b) \langle \vec{a}, c \rangle \notin R.$$

Then f, as a relation, is recursive. Hence f is a recursive partial function. Clearly its domain is Q. ■

Thus our indexing of the recursive partial functions induces an indexing of the recursively enumerable relations. Define

$$W_e = \operatorname{dom} [\![e]\!]_1.$$

Then $W_0, W_1, W_2, \ldots$ is a list of all recursively enumerable subsets of N. In Theorem 38E we showed that $\{e : W_e = N\}$ is not recursive. Similarly, Theorem 38F asserts that $\{e : e \in W_e\}$ is not recursive. Define a relation Q by

$$Q = \{\langle a, b\rangle : a \in W_b\}.$$

Then Q is recursively enumerable, since $\langle a, b\rangle \in Q \Leftrightarrow \exists k\, \langle b, a, k\rangle \in T_1$. Furthermore, Q is universal for recursively enumerable sets, in the sense that for any recursively enumerable $A \subseteq N$ there is some b such that $A = \{a : \langle a, b\rangle \in Q\}$. The unsolvability of the halting problem can be stated: Q is not recursive.

Reduction of decision problems

Suppose we have a two-place recursive partial function f. Then we claim that, for example, the function g defined by

$$g(a) = f(3, a)$$

is also a recursive partial function. On the basis of intuitive computability this is clear; one computes g by plugging in 3 for the first variable and then following the instructions for f. A proof can be found by formalizing this argument. There is some formula $\varphi = \varphi(\mathbf{v}_1, \mathbf{v}_2, \mathbf{v}_3)$ which weakly represents f (as a relation) in Cn A_E. Then g is weakly represented by $\varphi(\mathbf{S}^3\mathbf{0}, \mathbf{v}_1, \mathbf{v}_2)$, provided that $\mathbf{v}_1$ and $\mathbf{v}_2$ are substitutable in φ for $\mathbf{v}_2$ and $\mathbf{v}_3$. (If not, we can always use an alphabetic variant of φ).

Now all this is not very deep. But by standing back and looking at what was said, a more subtle fact is perceived. We were able to transform effectively the instructions for f into instructions for g. So there should be a recursive function which, given an index for f and the number 3, will produce an index for g.

Parameterization Theorem For each $m \geq 1$ and $n \geq 1$ there is a recursive function ρ such that for any e, $\vec{a}$, $\vec{b}$,

$$[\![e]\!]_{m+n}(a_1, \ldots, a_m, b_1, \ldots, b_n) = [\![\rho(e, a_1, \ldots, a_m)]\!]_n(b_1, \ldots, b_n).$$

(Equality here means of course that if one side is defined, then so also is the other side, and the values coincide.)

On the left side of the equation $\vec{a}$ consists of arguments for the function $[\![e]\!]_{m+n}$; on the right side $\vec{a}$ consists of parameters upon which the function $[\![\varrho(e, \vec{a})]\!]_n$ depends. In the example we had $m = n = 1$ and $a_1 = 3$. Since ϱ depends on m and n, the notation "ϱ_n^m" would be logically preferable. But, in fact, we will use just "ϱ."

Proof, for $m = n = 1$ It is possible to give a proof along the lines indicated by the discussion that preceded the theorem. But to avoid having to cope with alphabetic variants, we will adopt a slightly different strategy.

We know from the normal form theorem that the three-place partial function h defined by

$$h(e, a, b) = [\![e]\!]_2(a, b)$$

is a recursive partial function. Hence there is a formula ψ which weakly represents h (as a relation). We may suppose that in ψ the variables $\mathbf{v}_1$ and $\mathbf{v}_2$ are not quantified. We can then take

$$\begin{aligned}\varrho(e, a) &= \#\psi(\mathbf{S}^e\mathbf{0}, \mathbf{S}^a\mathbf{0}, \mathbf{v}_1, \mathbf{v}_2)\\ &= \mathrm{Sb}(\mathrm{Sb}(\mathrm{Sb}(\mathrm{Sb}(\#\psi, \#\mathbf{v}_1, \#\mathbf{S}^e\mathbf{0}), \#\mathbf{v}_2, \#\mathbf{S}^a\mathbf{0}), \#\mathbf{v}_3, \#\mathbf{v}_1), \#\mathbf{v}_4, \#\mathbf{v}_2).\end{aligned}$$

Then $\varrho(e, a)$ is the Gödel number of a formula weakly representing the function $g(b) = [\![e]\!]_2(a, b)$. Hence it is an index of g. ∎

We will utilize the parameterization theorem to show that certain sets are *not* recursive. We already know that $K = \{a : [\![a]\!]_1(a) \text{ is defined}\}$ is not recursive. For a given nonrecursive set A we can sometimes find a (total) recursive function g such that

$$a \in K \Leftrightarrow g(a) \in A$$

or a (total) recursive function g' such that

$$a \notin K \Leftrightarrow g'(a) \in A.$$

In either case it then follows at once that A cannot be recursive lest K be. (In the former case K is said to be *many–one reducible* to A; in the latter case the complement $\bar{K}$ of K is many–one reducible to A.) The function g or g' can often be obtained from the parameterization theorem.

EXAMPLE $\{a : W_a = \varnothing\}$ is not recursive.

Proof Call this set A. First, note that $A \in \Pi_1$, since $W_a = \varnothing$ iff $\forall b\, \forall k\, \langle a, b, k\rangle \notin T_1$. Consequently, K cannot be many–one reducible to A, but it is reasonable to hope that $\bar{K}$ might be. That is, we want a total recursive function g such that

$$[\![a]\!]_1(a) \text{ is undefined} \Leftrightarrow \text{dom } [\![g(a)]\!]_1 = \varnothing.$$

This will hold if for all b, $[\![g(a)]\!]_1(b) = [\![a]\!]_1(a)$. So start with the recursive partial function

$$f(a, b) = [\![a]\!]_1(a)$$

and let $g(a) = \varrho(\hat{f}, a)$, where $\hat{f}$ is an index for f. Then

$$[\![g(a)]\!]_1(b) = [\![\varrho(\hat{f}, a)]\!]_1(b) = f(a, b) = [\![a]\!]_1(a).$$

Thus this g shows that $\bar{K}$ is many–one reducible to A. ■

Theorem 38I (Rice, 1953) Let $\mathscr{C}$ be a set of one-place recursive partial functions. Then the set $\{e : [\![e]\!]_1 \in \mathscr{C}\}$ of indices of members of $\mathscr{C}$ is recursive iff either $\mathscr{C}$ is empty or $\mathscr{C}$ contains all one-place recursive partial functions.

Proof Let $I_{\mathscr{C}} = \{e : [\![e]\!]_1 \in \mathscr{C}\}$ be the set of indices of members of $\mathscr{C}$.

Case I: The empty function $\varnothing$ is not in $\mathscr{C}$. If nothing at all is in $\mathscr{C}$ we are done, but suppose some function ψ is in $\mathscr{C}$. We can show that K is many–one reducible to $I_{\mathscr{C}}$ if we have a recursive total function g such that

$$[\![g(a)]\!]_1 = \begin{cases} \psi & \text{if } a \in K, \\ \varnothing & \text{if } a \notin K. \end{cases}$$

For then $a \in K \Leftrightarrow g(a) \in I_{\mathscr{C}}$.

We can obtain g from the parameterization theorem by defining

$$g(a) = \varrho(e, a),$$

where

$$[\![e]\!]_1(a,\ b) = \begin{cases} \psi(b) & \text{if } a \in K, \\ \text{undefined} & \text{if } a \notin K. \end{cases}$$

The above *is* a recursive partial function, since

$$[\![e]\!]_1(a, b) = c \Leftrightarrow a \in K \;\&\; \psi(b) = c$$

and the right-hand side is recursively enumerable.

Case II: $\varnothing \in \mathscr{C}$. Then apply case I to the complement $\overline{\mathscr{C}}$ of $\mathscr{C}$. We can then conclude that $I_{\overline{\mathscr{C}}}$ is not recursive. But $I_{\overline{\mathscr{C}}}$ is the complement of $I_{\mathscr{C}}$, so $I_{\mathscr{C}}$ cannot be recursive. ■

EXAMPLES For any fixed e, the set $\{a : W_a = W_e\}$ is not recursive, as a consequence of Rice's theorem. In particular, $\{a : W_a = \varnothing\}$ is not recursive, a result proved in an earlier example. For two other applications of Rice's theorem, we can say that $\{a : W_a \text{ is infinite}\}$ and $\{a : W_a \text{ is recursive}\}$ are not recursive.

Register machines

There are many equivalent definitions of the class of recursive functions. Several of these definitions employ idealized computing devices. These computing devices are like digital computers but are free of any limitation on memory space. The first definition of this type was published by Alan Turing in 1936; similar work was done by Emil Post at roughly the same time. We will give here a variation on this theme, due to Shepherdson and Sturgis (1963).

A *register machine* will have a finite number of registers, numbered 1, 2, . . . , K. Each register is capable of storing a natural number of any magnitude. The operation of the machine will be determined by a *program*. A program is a finite sequence of *instructions*, drawn from the following list:

I r (where $1 \leq r \leq K$). "Increment r." The effect of this instruction is to increase the contents of register r by 1. The machine then proceeds to the next instruction in the program.

D r (where $1 \leq r \leq K$). "Decrement r." The effect of this instruction depends on the contents of register r. If that number is nonzero, it is decreased by 1 and the machine then proceeds, not to the next instruction, but the following one. But if the number in register r is zero, the machine just proceeds to the next instruction. In summary: The machine tries to decrement register r and skips an instruction if it is successful.

T q (where q is an integer—positive, negative, or zero). "Transfer q." All registers are left unchanged. The machine takes as its next instruction the qth instruction following this one in the program (if $q \geq 0$), or the $|q|$th instruction preceding this one (if $q < 0$). The machine halts if there is no such instruction in the program. An instruction of T 0 results in a loop, with the machine executing this one instruction over and over again.

EXAMPLES 1. Program to clear register 7.

D 7	Try to decrement 7.
T 2	
T −2	Go back and repeat.
	Halt.

2. Program to move a number from register r to register s.

Clear register s	(Use the program of the first example.)
D r	Take 1 from r.
T 3	Halt when zero.
I s	Add 1 to s.
T -3	Repeat.

This program has seven instructions altogher. It leaves a zero in register r.

3. Program to add register 1 to registers 2 and 3.

```
D 1
T 4
I 2
I 3
T −4
```

4. (Addition) Say that a and b are in registers 1 and 2. We want $a + b$ in register 3, and we want to leave a and b still in registers 1 and 2 at the end.

	Register contents			
Clear register 3.	a	b	0	
Move number from register 1 to register 4.	0	b	0	a
Add register 4 to registers 1 and 3.	a	b	a	0
Move number from register 2 to register 4.	a	0	a	b
Add register 4 to registers 2 and 3.	a	b	$a + b$	0

This program has twenty-seven instructions as it is written, but three of them are unnecessary. (In the fourth line we begin by clearing register 4, which is already clear.) At the end we have the number a back in register 1. But during the program register 1 must be cleared; this is the only way of determining the number a.

5. (Subtraction) Let $a \dot{-} b = \max(a - b, 0)$. We leave this program to the reader (Exercise 10).

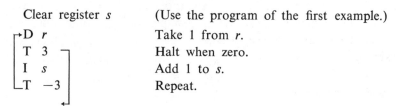

Now suppose f is an n-place partial function on $\mathbb{N}$. Possibly there will be a program P such that if we start a register machine (having all the registers to which P refers) with $a_1, \ldots, a_n$ in registers $1, \ldots, n$ and apply program P, then

(1) If $f(a_1, \ldots, a_n)$ is defined, then the calculation eventually terminates with $f(a_1, \ldots, a_n)$ is register $n + 1$. Furthermore, the calculation terminates by seeking a $(p + 1)$st instruction, where p is the length of P.

(2) If $f(a_1, \ldots, a_n)$ is undefined, then the calculation never terminates.

If there is such a program P, we say that P *calculates* f.

Theorem 38J Let f be a partial function. Then there is a program which calculates f iff f is a recursive partial function.

Thus by using register machines we arrive at exactly the class of recursive partial functions, a class we originally defined in terms of representability in consistent finitely axiomatizable theories. The fact that such different approaches produce the same class of partial functions is evidence that this is a significant class.

Outline of proof To show that the functions calculable by register machines are recursive partial functions, one "arithmetizes calculations" in the same spirit as we arithmetized deductions in Section 3.4. That is, one assigns Gödel numbers to programs and to sequences of memory configurations. One then verifies that the relevant concepts, translated into numerical relations by the Gödel numbering, are all recursive. (After going through this, one perceives that, from a sufficiently general point of view, deductions and calculations are really about the same sort of thing.)

Conversely, to show that the recursive partial functions are calculable by register machines, one can work through Sections 3.3 and 3.4 again, but where functions were previously shown to be representable in Cn A_E, they must now be shown to be calculable by register machines. This is not as hard as it might sound, since after the first few pages, the proofs are all the same as the ones used before. There is a reason for this similarity. It can be shown that the class of all recursive functions is generated from a certain handful of recursive functions by the operation of composition (in the sense of Theorem 33L) an the "least-zero" operator (Theorem 33M). Much of the work in Sections 3.3 and 3.4 amounts to a verification of this fact. Thus once one has shown that each function in this initial handful is calculable by register machine and that the class of functions calculable by register machines is closed under composition and the least-zero operator, then the earlier work can be carried over, yielding the calculability of all recursive functions.

EXERCISES

1. Define functions f and g by

$$f(n) = \begin{cases} 0 & \text{if Fermat's last theorem is true,} \\ 1 & \text{otherwise,} \end{cases}$$

$$g(n) = \begin{cases} 0 & \text{if in the decimal expansion of } \pi \text{ there} \\ & \text{is a run of at least } n \text{ consecutive 7's,} \\ 1 & \text{otherwise.} \end{cases}$$

Is f recursive? Is g recursive?

2. (a) Show that the range of any recursive partial function is recursively enumerable.

(b) Show that the range of a nondecreasing total recursive function is recursive.

3. (a) Let A be a nonempty recursively enumerable subset of $\mathbb{N}$. Show that A is the range of some total recursive function.

(b) Show that any infinite recursively enumerable subset of $\mathbb{N}$ includes an infinite recursive subset.

4. Show that every recursive partial function has infinitely many indices.

5. Give an example of a function f and a number e such that for all a,

$$f(a) = U(\mu k \, \langle e, a, k\rangle \in T_1)$$

but e is not the Gödel number of a formula weakly representing f in $\operatorname{Cn} A_E$.

6. Show that the parameterization theorem can be strengthened by requiring ϱ to be one-to-one.

7. Recall that the union of two recursively enumerable sets is recursively enumerable (Exercise 7 of Section 3.5). Show that there is a total recursive function g such that $W_{g(a,b)} = W_a \cup W_b$.

8. Show that $\{a : W_a$ has two or more members$\}$ is in Σ_1 but not in Π_1.

9. Show that there is no recursively enumerable set A such that $\{[\![a]\!]_1 : a \in A\}$ equals the class of total recursive functions on $\mathbb{N}$.

10. Give register machine programs which calculate the following functions:

(a) Subtraction, $a \mathbin{\dot{-}} b = \max(a - b, 0)$.
(b) Multiplication, $a \cdot b$.
(c) $\max(a, b)$.

11. Assume that there is a register machine program which calculates the n-place partial function f. Show that given any positive integers $r_1, \ldots, r_n$ (all distinct), p, and k, we can find a program Q such that whenever we start a register machine (having all the registers to which Q refers) with $a_1, \ldots, a_n$ in registers $r_1, \ldots, r_n$ and apply program Q, then (i) if $f(a_1, \ldots, a_n)$ is defined, then the calculation eventually terminates with $f(a_1, \ldots, a_n)$ in register p, with the contents of registers $1, 2, \ldots, k$ (except for register p) the same as their initial contents, and furthermore the calculation terminates by seeking a $(q + 1)$st instruction, where q is the length of Q; (ii) if $f(a_1, \ldots, a_n)$ is undefined, then the calculation never terminates.

12. Let $g : \mathcal{N}^{n+1} \to \mathcal{N}$ be a (total) function which is calculated by some register machine program. Let $f(a_1, \ldots, a_n) = \mu b[g(a_1, \ldots, a_n, b) = 0]$, where the right-hand side is undefined if no such b exists. Show that the partial function f can be calculated by some register machine program.

CHAPTER FOUR

Second-Order Logic

§ 4.1 SECOND-ORDER LANGUAGES

We can obtain richer, more expressive languages than the first-order languages considered thus far, by allowing quantification of predicate or function symbols. For example,

$$\exists x(Px \rightarrow \forall x\, Px)$$

is a valid formula having $\forall$ and P as its parameters. Because it is true no matter how P is interpreted,

$$\forall P\, \exists x(Px \rightarrow \forall x\, Px)$$

deserves to be called valid. (Now $\forall$ is the only parameter, since P is treated as a predicate variable.)

Suppose, then, that we have in addition to the symbols introduced at the beginning of Section 2.1, the further logical symbols:

4. Predicate variables: For each positive integer n we have the n-place predicate variables

$$\mathbf{X}_1^n, \mathbf{X}_2^n, \ldots .$$

5. Function variables: For each positive integer n, we have the n-place function variables

$$\boldsymbol{F}_1^n, \boldsymbol{F}_2^n, \ldots .$$

The usual variables $\boldsymbol{v}_1, \boldsymbol{v}_2, \ldots$ will now be called *individual* variables, to avoid confusion. The set of terms is as before defined as the set of expressions which can be built up from the constant symbols and the individual variables by applying the function symbols (both the function parameters and the function variables). Atomic formulas are again expressions $Pt_1 \cdots t_n$, where $t_1, \ldots, t_n$ are terms and P is an n-place predicate symbols (parameter or variable). The definition of wff is augmented by new formula-building operations: If φ is a wff, then so also are $\forall \boldsymbol{X}_i^n \varphi$ and $\forall \boldsymbol{F}_i^n \varphi$. The notion of a variable occurring free in φ is defined just as before. A sentence is a wff σ in which no variable (individual, predicate, or function) occurs free.

It should be remarked that the roles played by predicate parameters and free predicate variables are essentially the same. There is the same close relationship between constant symbols and free individual variables.

By a *structure* we continue to mean a function on the set of parameters meeting the conditions set forth in Section 2.2. We must extend the definition of satisfaction in the natural way. Let V now be the set of all variables, individual, predicate, or function. Let s be a function on V which assigns to each variable the suitable type of object. Thus $s(\boldsymbol{v}_1)$ is a member of the universe, $s(\boldsymbol{X}^n)$ is an n-ary relation on the universe, and $s(\boldsymbol{F}^n)$ is an n-ary operation. For a term t, $\bar{s}(t)$ is defined in the natural way. In particular, if $\boldsymbol{F}$ is a function variable, then $\bar{s}(\boldsymbol{F}t_1 \cdots t_n)$ is the result of applying the function $s(\boldsymbol{F})$ to $\bar{s}(t_1), \ldots, \bar{s}(t_n)$. Satisfaction of atomic formulas is also defined essentially as before. For a predicate variable $\boldsymbol{X}$,

$$\models_{\mathfrak{A}} \boldsymbol{X}t_1 \cdots t_n [s] \quad \text{iff } \langle \bar{s}(t_1), \ldots, \bar{s}(t_n) \rangle \in s(\boldsymbol{X}).$$

The only new features in the definition of satisfaction arise from our new quantifiers.

5. $\models_{\mathfrak{A}} \forall \boldsymbol{X}_i^n \varphi [s]$ iff for every n-ary relation R on $|\mathfrak{A}|$, $\models_{\mathfrak{A}} \varphi [s(\boldsymbol{X}_i^n | R)]$.
6. $\models_{\mathfrak{A}} \forall \boldsymbol{F}_i^n \varphi [s]$ iff for every function $f : |\mathfrak{A}|^n \to |\mathfrak{A}|$, $\models_{\mathfrak{A}} \varphi [s(\boldsymbol{F}_i^n | f)]$.

Again it is easy to see that only the values of s at variables occurring free in the formula are significant. For a sentence σ, we may unambiguously speak of its being true or false in $\mathfrak{A}$. Logical (semantical) implication is defined exactly as before.

EXAMPLE 1 A well-ordering is an ordering relation such that any nonempty set has a least (with respect to the ordering) element. This last condition can be translated into the second-order sentence

$$\forall X(\exists y\, Xy \rightarrow \exists y(Xy \wedge \forall z(Xz \rightarrow y \leq z))).$$

Here, as elsewhere, we omit the subscripts on X and F when they are immaterial, and we omit the superscripts if they are clear from the context.

EXAMPLE 2 One of Peano's postulates (the induction postulate) states that any set of natural numbers which contains 0 and is closed under the successor function is, in fact, the set of all natural numbers. This can be translated into the second-order language for number theory as

$$\forall X(X\mathbf{0} \wedge \forall y(Xy \rightarrow X\mathrm{S}y) \rightarrow \forall y\, Xy).$$

Any model of S1, S2, and the above Peano induction postulate is isomorphic to $(\mathbb{N}, 0, S)$; see Exercise 1. Thus this set of sentences is categorical; i.e., all its models are isomorphic.

EXAMPLE 3 For any formula φ in which the predicate variable X^n does not occur free, the formula

$$\exists X^n \forall v_1 \cdots \forall v_n[X^n v_1 \cdots v_n \leftrightarrow \varphi]$$

is valid. (Here other variables may occur free in φ in addition to $v_1, \ldots, v_n$.) Formulas of this form are called *relation comprehension formulas*. There are also the analogous *function comprehension formulas*. If ψ is a formula in which the variable F^n does not occur free, then

$$\forall v_1 \cdots \forall v_n \exists! v_{n+1} \psi \rightarrow \exists F^n \forall v_1 \cdots \forall v_n(F^n v_1 \cdots v_n \approx v_{n+1} \leftrightarrow \psi)$$

is valid. (Here "$\exists! v_{n+1} \psi$" is an abbreviation for a formula obtained from Exercise 21 of Section 2.2.)

EXAMPLE 4 In the ordered field of real numbers, any bounded nonempty set has a least upper bound. We can translate this by the second-order sentence

$$\forall X[\exists y \forall z(Xz \rightarrow z < y) \wedge \exists z\, Xz \rightarrow \exists y \forall y'(\forall z(Xz \rightarrow z < y') \leftrightarrow y \leq y')].$$

It is known that any ordered field which satisfies this second-order sentence is isomorphic to the field of reals.

EXAMPLE 5 For each $n \geq 2$, we have a first-order sentence λ_n which translates, "There are at least n things." For example, λ_3 is

$$\exists x\, \exists y\, \exists z(x \not\approx y \wedge x \not\approx z \wedge y \not\approx z).$$

The set $\{\lambda_2, \lambda_3, \ldots\}$ has for its class of models the EC_Δ class consisting of the infinite structures. There is a single second-order sentence which is equivalent. A set is infinite iff there is an ordering on it having no last element. Or more simply, a set is infinite iff there is a transitive irreflexive relation R on the set whose domain is the entire set. This condition can be translated into a second-order sentence λ_∞:

$$\exists X[\forall u\, \forall v\, \forall w(Xuv \rightarrow Xvw \rightarrow Xuw) \wedge \forall u\, \neg Xuu \wedge \forall u\, \exists v\, Xuv].$$

Another sentence (using a function variable) which defines the class of infinite structures is

$$\exists F[\forall x\, \forall y(Fx \approx Fy \rightarrow x \approx y) \wedge \exists z\, \forall x\, Fx \not\approx z].$$

The preceding example shows that the compactness theorem fails for second-order logic.

Theorem 41A There is an unsatisfiable set of second-order sentences every finite subset of which is satisfiable.

Proof The set is, in the notation of the above example,

$$\{\neg\lambda_\infty, \lambda_2, \lambda_3, \ldots\}. \quad \blacksquare$$

The Löwenheim–Skolem theorem also fails for second-order logic. By the *languange of equality* we mean the language (with $\approx$) having no parameters other than $\forall$. A structure for this language can be viewed as being just a nonempty set. In particular, a structure is determined to within isomorphism by its cardinality. A sentence in this language is therefore determined to within logical equivalence by the set of cardinalities of its models (called its *spectrum*).

Theorem 41B There is a sentence in the second-order language of equality which is true in a set iff its cardinality is $2^{\aleph_0}$.

Proof, using concepts from algebra Consider first the conjunction of the (first-order) axioms for an ordered field, further conjoined with the second-order sentence expressing the least-upper-bound property (see Example 4 of this section). This is a sentence whose models are exactly the isomorphs of the real ordered field (i.e., the structures isomorphic to the field of real numbers). We now convert the parameters $\mathbf{0}, \mathbf{1}, +, \cdot, <$ to variables (individual, function, or predicate as appropriate) which we existentially quantify. The resulting sentence has the desired properties. ■

Theorem 41C The set of Gödel numbers of valid second-order sentences is not definable in $\mathfrak{N}$ by any second-order formula.

Here we assume that Gödel numbers have been assigned to second-order expressions in a manner like that used before. Although our proof applies to the second-order language of number theory, the theorem is true for any recursively numbered language having at least a two-place predicate symbol.

Proof Let T^2 be the second-order theory of $\mathfrak{N}$, i.e., the set of second-order sentences true in $\mathfrak{N}$. The same argument used to prove Tarski's theorem shows that $\sharp T^2$ is not definable in $\mathfrak{N}$ by any second-order formula.

Now let α be the conjunction of the members of A_E with the second-order Peano induction postulate (Example 2). Any model of α is isomorphic to $\mathfrak{N}$; cf. Exercise 1. Consequently, for any sentence σ,

$$\sigma \in T^2 \qquad \text{iff } (\alpha \to \sigma) \text{ is valid.}$$

Consequently, the set of (Gödel numbers of) validities cannot be definable lest $\sharp T^2$ be. ■

A fortiori, the set of Gödel numbers of second-order validities is not arithmetical and not recursively enumerable. That is, the enumerability theorem fails for second-order logic. (In the other direction, one can show that this set is not definable in number theory of order three, or even of order ω. But these are topics we will not enter into here.)

It is interesting to compare the effect of a second-order universal sentence, such as the Peano induction postulate

$$\forall X(X\mathbf{0} \wedge \forall y(Xy \to X\mathrm{S}y) \to \forall y\, Xy)$$

and the corresponding first-order schema, i.e., the set of all sentences

$$\varphi(\mathbf{0}) \wedge \forall y(\varphi(y) \to \varphi(\mathrm{S}y)) \to \forall y\, \varphi(y),$$

where φ is a first-order formula having just $\boldsymbol{v}_1$ free. If $\mathfrak{A}$ is a model of the Peano induction postulate, then any subset of $|\mathfrak{A}|$ containing $\mathbf{0}^{\mathfrak{A}}$ and closed under $\mathbf{S}^{\mathfrak{A}}$ is in fact all of $|\mathfrak{A}|$. On the other hand, if $\mathfrak{A}$ is a model of the corresponding axiom schema, we can only say that every *definable* subset of $|\mathfrak{A}|$ containing $\mathbf{0}^{\mathfrak{A}}$ and closed under $\mathbf{S}^{\mathfrak{A}}$ is all of $|\mathfrak{A}|$. There may well be undefinable subsets for which this fails. (For example, take any model $\mathfrak{A}$ of Th $(N, 0, S)$ having Z-chains. Then $\mathfrak{A}$ satisfies the above first-order schema, but it does not satisfy the second-order induction postulate. The set of standard points is simply not definable in $\mathfrak{A}$.)

EXERCISES

1. Show that any structure for the language with parameters $\forall$, $\mathbf{0}$, and $\mathbf{S}$ that satisfies the sentences

S1. $$\forall x\, \mathbf{S}x \not\approx \mathbf{0},$$

S2. $$\forall x\, \forall y(\mathbf{S}x \approx \mathbf{S}y \rightarrow x \approx y),$$

and the Peano induction postulate

$$\forall X(X\mathbf{0} \wedge \forall y(Xy \rightarrow X\mathbf{S}y) \rightarrow \forall y\, Xy)$$

is isomorphic to $\mathfrak{N}_S = (N, 0, S)$.

2. Give a sentence in the second-order language of equality which is true in a set iff its cardinality is $\aleph_0$. Similarly for $\aleph_1$.

3. Let φ be a formula in which only the n-place predicate variable $\boldsymbol{X}$ occurs free. Say that an n-ary relation R on $|\mathfrak{A}|$ is *implicitly defined* in $\mathfrak{A}$ by φ iff $\mathfrak{A}$ satisfies φ with an assignment of R to $\boldsymbol{X}$ but does not satisfy φ with an assignment of any other relation to $\boldsymbol{X}$. Show that $\sharp$Th $\mathfrak{N}$, the set of Gödel numbers of first-order sentences true in $\mathfrak{N}$, is implicitly definable in $\mathfrak{N}$ by a formula without quantified predicate or function variables.

4. Consider a language (with equality) having the one-place predicate symbols I and S and the two-place predicate symbol E. Find a second-order sentence σ such that (i) if A is a set for which $A \cap \mathscr{P}A = \varnothing$ and if $|\mathfrak{A}| = A \cup \mathscr{P}A$, $I^{\mathfrak{A}} = A$, $S^{\mathfrak{A}} = \mathscr{P}A$, $E^{\mathfrak{A}} = \{\langle a, b\rangle : a \in b \subseteq A\}$, then $\mathfrak{A}$ is a model of σ; and (ii) any model of σ is isomorphic to one of the sort described in (i).

§ 4.2 SKOLEM FUNCTIONS

We want to show how, given any *first*-order formula, one can find a logically equivalent prenex second-order formula of a very special form:

existential quantifiers	universal individual quantifiers	quantifier-free formula

This is a prenex formula wherein all universal quantifiers are individual ones which follow a string of existential individual and function quantifiers.

In the simplest example, observe that

$$\forall x\, \exists y\, \varphi(x, y) \models\!\!\!\dashv \exists F\, \forall x\, \varphi(x, Fx).$$

In the "$\dashv$" direction this is easy to see. For the "$\models$" direction, consider a structure $\mathfrak{A}$ and an assignment function s satisfying $\forall x\, \exists y\, \varphi(x, y)$. We know that for any $a \in |\mathfrak{A}|$ there is at least one $b \in |\mathfrak{A}|$ such that

$$\models_{\mathfrak{A}} \varphi(x, y)\, [s(x|a)(y|b)].$$

We obtain a function f on $|\mathfrak{A}|$ by choosing one such b for each a and taking $f(a) = b$. (The axiom of choice is used here.) Then

$$\models_{\mathfrak{A}} \forall x\, \varphi(x, Fx)\, [s(F|f)].$$

This function f is called a *Skolem function* for the formula $\forall x\, \exists y\, \varphi$ in the structure $\mathfrak{A}$.

The same argument applies more generally. As a second example, suppose that we begin with the formula

$$\exists y_1\, \forall x_1\, \exists y_2\, \forall x_2\, \forall x_3\, \exists y_3\, \psi(y_1, y_2, y_3).$$

(We have listed only y_1, y_2, and y_3, but presumably other variables occur free in ψ as well.) Here we already have the existential quantifier $\exists y_1$ at the left. What remains is

$$\forall x_1\, \exists y_2\, \forall x_2\, \forall x_3\, \exists y_3\, \psi(y_1, y_2, y_3).$$

This is a special case of the first example (with $\varphi(x_1, y_2) = \forall x_2\, \forall x_3\, \exists y_3\, \psi(y_1, y_2, y_3)$). It is logically equivalent, as before, to

$$\exists F_2\, \forall x_1\, \forall x_2\, \forall x_3\, \exists y_3\, \psi(y_1, F_2 x_1, y_3).$$

Now we have the existential quantifiers $\exists y_1\ \exists \boldsymbol{F}_2$ at the left; what remains is

$$\forall x_1\ \forall x_2\ \forall x_3\ \exists y_3\ \psi(y_1, \boldsymbol{F}_2 x_1, y_3).$$

By the same reasoning as before, this is logically equivalent to

$$\exists \boldsymbol{F}_3\ \forall x_1\ \forall x_2\ \forall x_3\ \psi(y_1, \boldsymbol{F}_2 x_1, \boldsymbol{F}_3 x_1 x_2 x_3),$$

where $\boldsymbol{F}_3$ is a three-place function variable. Thus the original formula is equivalent to

$$\exists y_1\ \exists \boldsymbol{F}_2\ \exists \boldsymbol{F}_3\ \forall x_1\ \forall x_2\ \forall x_3\ \psi(y_1, \boldsymbol{F}_2 x_1, \boldsymbol{F}_3 x_1 x_2 x_3).$$

For quantifier-free ψ, this is in the form we desire.

Skolem Normal Form Theorem For any first-order formula we can find a logically equivalent second-order formula consisting of

(a) first a string (possibly empty) of existential individual and function quantifiers, followed by

(b) a string (possibly empty) of universal individual variables, followed by

(c) a quantifier-free formula.

A formal proof could be given using induction, but the preceding example illustrates the general method.

Recall that a universal ($\forall_1$) formula is a first-order prenex formula all of whose quantifiers are universal: $\forall x_1\ \forall x_2 \cdots \forall x_k\ \alpha$, where α is quantifier-free. Similarly, an existential ($\exists_1$) formula is a first-order prenex formula all of whose quantifiers are existential.

Corollary 42A For any first-order φ, we can find a universal formula θ in an expanded language containing function symbols, such that φ is satisfiable iff θ is satisfiable.

By applying this corollary to $\neg\varphi$, we obtain an existential formula (with function symbols) which is valid iff φ is valid.

Proof Again we will only illustrate the situation by an example. Say that φ is

$$\exists y_1\ \forall x_1\ \exists y_2\ \forall x_2\ \forall x_3\ \exists y_3\ \psi(y_1, y_2, y_3).$$

First we replace φ by the logically equivalent formula in Skolem form:

$$\exists y_1\ \exists \boldsymbol{F}_2\ \exists \boldsymbol{F}_3\ \forall x_1\ \forall x_2\ \forall x_3\ \psi(y_1, \boldsymbol{F}_2 x_1, \boldsymbol{F}_3 x_1 x_2 x_3).$$

Then for θ we take

$$\forall x_1 \, \forall x_2 \, \forall x_3 \, \psi(c, fx_1, gx_1x_2x_3),$$

where c, f, and g are new function symbols having zero, one, and three places, respectively. ■

Corollary 42B In a recursively numbered language having for each $k \geq 0$ infinitely many k-place function symbols and a two-place predicate symbol, the set of Gödel numbers of valid existential (first-order) sentences is not recursive.

Proof Given any sentence σ we can, by applying Corollary 42A to $\neg\sigma$, effectively find an existential sentence which is valid iff σ is valid. Hence a decision procedure for the existential validities would yield a decision procedure for arbitrary validities, in contradiction to Church's theorem. ■

It is possible to strengthen this corollary by weakening the hypotheses. It is not too hard to see that the two-place predicate symbol is unnecessary, and that it suffices that there be k-place function symbols for arbitrarily large k. (The proof utilizes faithful interpretations.)

We can use predicate variables instead of function variables in these results, but at a price. Suppose we begin with a first-order formula. It is equivalent to a formula ψ in Skolem normal form; suppose for simplicity that $\psi = \exists \mathbf{F} \varphi$, where φ has only individual quantifiers and $\mathbf{F}$ is a one-place function variable. We can choose φ in such a way that $\mathbf{F}$ occurs only in equations of the form $u \approx \mathbf{F}t$ (for terms t and u not containing $\mathbf{F}$). This can be done by replacing, for example, an atomic formula $\alpha(\mathbf{F}t)$ by either $\forall x(x \approx \mathbf{F}t \rightarrow \alpha(x))$ or $\exists x(x \approx \mathbf{F}t \wedge \alpha(x))$.

Next observe that a formula

$$\exists \mathbf{F} \underline{\quad} u \approx \mathbf{F}t \underline{\quad},$$

wherein $\mathbf{F}$ occurs only in the form shown, is equivalent to

$$\exists \mathbf{X}(\forall y \, \exists! z \, \mathbf{X}yz \wedge \underline{\quad} \mathbf{X}tu \underline{\quad}).$$

If one pursues this question (as we will not do here) one finds that any first-order formula is logically equivalent to a second-order formula consisting of

(a) a string of existential predicate quantifiers, followed by
(b) a string of universal individual quantifiers, followed by

(c) a string of existential individual quantifiers, followed by
(d) a quantifier-free formula.

There are corresponding versions (which the reader is invited to formulate) of Corollaries 42A and 42B.

EXERCISES

1. Prove the Löwenheim–Skolem theorem in the following improved form: Let $\mathfrak{A}$ be a structure for a language of cardinality $\varkappa$. Let S be a subset of $|\mathfrak{A}|$ having cardinality λ. Then there is a *substructure* $\mathfrak{B}$ of $\mathfrak{A}$ of cardinality $\lambda + \varkappa$ such that for any function s mapping the variables into $|\mathfrak{B}|$ and any (first-order) φ,

$$\models_{\mathfrak{A}} \varphi\,[s] \qquad \text{iff } \models_{\mathfrak{B}} \varphi\,[s].$$

Suggestion: Choose Skolem functions for all formulas. Close S under the functions.

2. State the two corollaries referred to in the last sentence of Section 4.2. (The second of these can be compared with Exercise 5(b) of Section 2.5, where it is shown that the set of $\forall_2$ validities without function symbols is decidable.)

§ 4.3 MANY-SORTED LOGIC

We now return to first-order languages, but with many sorts of variables, ranging over different universes. (In the next section this will be applied to the case in which one sort of variable is for elements of a universe, another for subsets of that universe, yet another for binary relations, and so forth.)

In informal mathematics one sometimes says things like, "We use Greek letters for ordinals, capital script letters for sets of integers, ..." In effect, one thereby adopts several sorts of variables, each sort having its own universe. We now undertake to examine this situation precisely. As might be expected, nothing is drastically different from the usual one-sorted situation. None of the results of this section are at all deep, and most of the proofs are omitted.

Assume that we have a nonempty set I, whose members are called *sorts*, and symbols arranged as follows:

A. *Logical symbols*

0. Parentheses: (,).
1. Sentential connective symbols: $\neg$, $\rightarrow$.

2. Variables: For each sort i, there are variables $\boldsymbol{v}_1^i, \boldsymbol{v}_2^i, \ldots$ of sort i.
3. Equality symbols: For some $i \in I$ there may be the symbol $\approx_i$, said to be a predicate symbol of sort $\langle i, i \rangle$.

B. *Parameters*

0. Quantifier symbols: For each sort i there is a universal quantifier symbol $\forall_i$.
1. Predicate symbols: For each $n > 0$ and each n-tuple $\langle i_1, \ldots, i_n \rangle$ of sorts, there is a set (possibly empty) of n-place predicate symbols, each of which is said to be of sort $\langle i_1, \ldots, i_n \rangle$.
2. Constant symbols: For each sort i there is a set (possibly empty) of constant symbols each of which is said to be of sort i.
3. Function symbols: For each $n > 0$ and each $(n+1)$-tuple $\langle i_1, \ldots, i_n, i_{n+1} \rangle$ of sorts, there is a set (possibly empty) of n-place function symbols, each of which is said to be of sort $\langle i_1, \ldots, i_n, i_{n+1} \rangle$.

As usual, we must assume that these categories of symbols are disjoint, and further that no symbol is a finite sequence of other symbols.

Each term will be assigned a unique sort. We define the set of terms of sort i inductively, simultaneously for all i:

1. Any variable of sort i or constant symbol of sort i is a term of sort i.

2. If $t_1, \ldots, t_n$ are terms of sort $i_1, \ldots, i_n$, respectively, and f is a function symbol of sort $\langle i_1, \ldots, i_n, i_{n+1} \rangle$, then $ft_1 \cdots t_n$ is a term of sort i_{n+1}.

This definition can be recast into a more familiar form. The set of pairs $\langle t, i \rangle$ such that t is a term of sort i is generated from the basic set

$$\{\langle \boldsymbol{v}_n^i, i \rangle : n \geq 1 \ \&\ i \in I\} \cup \{\langle c, i \rangle : c \text{ is a constant symbol of sort } i\}$$

by the operations which, for a function symbol f of sort $\langle i_1, \ldots, i_n, i_{n+1} \rangle$, produce the pair $\langle ft_1 \cdots t_n, i_{n+1} \rangle$ from the pairs $\langle t_1, i_1 \rangle, \ldots, \langle t_n, i_n \rangle$.

An atomic formula is a sequence $Pt_1 \cdots t_n$ consisting of a predicate symbol of sort $\langle i_1, \ldots, i_n \rangle$ and terms $t_1, \ldots, t_n$ of sort $i_1, \ldots, i_n$, respectively. The nonatomic formulas are then formed using the connectives $\neg, \rightarrow$, and the quantifiers $\forall_i \boldsymbol{v}_n^i$.

A many-sorted structure $\mathfrak{A}$ is a function on the set of parameters which assigns to each the correct type of object:

1. To the quantifier symbol $\forall_i$, $\mathfrak{A}$ assigns a nonempty set $|\mathfrak{A}|_i$, called the *universe* of $\mathfrak{A}$ of sort i.

2. To each predicate symbol P of sort $\langle i_1, \ldots, i_n \rangle$, $\mathfrak{A}$ assigns the relation

$$P^{\mathfrak{A}} \subseteq |\mathfrak{A}|_{i_1} \times \cdots \times |\mathfrak{A}|_{i_n}.$$

3. To each constant symbol c of sort i, $\mathfrak{A}$ assigns a point $c^{\mathfrak{A}}$ in $|\mathfrak{A}|_i$.

4. To each function symbol f of sort $\langle i_1, \ldots, i_n, i_{n+1} \rangle$, $\mathfrak{A}$ assigns a function

$$f^{\mathfrak{A}} : |\mathfrak{A}|_{i_1} \times \cdots \times |\mathfrak{A}|_{i_n} \to |\mathfrak{A}|_{i_{n+1}}.$$

The definitions of truth and satisfaction are the obvious ones, given that $\forall_i$ is to mean "for all members of the universe $|\mathfrak{A}|_i$ of sort i."

In a many-sorted structure, the universes of the various sorts might or might not be disjoint. But since we have no equality symbols *between* sorts, any nondisjointness must be regarded as accidental. In particular, there will always be an elementarily equivalent structure whose universes are disjoint.

Reduction to one-sorted logic

Many-sorted languages may at times be convenient (as in the following section). But there is nothing essential that can be done with them that cannot already be done without them. We now proceed to make this assertion in a more precise form.

We will consider a one-sorted language having all the predicate, constant, and function symbols of our assumed many-sorted language. In addition, it will have a one-place predicate symbol Q_i for each i in I. There is a syntactical translation taking each many-sorted sentence σ into a one-sorted sentence σ^*. In this translation all equality symbols are replaced by $\approx$. The only other change is in the quantifiers (the quantifier symbols and the quantified variables): We replace

$$\forall_i \mathbf{v}_n^i \_\_\_ \mathbf{v}_n^i \_\_\_$$

by

$$\forall v (Q_i v \to \_\_\_ v \_\_\_),$$

where v is a variable chosen not to conflict with the others. Thus the quantifiers of sort i are relativized to Q_i.

Turning now to semantics, we can convert a many-sorted structure $\mathfrak{A}$ into a structure $\mathfrak{A}^*$ for the above one-sorted language. The universe $|\mathfrak{A}^*|$ is the union $\bigcup_{i \in I} |\mathfrak{A}|_i$ of all the universes of $\mathfrak{A}$. To Q_i is assigned the set $|\mathfrak{A}|_i$. On the predicate and constant symbols $\mathfrak{A}^*$ agrees with $\mathfrak{A}$. For a function symbol f, the function $f^{\mathfrak{A}^*}$ is an arbitrary extension of $f^{\mathfrak{A}}$. (Of course this last sentence does not completely specify $f^{\mathfrak{A}^*}$. The results we give for $\mathfrak{A}^*$ hold for any structure obtained in the manner just described.)

Lemma 43A A many-sorted sentence σ is true in $\mathfrak{A}$ iff σ^* is true in $\mathfrak{A}^*$.

To prove this, one makes a stronger statement concerning formulas. The stronger statement is then proved by induction.

Consider now the other direction. A one-sorted structure is not always convertible into a many-sorted structure. So we will impose some conditions. Let Φ be the set consisting of the following one-sorted sentences:

1. $\exists v\, Q_i v$, for each i in I.
2. $\forall \mathbf{v}_1 \cdots \forall \mathbf{v}_n (Q_{i_1}\mathbf{v}_1 \to \cdots \to Q_{i_n}\mathbf{v}_n \to Q_{i_{n+1}} f\mathbf{v}_1 \cdots \mathbf{v}_n)$, for each function symbol f of sort $\langle i_1, \ldots, i_n, i_{n+1}\rangle$. We include the case $n = 0$, in which case the above becomes the sentence $Q_i c$ for a constant symbol c of sort i.

Notice that the above $\mathfrak{A}^*$ was a model of Φ. A one-sorted model $\mathfrak{B}$ of Φ does convert into a many-sorted $\mathfrak{B}^\#$. The conversion is performed in the natural way:

$$|\mathfrak{B}^\#|_i = Q_i^{\mathfrak{B}};$$

$$P^{\mathfrak{B}^\#} = P^{\mathfrak{B}} \cap (Q_{i_1}^{\mathfrak{B}} \times \cdots \times Q_{i_n}^{\mathfrak{B}}),$$ where P is a predicate symbol of sort $\langle i_1, \ldots, i_n\rangle$;

$$c^{\mathfrak{B}^\#} = c^{\mathfrak{B}};$$

$$f^{\mathfrak{B}^\#} = f^{\mathfrak{B}} \cap (Q_{i_1}^{\mathfrak{B}} \times \cdots \times Q_{i_n}^{\mathfrak{B}} \times Q_{i_{n+1}}^{\mathfrak{B}}),$$ the restriction of f to $Q_{i_1}^{\mathfrak{B}} \times \cdots \times Q_{i_n}^{\mathfrak{B}}$, where f is a function symbol of sort $\langle i_1, \ldots, i_n, i_{n+1}\rangle$.

Lemma 43B If $\mathfrak{B}$ is a model of Φ, then $\mathfrak{B}^\#$ is a many-sorted structure. Furthermore, a many-sorted sentence σ is true in $\mathfrak{B}^\#$ iff σ^* is true in $\mathfrak{B}$.

The proof is similar to the proof of Lemma 43A.

Notice that $\mathfrak{B}^{\#*}$ is not in general equal to $\mathfrak{B}$. (For example, $|\mathfrak{B}|$ may contain points not belonging to any $Q_i^{\mathfrak{B}}$.) On the other hand, $\mathfrak{A}^{*\#}$ is equal to $\mathfrak{A}$.

Theorem 43C In the many-sorted language

$$\Sigma \models \sigma$$

iff in the one-sorted language

$$\Sigma^* \cup \Phi \models \sigma^*.$$

Proof ($\Rightarrow$) Let $\mathfrak{B}$ be a one-sorted model of $\Sigma^* \cup \Phi$. Then $\mathfrak{B}^\#$ is a model of Σ by Lemma 43B. Hence $\mathfrak{B}^\#$ is a model of σ. So by Lemma 43B again, $\mathfrak{B}$ is a model of σ^*.

($\Leftarrow$) Similar, with Lemma 43A. ■

By using Theorem 43C, we can now infer the following three theorems from the corresponding one-sorted results.

Compactness Theorem If every finite subset of a set Σ of many-sorted sentences has a model, then Σ has a model.

Proof Assume that every finite subset Σ_0 of Σ has a many-sorted model $\mathfrak{A}_0$. Then a finite subset Σ_0^* of Σ^* has the model $\mathfrak{A}_0^*$. Hence by the ordinary compactness theorem, Σ^* has a model $\mathfrak{B}$. $\mathfrak{B}^\#$ is then a model of Σ. ■

Enumerability Theorem For a recursively numbered many-sorted language, the set of Gödel numbers of valid sentences is recursively enumerable.

Proof For a many-sorted σ, we have by Theorem 43C,

$$\models \sigma \qquad \text{iff } \Phi \models \sigma^*.$$

Since Φ is recursive, Cn Φ is recursively enumerable. And σ^* depends recursively on σ, so we can apply Exercise 7(*b*) of Section 3.5. ■

Löwenheim–Skolem Theorem For any many-sorted structure (for a countable language) there is an elementarily equivalent countable structure.

Proof Say that the given structure is $\mathfrak{A}$. Then $\mathfrak{A}^*$ is a one-sorted model of $(\text{Th}\,\mathfrak{A})^* \cup \Phi$. Hence by the ordinary Löwenheim–Skolem theorem, $(\text{Th}\,\mathfrak{A})^* \cup \Phi$ has a countable model $\mathfrak{B}$. $\mathfrak{B}^\#$ is a model of Th $\mathfrak{A}$ and so is elementarily equivalent to $\mathfrak{A}$. ■

§ 4.4 GENERAL STRUCTURES

We now return to the discussion of second-order logic begun in Section 4.1. We discussed there (a) the grammar, i.e., the set of wffs for second order, and (b) the semantics, i.e., the notion of structure (which was the same as for first order) and the definition of satisfaction and truth.

In this section we want to leave (a) unchanged, but we want to present an alternative to (b). The idea can be stated very briefly: We view the language (previously thought of as second-order) now as being a many-sorted elementary (i.e., first-order) language. The result is to make open to interpretation not only the universe over which individual variables range but also the universes for the predicate and function variables. This approach is particularly suited to number theory; that case is examined briefly at the end of this section.

The many-sorted language

Despite the fact that we want ultimately to consider the grammar of Section 4.1, it will be expedient to consider also a many-sorted language constructed from the second-order language of Section 4.1. We take $\aleph_0$ sorts: the one individual sort (with variables $\mathbf{v}_1, \mathbf{v}_2, \ldots$); for each $n > 0$, the n-place predicate sort (with variables $\mathbf{X}_1^n, \mathbf{X}_2^n, \ldots$); and for each $n > 0$ the n-place function sort (with variables $\mathbf{F}_1^n, \mathbf{F}_2^n, \ldots$). We will use equality ($\approx$) only between terms of the individual sort. The predicate and function parameters of our assumed second-order language will also be parameters of the many-sorted language, and will take as arguments terms of the individual sort.

In addition, we now use two new classes of parameters. For each $n > 0$ there is a *membership* predicate parameter ε_n which takes as arguments one term of the n-place predicate sort and n terms of the individual sort. Thus, for example,

$$\varepsilon_3 \mathbf{X}^3 \mathbf{v}_2 \mathbf{v}_1 \mathbf{v}_8$$

is a wff. Its intended interpretation is that the triple denoted by $(\mathbf{v}_2, \mathbf{v}_1, \mathbf{v}_8)$ is to belong to the relation denoted by $\mathbf{X}^3$. This is exactly the interpretation assigned previously to the second-order formula

$$\mathbf{X}^3 \mathbf{v}_2 \mathbf{v}_1 \mathbf{v}_8,$$

and, in fact, the reader is advised to identify these two formulas closely in his mind.

For each $n > 0$, there is also the *evaluation* function parameter E_n. E_n takes as arguments one term of the n-place function sort and n terms of the individual sort. The resulting term,

$$E_n \mathbf{F}^n t_1 \cdots t_n,$$

is itself of the individual sort. Again the reader is advised to identify closely the term $E_n \mathbf{F}^n t_1 \cdots t_n$ with the previous $\mathbf{F}^n t_1 \cdots t_n$.

There is an obvious way of translating between the second-order language of Section 4.1 and the present many-sorted language. In one direction we stick on the ε_n and E_n symbols; in the other direction we take them off. The purpose of these symbols is to make the language conform to Section 4.3.

A many-sorted structure has universes for each sort and assigns suitable objects to the various parameters (as described in the preceding section). First, we want to show that without loss of generality, we may suppose that ε_n is interpreted as genuine membership and E_n as genuine evaluation.

Theorem 44A Let $\mathfrak{A}$ be a structure for the above many-sorted language such that the different universes of $\mathfrak{A}$ are disjoint. Then there is a homomorphism h of $\mathfrak{A}$ onto a structure $\mathfrak{B}$ such that

(a) h is one-to-one, in fact the identity, on the individual universe (from which it follows that

$$\models_{\mathfrak{A}} \varphi\,[s] \qquad \text{iff } \models_{\mathfrak{B}} \varphi\,[h \circ s]$$

for each formula φ).

(b) The n-place predicate universe of $\mathfrak{B}$ consists of certain n-ary relations over the individual universe, and $\langle R, a_1, \ldots, a_n\rangle$ is in $\varepsilon_n^{\mathfrak{B}}$ iff $\langle a_1, \ldots, a_n\rangle \in R$.

(c) The n-place function universe of $\mathfrak{B}$ consists of certain n-place functions on the individual universe, and $E_n^{\mathfrak{B}}(f, a_1, \ldots, a_n) = f(a_1, \ldots, a_n)$.

Proof Since the universes of $\mathfrak{A}$ are disjoint, we can define h on one universe at a time. On the individual universe U, h is the identity. On the universe of the n-place predicate sort,

$$h(Q) = \{\langle a_1, \ldots, a_n\rangle\text{: Each } a_i \text{ is in } U \text{ and } \langle Q, a_1, \ldots, a_n\rangle \text{ is in } \varepsilon_n^{\mathfrak{A}}\}.$$

Thus

(1) $$\langle a_1, \ldots, a_n\rangle \in h(Q) \qquad \text{iff } \langle Q, a_1, \ldots, a_n\rangle \text{ is in } \varepsilon_n^{\mathfrak{A}}.$$

Similarly, on the universe of the n-place function sort,

$h(g)$ is the n-place function on U whose value at $\langle a_1, \ldots, a_n\rangle$ is $E_n^{\mathfrak{A}}(g, a_1, \ldots, a_n)$.

Thus

(2) $$h(g)(a_1, \ldots, a_n) = E_n^{\mathfrak{A}}(g, a_1, \ldots, a_n).$$

For $\varepsilon_n^{\mathfrak{B}}$ we take simply the membership relation,

(3) $$\langle R, a_1, \ldots, a_n\rangle \text{ is in } \varepsilon_n^{\mathfrak{B}} \qquad \text{iff } \langle a_1, \ldots, a_n\rangle \in R.$$

For $E_n^{\mathfrak{B}}$ we take the evaluation function,

$$E_n^{\mathfrak{B}}(f, a_1, \ldots, a_n) = f(a_1, \ldots, a_n). \tag{4}$$

On the other parameters (inherited from the second-order language) $\mathfrak{B}$ agrees with $\mathfrak{A}$.

Then it is clear upon reflection that h is a homomorphism of $\mathfrak{A}$ onto $\mathfrak{B}$. That h preserves ε_n follows from (1) and (3), where in (3) we take $R = h(Q)$. Similarly, from (2) and (4) it follows that h preserves E_n.

Finally, we have to verify the parenthetical remark of part (a). This follows from the many-sorted analog of the homomorphism theorem of Section 2.2, by using the fact that we have equality only for the individual sort, where h is one-to-one. ■

By the above theorem, we can restrict attention to structures $\mathfrak{B}$ in which ε_n and E_n are fixed by (b) and (c) of the theorem. But since $\varepsilon_n^{\mathfrak{B}}$ and $E_n^{\mathfrak{B}}$ are determined by the rest of $\mathfrak{B}$, we really do not need them at all. When we discard them, we have a general prestructure for our second-order grammar.

General structures for second-order languages

These structures provide the alternative semantics mentioned at the beginning of this section.

Definition A *general pre-structure* $\mathfrak{A}$ for our second-order language consists of a structure (in the original sense), together with the additional sets:

(a) for each $n > 0$, an n-place *relation universe*, which is a set of n-ary relations on $|\mathfrak{A}|$;

(b) for each $n > 0$, an n-place *function universe*, which is a set of functions from $|\mathfrak{A}|^n$ into $|\mathfrak{A}|$.

$\mathfrak{A}$ is a *general structure* if, in addition, all comprehension sentences are true in $\mathfrak{A}$.

The last sentence of the definition requires explanation. First, a comprehension sentence is a sentence obtained as a generalization of a comprehension formula (see Example 3 of Section 4.1). Thus it is a sentence of the form

$$\forall y_1 \cdots \forall y_m \exists X^n \forall \mathbf{v}_1 \cdots \forall \mathbf{v}_n (X^n \mathbf{v}_1 \cdots \mathbf{v}_n \leftrightarrow \varphi),$$

where X^n does not occur free in φ, or

$$\forall y_1 \cdots \forall y_m[\forall v_1 \cdots \forall v_n \, \exists! v_{n+1} \, \varphi \rightarrow \exists F^n \, \forall v_1 \cdots \forall v_{n+1}(F^n v_1 \cdots v_n \approx v_{n+1} \leftrightarrow \varphi)],$$

where F^n does not occur free in φ.

Next we must say what it means for a comprehension sentence (or any second-order sentence for that matter) to be true in $\mathfrak{A}$. Assume then that $\mathfrak{A}$ is a general pre-structure. Then a sentence σ is true in $\mathfrak{A}$ iff the result of converting σ into a many-sorted sentence (by adding ε_n and E_n) is true in $\mathfrak{A}$, with ε_n interpreted as membership and E_n as evaluation.

More generally, let φ be a second-order formula, and let s be a function which assigns to each individual variable a member of $|\mathfrak{A}|$, to each predicate variable a member of the relation universe of $\mathfrak{A}$, and to each function variable a member of the function universe of $\mathfrak{A}$. Then we say that $\mathfrak{A}$ satisfies φ with s, $\models^{G}_{\mathfrak{A}} \varphi\,[s]$, iff the many-sorted version of φ is satisfied with s in the structure $\mathfrak{A}$, where ε_n is interpreted as membership and E_n as evaluation.

The essential consequences of this definition of satisfaction are the following, which should be compared with 5 and 6 of page 269.

$\models^{G}_{\mathfrak{A}} \forall X^n \varphi\,[s]$	iff for every R in the n-place relation universe of $\mathfrak{A}$, $\models^{G}_{\mathfrak{A}} \varphi\,[s(X^n \mid R)]$.
$\models^{G}_{\mathfrak{A}} \forall F^n \varphi\,[s]$	iff for every f in the n-place function universe of $\mathfrak{A}$, $\models^{G}_{\mathfrak{A}} \varphi\,[s(F^n \mid f)]$.

This, then, is the alternative approach mentioned at the beginning of the section. It involves treating the second-order grammar as being a many-sorted first-order grammar in disguise. Because this approach is basically first-order, we have the Löwenheim–Skolem theorem, the compactness theorem, and the enumerability theorem.

Löwenheim–Skolem Theorem If the set Σ of sentences in a countable second-order language has a general model, then it has a countable general model.

Here a countable general model is one in which every universe is countable (or equivalently, the union of all the universes is countable).

Proof Let Γ be the set of comprehension sentences. Then $\Sigma \cup \Gamma$, viewed as a set of many-sorted sentences, has a countable many-sorted model by the Löwenheim–Skolem theorem of the preceding section. By Theorem 44A, a homomorphic image of that model is a general pre-structure satisfying $\Sigma \cup \Gamma$, and hence is a general model of Σ. ■

Compactness Theorem If every finite subset of a set Σ of second-order sentences has a general model, then Σ has a general model.

Proof The proof is exactly as above. Every finite subset of $\Sigma \cup \Gamma$ has a many-sorted model, so we can apply the compactness theorem of the preceding section. ■

Enumerability Theorem Assume that the language is recursively numbered. Then the set of Gödel numbers of second-order sentences which are true in every general structure is recursively enumerable.

Proof A sentence σ is true in every general structure iff it is a many-sorted consequence of Γ. And $\sharp\Gamma$ is recursive. ■

The above two theorems assure us that there is an acceptable deductive calculus such that τ is deducible from Σ iff τ is true in every general model of Σ (see the remarks at the beginning of Section 2.4). But now that we know there is such a complete deductive calculus, there is no reason to go into the detailed development of one.

We can compare the two approaches to second-order semantics as follows: The version of Section 4.1 (which we will call *absolute* second-order logic) is a hybrid creature, in which the meaning of the parameters is left open to interpretation by structures, but the notion of being (for example) a subset is not left open, but is treated as having a fixed meaning. The version of the present section (*general* second-order logic) avoids appealing to a fixed notion of subset, and consequently is reducible to first-order logic. In this respect it is like axiomatic set theory, where one speaks of sets and sets of sets and so forth, but the theory is a first-order theory.

By enlarging the class of structures, general second-order logic diminishes the cases in which logical implication holds. That is, if every general model of Σ is a general model of σ, it then follows that $\Sigma \models \sigma$ in absolute second-order logic. But the converse fails. For example, take $\Sigma = \varnothing$: The set of sentences true in all general models is a recursively enumerable subset of the nonarithmetical set of valid sentences of absolute second-order logic.

Models of analysis

We can illustrate the ideas of this section by focusing attention on the most interesting special case: general models of second-order number theory. Consider then the second-order language for number theory, with the parameters **0**, **S**, $<$, $\cdot$, and **E**. We take as our set of axioms the set A_E^2 obtained from A_E by adding as a twelfth member the Peano induction postulate (Example 2, Section 4.1). From Exercise 1 of Section 4.1, we can conclude that any model (in the semantics of that section) of A_E^2 is isomorphic to $\mathfrak{N}$.

But what of the general models of our axiom set? They can differ from $\mathfrak{N}$ in either (or both) of two ways. We can employ the compactness theorem as before to construct (nonstandard) general models of the axioms having infinite numbers (i.e., models $\mathfrak{A}$ with an element a which is larger than the denotation of $\mathbf{S}^n\mathbf{0}$ in the ordering $<^{\mathfrak{A}}$). We can also find (nonabsolute) general models in which, for example, the set universe (the unary relation universe) is less than the full power set of the individual universe. Indeed, any countable general model must be of this kind.

It is traditional for logicians to refer to second-order number theory as *analysis*. The name derives from the fact that it is possible to identify real numbers with sets of natural numbers. In second-order number theory we have quantifiers over sets of natural numbers, which we can view as quantifiers over real numbers. The appropriateness of the name is nevertheless open to question, but its usage is well established. By a *model of analysis* we will mean a general model of the above axiom set A_E^2.

Define an *ω-model of analysis* to be a model of analysis in which the individual universe is N and the denotations of **0** and **S** are the genuine 0 and S. (Consequently, the denotations of $<$, $+$, $\cdot$, and **E** are also standard.) The motivation for studying ω-models can be stated as follows: We have a clear understanding of the set N. But we do not have anything like the same understanding of its power set $\mathscr{P}N$. For example, we may be uncertain whether its cardinality is $\aleph_1$ or $\aleph_2$ or more. So it is reasonable to hold fixed that which we are sure of (N), but to leave open to interpretation by a structure that which we are not sure of ($\mathscr{P}N$).

Among the ω-models of analysis there is the one *absolute* model, whose n-place relation universe consists of all n-ary relations on N (and whose function universes consist of all possible functions). A first-order sentence is true in an arbitrary ω-model of analysis iff it is true in $\mathfrak{N}$. But the ω-model may disagree with the absolute model on second-order sentences.

In the next theorem we assert that an ω-model of analysis is completely determined by its set universe (i.e., its one-place relation universe).

Theorem 44B If $\mathfrak{A}$ and $\mathfrak{B}$ are ω-models of analysis having the same one-place relation universe, then $\mathfrak{A} = \mathfrak{B}$.

Proof Suppose R belongs to the three-place relation universe of $\mathfrak{A}$. Let $\langle R\rangle$ be the "compression" of R into a unary relation:

$$\langle R\rangle = \{\langle a, b, c\rangle : \langle a, b, c\rangle \in R\}.$$

Our sequence-encoding function is recursive and hence is first-order definable in number theory by a formula φ. $\langle R\rangle$ is in the set universe of $\mathfrak{A}$ by virtue of the comprehension sentence

$$\forall \mathbf{X}^3\, \exists \mathbf{X}^1\, \forall u[\mathbf{X}^1 u \leftrightarrow \exists \mathbf{v}_1\, \exists \mathbf{v}_2\, \exists \mathbf{v}_3(\varphi(\mathbf{v}_1, \mathbf{v}_2, \mathbf{v}_3, u) \wedge \mathbf{X}^3\mathbf{v}_1\mathbf{v}_2\mathbf{v}_3)].$$

Thus $\langle R\rangle$ is in the set universe of $\mathfrak{B}$; we unpack it by a similar argument. R is in the three-place relation universe of $\mathfrak{B}$ by virtue of the comprehension sentence

$$\forall \mathbf{X}^1\, \exists \mathbf{X}^3\, \forall \mathbf{v}_1\, \forall \mathbf{v}_2\, \forall \mathbf{v}_3[\mathbf{X}^3\mathbf{v}_1\mathbf{v}_2\mathbf{v}_3 \leftrightarrow \exists u(\varphi(\mathbf{v}_1, \mathbf{v}_2, \mathbf{v}_3, u) \wedge \mathbf{X}^1 u)].$$

A similar argument applies to the function universes. ■

Consequently, we can identify an ω-model of analysis with its set universe (which is included in $\mathscr{P}N$). Not every subclass of $\mathscr{P}N$ is then an ω-model of analysis, but only those for which the comprehension sentences are satisfied.

EXAMPLES of ω-models We need only specify the set universe.

1. $\mathscr{P}N$ is the absolute model.
2. Let $(A, \in_A)$ be a model of the usual axioms for set theory such that (i) the relation $\in_A$ is the genuine membership relation $\{\langle a, b\rangle : a \in A, b \in A, \text{and } a \in b\}$ on the universe A, and (ii) A is transitive, i.e., if $a \in b \in A$, then $a \in A$. Then the collection of all those subsets of N that belong to A is an ω-model of analysis.
3. For a class $\mathscr{A} \subseteq \mathscr{P}N$, define $D\mathscr{A}$ to be the class of all sets $B \subseteq N$ which are definable in the ω pre-structure with set universe $\mathscr{A}$, by a formula of the language of second-order number theory, augmented by parameters for each set in $\mathscr{A}$. Then define by transfinite recursion on the ordinals:

$$\mathscr{A}_0 = \varnothing,$$

$$\mathscr{A}_{\alpha+1} = D\mathscr{A}_\alpha,$$

$$\mathscr{A}_\lambda = \bigcup_{\alpha<\lambda} \mathscr{A}_\alpha \quad \text{for limit } \lambda.$$

By cardinality considerations we see that this stops growing at some ordinal

β for which $\mathscr{A}_{\beta+1} = \mathscr{A}_{\beta}$. Let β_0 be the least such β; it can be shown (from the Löwenheim–Skolem theorem) that β_0 is a countable ordinal. $\mathscr{A}_{\beta_0}$ coincides with $\bigcup_{\alpha} \mathscr{A}_{\alpha}$ (the union being over all ordinals α), and is called the class of *ramified analytical* sets. It is an ω-model of analysis; the truth of the comprehension sentences follows from the fact that $\mathbb{D}\mathscr{A}_{\beta_0} \subseteq \mathscr{A}_{\beta_0}$.

Index

F
G
H 3
I 4
J 5